KU-526-192

The Complete Encyclopedia of

VEGETABLES AND

VEGETARIAN

COOKING

The Complete Encyclopedia of

VEGETABLES AND VEGETARIAN COOKING

CHRISTINE INGRAM
WITH
ROZ DENNY AND KATHERINE RICHMOND

LORENZ BOOKS

This edition published by Lorenz Books
an imprint of
Anness Publishing Limited
Hermes House
88-89 Blackfriars Road
London SE1 8HA

All rights reserved. No part of this publication may be reproduced, stored in a retrieval system, or transmitted in any way or by any means, electronic, mechanical, photocopying, recording or otherwise, without the prior written permission of the copyright holder.

A CIP catalogue record for this book is available from the British Library

ISBN 0 7548 0294 9

Publisher: Joanna Lorenz
Editors: Christopher Fagg and Lydia Darbyshire
Guide to Vegetables by: Christine Ingram
Recipes by: Christine Ingram, Roz Denny and Katherine Richmond
Photographs by: Patrick McLeavey, Michael Michaels and Michelle Garrett
Home Economists: Jane Stevenson, Wendy Lee and Liz Trigg
Design Styling by: Patrick McLeavey
Setting by: SX Composing DTP
Index by: Lydia Darbyshire

Printed and bound in Germany

© Anness Publishing Limited 1997
Updated © 1999

1 3 5 7 9 10 8 6 4 2

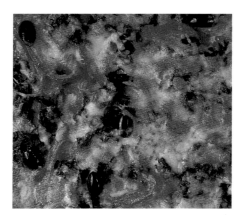

NOTES

....

For all recipes, quantities are given in both metric and imperial measures and, where appropriate, measures are also given in standard cups and spoons. Follow one set, but not a mixture, because they are not interchangeable.

Size 3 (standard) eggs should be used unless otherwise stated.

CONTENTS

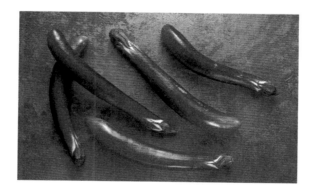

INTRODUCTION

VEGETABLES CAN PLAY A STARRING ROLE in a recipe or they may be combined with other ingredients in a harmony of flavours. Some of the best known vegetable soups are examples of well-tuned mixtures – minestrone, for instance, is a blend of carrots, tomatoes and beans with pasta; or a good old-fashioned vegetable soup brings together simple ingredients such as carrots, turnips and leeks with a grain, such as barley. These, and many other dishes, are loved for the sum of their parts, rather than for the taste of the different vegetables from which they have been made. In general, however, in this book you will find recipes that make the most of individual vegetable, so that the particular virtues of each one may be appreciated to the full.

The recipes in the second half of this book are an eclectic mix of classic dishes from around the world, together with others that have been devised to make the most of individual ingredients or combinations of ingredients. By far the majority of recipes in this book are designed with vegetarians in mind, and most of them are

vegetables that can delight our palates. So this book has been written not just for those converted to the virtues of vegetarianism; it is also intended to tempt "omnivores" to the vegetarian way of eating and to show that it is possible to produce a delicious and attractive meal, free from the tyranny of the "meat and two veg" way of menu planning.

THE VEGETARIAN DIET

The golden rule that nutritionists and doctors ask us to practise is to eat all foods in moderation and to eat a great variety of foods. This variety is especially important for vegetarians.

Because vegetarians eat more grains, vegetables, pulses and fruit in their diet than meat eaters, they seem to obtain a greater amount of dietary fibre or, as it is now known, non-starch polysaccharides (NSP). However, they have to be careful

problem for vegetarians if they are not aware of the sources from which it can be obtained. In addition, iron from vegetable sources cannot be utilized by the body unless there is vitamin C present in the same meal to act as a catalyst. But a small piece of fruit, fresh salad or even a good squeeze of lemon juice will soon redress that problem.

Vegans (those who exclude dairy products from their diet) need to make sure that they take in sufficient calcium, either in the form of calcium tablets or calcium-enriched soya milk. They may also need to supplement their diet with other vitamins and with minerals. A vegan diet can be just as healthy as a well-balanced omnivorous diet, just as long as followers are well informed about suitable foods.

THE VEGETARIAN LARDER

The basis of the vegetarian diet is, of course, formed by vegetables of all kinds, and on pages 14–139 we look at the enormous variety of vegetables that is available today and assess their nutritional value and ways of preparing them. First,

suitable for vegans. There is, however, a small section of "virtually vegetarian" recipes, which include seafood and fish.

The great thing about cooking with vegetables is that, once you have got the hang of using them, recipes become more or less unnecessary. As you experiment, perhaps substituting seasonal produce for the ingredients listed, you will discover how you like to enjoy carrots, asparagus or any of the less often seen and used

not to increase their intake of high-fat dairy products such as cheese, butter and cream. Just like meat eaters, they should watch their intake of these potentially high cholesterol products. Whenever possible, choose lower fat versions, which are usually well labelled. Changing from full-fat milk to skimmed or semi-skimmed milk helps, as does eating plenty of low-fat yogurt, fromage frais and skimmed milk soft cheeses.

Iron can also be a particular

however, we will review in nutritional terms the main ingredients that are used in the recipes in the second half of this book.

THE STARCHY FOODS

The total contribution of carbohydrate (starchy) foods and vegetables in a well-balanced diet may not be entirely understood, but the benefits and protective role they play in ensuring good health is widely recognized. Expert opinions agree that we should all eat a high proportion of vegetables in our everyday diets, more of complex carbohydrate foods, which is great news for vegetarians and all creative cooks, as these foodstuffs are all so versatile, nourishing and, best of all, cheap. They also store well without refrigeration and can be cooked with the minimum of preparation. Starchy foods have reasonable amounts of protein plus vitamins from the B group and minerals such as phosphorus, zinc, irons, potassium and, in the case of bread, added calcium. The other main advantage is that they are good sources of dietary fibre (NSP).

FLOURS

Have a selection of different flours to ring the changes. Often it is a good idea to mix two types together in baking for flavour and texture. Use half wholemeal and half plain white flours for a lighter brown pastry crust or bread loaf. Mix buckwheat flour with plain white for pancakes and so on. Flour is a good source of protein as well as complex starchy carbohydrate, and it is indispensable in cooking. Brown/wholemeal flours have a shorter shelf life than the more refined types of white flours. Remember, too, that self-raising flours, with their added raising agent, lose their lightening ability after about six months.

RICE

Top of the rice range is basmati, an elegant, fragrant, long-grain rice, grown in the foothills of the Himalayas. Traditionally eaten with curries, basmati is marvellous in almost all dishes, sweet and savoury, especially pilaffs. Brown

basmati is a lighter wholegrain rice with higher levels of dietary fibre. Thai rices are delicate and lightly sticky, and they are particularly good in stir-fries and wonderful in milk puddings. Wild rice (which is not a real rice but an aquatic grass) has good levels of proteins. Pre-soaking shortens the cooking time.

PASTA

The mainstay of many a cook in a hurry, pasta is, again, a good source of starchy complex carbohydrates, and it is available in a multitude of shapes, colours and now even flavours. Good pasta should be cooked to a tender texture but retain a firm bite, which the Italians call *al dente* ("to the tooth"). For this, choose pasta that is made with durum wheat or durum semolina. Cook pasta in

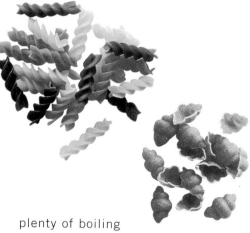

plenty of boiling salted water according to the instructions on the pack, then drain, rinse in cold water and shake lightly. Italians serve pasta slightly wet. Return it to the pan with some olive oil, seasoning and a grating of fresh nutmeg.

FATS AND OILS

For general frying, choose oils high in polyunsaturates. Sunflower, rape seed and groundnut have the lightest flavours, and these are the ones preferred by gourmet cooks.

Corn oil and blended vegetable oils are stronger in flavour. Increasingly popular is olive oil. Not only highly prized for flavour, it also has properties beneficial to health, being high in monosaturates, which are thought to help reduce blood cholesterol levels. Two main qualities

are available – pure olive oil, which is excellent for general cooking, and extra virgin oil, which is made from the first cold pressing of the olives, producing a full-flavoured, almost peppery taste. It is ideal for dressing salads and as a healthier substitute for butter. Aromatic seed and nut oils, such as sesame, walnut and hazelnut, are too heavy and expensive for general use, but they are ideal for trickling on hot vegetables, pulses or pasta.

All fats, unless labelled specifically "lower" or "reduced fat" contain approximately the same amount of calories. It is the type of fat within them that counts when it comes to our health. Sunflower and olive oil spreads are lower in harmful saturates and higher in healthier polyunsaturates and monosaturates. Spreads labelled "low fat" or "reduced fat" will have more added water, which helps to reduce the calories but almost makes them more difficult for frying and baking.

Fats and oils are important in our diets, contributing vital vitamins, such as A, D and E, so do not cut them out altogether. Include them in moderate amounts. Check pack labels for full nutritional details, and remember to restrict your fat intake to no more than one-third of your total daily intake of calories.

CHEESE

A popular high-protein food with vegetarians, cheese is also high in calories, having twice the number of many carbohydrate and protein foods. For fuller flavour choose well-matured varieties of cheese such as mature farmhouse Cheddar or fresh Parmesan – you will then not need to use as much.

For cooking, choose mature cheeses. Leave some full-flavoured cheeses unwrapped in the refrigerator to dry out: this concentrates the flavour and makes them go further when they are finely grated. Among the most useful cheese are fully mature Cheddar, fresh Parmesan, mature Gruyère and Pecorino (an Italian sheep's cheese). Lower fat soft cheeses and goats' cheeses are ideal for stirring into hot food to make an instant, tasty creamy sauce.

DAIRY PRODUCTS

Supermarket chiller cabinets carry a wide range of cultured dairy goods, which present many exciting opportunities for the home cook. Crème fraîche is a French-style soured cream, which does not curdle when boiled, so it is ideal stirred into hot dishes. It will also whip, adding a light piquancy to desserts. However, like double cream, it is quite high in fat (40 per cent) so use it sparingly.

Fromage frais is a smooth, lightly tangy, lower fat to virtually fat-free "cream", ideal for use in dressings, baked potatoes and desserts. Quark is a soft cheese made with skimmed milk and it is, therefore, very low in fat. It is a traditional cheese for using in cheesecakes,

but it is ideal for savoury dishes, too. Cream and cottage cheeses are long-time favourites, the latter now available in very low-fat versions for even healthier eating.

DAIRY-FREE PRODUCTS

The unassuming soy bean is one of the best sources of high vegetable protein foods. As such, it is ideal as a base for dairy-free milks, creams, fat spreads, ice creams and cheeses, making it perfect for vegans and those with dairy product allergies. Use these products in the same way as their dairy counterparts, although those

changing over will find that the soy products taste slightly sweeter.

Tofu, or bean curd, is made with soy milk and is particularly versatile in vegetarian cooking, both as a main ingredient in recipes or to add creamy, firm texture. On its own, tofu has little flavour, making it ideal to use as an absorber of other flavours, which is why it is so popular in Oriental cooking. Firm tofu or bean curd can be cut into cubes, marinated or smoked. It is very good fried in oil or grilled to a

crisp, golden crust. A softer set tofu, called silken tofu, is a good substitute for cream in cooking, and as such can be stirred into hot soups or used as a base for baked flans. Indeed, at any time when cream or milk is called for in a recipe, tofu can be used. Not only high in protein, tofu is a good source of vitamins of the B group and of iron, although, as a vegetable source of iron, you will need to serve some vitamin C at the same meal to utilize it.

Mycoprotein (sold under the brand name Quorn), is a new man-made food, which is a distant relation to the mushroom. Low in fat and calories, it is high in protein, with as much fibre as green vegetables. It cooks quickly, absorbing flavours as easily as soy bean curd, and it has a firmer texture. It is good for stir-frying, stews and casseroles.

NUTS AND SEEDS

Not only are they full of flavour, texture and colour, but nuts and seeds are great nutritional power packs. Like cheese, however, they can be high in fats as well as proteins, so go easy.

Cheapest and most versatile are peanuts, which are best bought unsalted or, even better, roasted unsalted. Almonds (blanched or flaked) are also very useful, as are walnuts, pine nuts, hazelnuts and the more expensive cashews. Often it is nice to mix two or three together. But nuts can go rancid if they are stored for too long (over about six months), so if you are not a regular user, buy in small quantities.

For maximum flavour, lightly roast nuts first before chopping or crushing. Use them as crisp coatings, too, but watch that they don't burn.

There is an increasing range of colourful and exciting seeds now in wholefood stores. Most useful are sunflower and sesame seeds, while pumpkin and melon seeds are most attractive scattered into salads or simply nibbled as a snack. Seeds for attractive garnish as well as flavour include poppy seeds, black mustard, fenugreek and caraway seeds. Most of all, nuts and seeds look simply stunning lined up in clean storage jars on kitchen shelves, tempting you to toss them into a whole variety of dishes, hot and cold.

HERBS

Wherever possible, try to use fresh herbs. There are many that will grow easily and obligingly in pots and small back gardens, as well as surviving winter cold. Good candidates include shrubby rosemary, thyme, bay and sage. Even chives and marjoram can survive well into late autumn and return obligingly in early spring. However, more and more food shops are selling packs and bunches of fresh herbs, which are grown commercially.

Ethnic shops and stores are good sources of a wide variety of unusual and good flavoured herbs. Most useful are flat-leaved parsley, coriander, dill, basil, chives and mint. Do not use one herb per dish: mix and match, experimenting with

different combinations. Although you should not skimp on herbs, the more pungent ones, such as tarragon, rosemary and sage, still need a cautious touch.

Handfuls of fresh, leafy herbs are wonderfully exciting additions to green salads. There is no need to chop them finely. Fill a mug with washed sprigs of herbs, then snip them roughly with scissors. Store leafy herbs loosely in polythene food bags in the fridge, spraying them lightly with water if they look limp. They spring back almost magically.

If you use dried herbs, buy small amounts and store them in a dry,

cool cupboard so that they retain their flavour. Replace dried herbs regularly, as they soon lose their colour and flavour, and end up tasting rather like dried grass.

SPICES

The vegetarian's best friend! Warm, aromatic, colourful and easy to use, spices can make the simplest dish

supreme. It is a misconception to think that spices are pungently hot. Most are not; it is really only those of the chilli family, including cayenne, that are. Some spices are useful for both sweet and savoury dishes, such as nutmeg, cinnamon, mace, cloves, cardamom and ginger. Others, such as fenugreek, turmeric, paprika, cumin, coriander berries and chilli, are used for

savoury dishes.

Spices are made for experimenting. Gradually you will learn which are the most pungent and to your liking. Some have glorious colours, like turmeric and paprika; others have a distinctive flavour. Add them in cautious pinches at first, until you decide what you most enjoy.

Spices are best roasted or fired first, to bring out the aromatic oils. This can be done either in a hot oven or a frying pan. Where possible, use the seeds or grains of spices first, and grind them down in a small electric spice mill or a pestle and mortar. Saffron (a most expensive spice) is best soaked briefly in a little warm water or milk to bring out its true flavour and pretty colour.

And don't keep spices just for ethnic and exotic dishes. Add them to home-cooked favourites – try macaroni cheese with paprika and cumin or fried eggs sprinkled with black mustard seeds.

The Vegetables

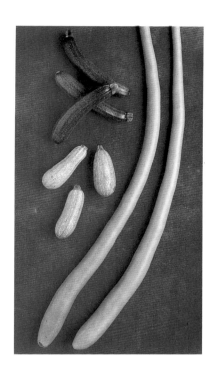

Plants have formed an essential part of man's diet since the earliest times. Archaeological evidence indicates that forms of wheat and barley, believed to be the first crops, were farmed in the Middle East as early as 8000BC. Many edible plants are native to this area, so it can be assumed that vegetables were also eaten, if only to relieve what must have been a fairly monotonous diet.

Beans and peas were among the first vegetables to be farmed in Thessaly and Macedonia, and these legumes would have been enormously important to early societies because they grew easily to provide seeds, which were a starchy food, high in protein, that dried well for long storage.

Many of our familiar vegetables were cultivated in historic times: the Egyptians grew onions, garlic,

radishes, lettuce and broad beans; the Greeks and Romans farmed produce that was native to their own countries, and they also discovered a wide range of plants through their contacts with other cultures. Not only did the Romans discover the fruits of other lands, but their expansion introduced ingredients to the countries they conquered. At the beginning of the first century AD, for example, beans, peas, leeks, parsnips and turnips were widely grown in Britain. By the Middle Ages a wealth of vegetables was available, and recipes for them were recorded in the first cookery books.

Early explorers displayed exotic ingredients, the spoils of their

Left: Italian Roast Peppers.
Above: Roast Asparagus Crêpes.

opportunity to buy almost anything we want, when we want it. This is a luxury we have come to expect, and we are generally ready to pay for international variety all year round. Gluts of vegetables, once common seasonal occurrences when food was cultivated and marketed on a local scale, are not features of modern food stores, although the availability of home-grown produce, of course, is still subject to the seasons. Therefore, in addition to taking advantage of the fantastic multi-national displays we find in most large stores, we should also seek local growers and try to buy freshly harvested produce.

We should learn to enjoy again summer vegetables during the season to which they traditionally belong and to savour winter produce in dishes like wholesome stews and broths, which are so well suited to cooler weather.

Although the nutritional value of vegetables varies according to type, freshness, preparation and cooking method, they are a main source of many vitamins, especially vitamin C. Some of the B-group vitamins are also found in vegetables, particularly in green vegetables and pulses. Carrots and dark green vegetables also contain carotene, which is used by the body to manufacture vitamin A. Vegetable oils are a useful source of vitamin E. Vegetables also contain calcium, iron, potassium and magnesium, as well as some trace elements, which we all need in small quantities.

Starchy vegetables are an important source of energy-giving carbohydrate, and they may include useful quantities of fibre.

travels, in their native countries and created a huge appetite for new flavours among the wealthy classes of Europe. Marco Polo travelled to China and carried aromatic spices on his return to Europe. Christopher Columbus and subsequent explorers found potatoes, tomatoes, peppers, squashes and maize in the New World. Such produce received a somewhat lukewarm reception

when it first appeared – potatoes and tomatoes, for example, were viewed with grave suspicion – but today there is great interest in vegetables cultivated all over the world.

In contrast to most store-cupboard ingredients, the great characteristic of many vegetables is that they still have natural seasons. When we go into supermarkets we have the

Above: Spinach.

Potatoes are beginning to enjoy something of a culinary renaissance as more cooks realize that choosing the right variety for a dish is one secret of success. Also, more potatoes are being grown for flavour and are often delicious served simply boiled in their skins and lightly dressed with a little olive oil or butter. Choose small, waxy, firm potatoes for salads, larger firm ones for roasting and floury varieties for mashing. More and more producers are printing suitable uses on the bags, so check these first.

In a vegetarian diet, ingredients such as pulses, beans and sprouting seeds make a valuable contribution to the overall intake of protein. It has been estimated that over half the world's main source of protein comes from pulses in one form or another. However, although they are high in protein, pulses are not complete in all amino acids. In particular, they lack one of the amino acids called methionine. Grain foods, on the other hand, lack lysine and tryptophan, which pulses do have. But, put them together and you have completed the usual protein circle. So, when you eat any of the pulses, try to include starches in the same meal, such as lentils with rice, hummus with bread, beans with pasta, and so on. In addition, include some fresh vitamin C in the meal (from fruits or leafy vegetables) so that your body can utilize the iron in the grains and pulses.

The variety of pulses is exciting and seemingly endless. Dried pulses benefit from soaking, preferably overnight. Older pulses may need longer. To shorten soaking time, cover with boiling water and leave for 2 hours. Drain and boil in fresh water. Boil pulses fast for the first 10 minutes of cooking to destroy any potential mild toxins present. Then lower the heat and gently simmer. Do not add salt or lemon juice during cooking, because this toughens the skins, although fresh herbs and onion slices add flavour. Like pasta, don't overdrain. Leave wet, season and perhaps dress with extra virgin olive oil.

Certain lentils can be cooked without pre-soaking. The small split red lentils (or masoor dhal) are marvellous for sprinkling as thickeners for soups and stews, and they take just 20 minutes to cook. Beans with a good creamy texture, which is perfect for soups, pâtés and purées, are butter beans, kidney beans (red or black), cannellini, haricots, borlotti, pinto beans and flageolets. Split peas and red lentils make marvellous dips. Chick-peas and aduki beans hold their texture well and make a good base for burgers and stews.

Vegetables have the highest nutritional content when they are freshly picked. The vitamin content diminishes with staleness and exposure to sunlight. Use fresh vegetables as soon as possible after purchase, and always avoid stale, limp specimens. The peel and the layer directly beneath it contain a high concentration of nutrients, so it is best to avoid peeling vegetables or, when this is necessary, to

Above: Tomato and Basil Tart.
Left: Red, orange and green peppers.

remove the thinnest possible layer for maximum nutrient retention.

Minerals and vitamins C and B are water soluble, and they are lost by seepage into cooking water or the liquid over which vegetables are steamed. To minimize loss of nutrients, do not cut up vegetables finely, because this creates a greater surface area for seepage. Vitamin C is also destroyed by long cooking and exposure to alkalines.

Raw and lightly cooked vegetables provide the best nutritional value and source of fibre. Any cooking liquid should, whenever possible, be used in stocks, gravies or sauces.

— ONE —

ONIONS
AND
LEEKS

Onions

Shallots

Chives

Garlic

Leeks

ONIONS

There are bound to be vegetables you like better than others but a cook would be lost without onions. There are many classic recipes specifically for onion dishes so they can be appreciated in their own right. Onion tarts or French onion soup, for instance, have a sublime flavour and only onions are appropriate. But also, there is hardly a recipe where onions, or their cousins – garlic, leeks or shallots – are not used. Gently fried until soft, or fried more fiercely until golden brown, they add a unique, savoury flavour to dishes.

History

Onions, along with shallots, leeks, chives and garlic, belong to the *Allium* family which, including wild varieties, has some 325 members. All have the characteristic onion smell which is caused by volatile acids beneath the skin.

Archaeological and historical records show that onions have been eaten for thousands of years. They are believed to have originally come from the Middle East and their easy cultivation suggests that their use spread quickly. There are references to the onion in the Bible and it was widely eaten in Egypt. There was, we are told, an inscription on the Great Pyramid stating that the slaves who built the tomb ate their way through 1,600 talents worth of onions, radishes and garlic – presumably a lot, given that the Great Pyramid was made using more than two million 2¹/₂-tonne blocks of stone.

By the Middle Ages, onions were a common vegetable throughout Europe and would have been used in soups, stews and sauces when strong flavouring was preferred.

Varieties

As they keep well in a cool place, most people keep a handy stock of onions, usually a general purpose type that can be sautéed or browned. However, onions come in a variety of different colours and strengths, and for certain recipes particular onions are needed.

Right: Spanish onions.
Far right top: Yellow onions.
Far right below: Red onions.

Spanish Onions: Onions raised in warm areas are milder in taste than onions from cooler regions, and Spanish onions are among the mildest cultivated onions. They are a beautiful pale copper colour and are noticeably larger than yellow onions. They have a delicate, sweet flavour which makes them ideal for serving raw in salads, thinly sliced, while their size makes them suitable for stuffing and baking whole.

Yellow Onions: These are the widely available onions you find everywhere and, though called yellow onions, their skins are more golden brown. They are the most pungent of all the onions and are a good, all-purpose variety. The smallest ones, referred to as baby, button or pickling onions, are excellent for pickling but can also be added whole to a casserole or sautéed in butter to make a delicious vegetable accompaniment.

Red Onions: Sometimes called Italian onions, these mild onions have an attractive appearance and are now widely available from most good greengrocers and supermarkets. Below their ruby red skins the flesh is blushed with red. They have a mild, sweet flavour and are excellent thinly sliced and used raw in salads and *antipasti* dishes.

White Onions: These come in all sorts of interesting shapes and sizes – squat, round and oval, big and small. The very small white onions, with shimmery silver skins, are mild and best added whole to stews or served in a creamy sauce. Larger white onions can be mild or strong – there is no way of telling. Like yellow onions, white onions are extremely versatile whether used raw or cooked. The very small white onions, called Paris Silverskin, are the ones used for dry martinis and for commercial pickling.

Vidalia Onions: These popular American onions are a speciality of and named after a town in Georgia, USA. They are a large, pale yellow onion and are deliciously sweet and juicy. Used in salads, or roasted with meat or with other vegetables, they are superb.

Bermuda Onions: These are similar in size to Spanish onions but are rather more squat. They have a mild flavour and are good thinly sliced, fried until golden and served with steaks or burgers.

remove the next layer of onion, as it is often dry or damaged. Unless slicing onions for stir-fries, for which it is customary to slice the onion into wedges, always slice the onion through the rings, widthways. Make whole rings, or for half-rings, cut in half lengthways through the root before slicing (*below*). For finely chopped onion, slice again lengthways.

Spring Onions or Scallions: These are also true onions but harvested very young while their shoots are still green and fresh. They have a mild, delicate taste and both the small white bulb and the green tops can be used in salads, omelettes and stir-fries, or indeed any dish which requires a mild onion flavour.

Nutrition

As well as tasting good, onions are good for you. They contain vitamins B and C together with calcium, iron and potassium. Like garlic, they also contain cycloallin, an anticoagulant which helps protect against heart disease.

Buying and Storing

It used to be a common sight to see an onion seller travelling around the streets on a bicycle with strings of onions hanging from every available support including his own neck.

Nowadays strings of onions are hard to come by although, if you do find them in shops, they are a good way of buying and storing the vegetable.

Onions, more than almost any other vegetable, keep well provided they are stored in a cool, dry place, such as a larder or outhouse. Do not store them in the fridge as they will go soft, and never keep cut onions in the fridge – or anywhere else – unless you want onion-flavoured milk and an onion-scented home. Onions do not keep well once cut and it is worth buying onions in assorted sizes so that you do not end up having bits left over. Unused bits of onion can be added to stocks; otherwise throw them away.

Preparing

Onions contain a substance which is released when they are cut and causes the eyes to water, quite painfully sometimes. There are all sorts of ways which are supposed to prevent this, including cutting onions under running water, holding a piece of bread between your teeth or wearing goggles!

As well as the outer brown leaves,

Cooking

The volatile acids in onions are driven off during cooking, which is why cooked onion is never as strong as raw onion. The method of cooking, even the way of frying an onion, affects its eventual taste. Boiled onion or chopped onion added neat to soups or casseroles has a stronger, more raw taste. Frying or sautéing briefly, or sweating (frying in a little fat with the lid on) until soft and translucent gives a mild flavour. When fried until golden brown, onions develop a distinct flavour, both sweet and savoury, that is superb in curries and with grilled meats and is essential for French onion soup.

Above far left: Vidalia onions.
Left: White onions.
Above left: Large and small Spring onions.

SHALLOTS

Shallots are not baby onions but a separate member of the onion family. They have a delicate flavour, less intense than most onions and they also dissolve easily into liquids, which is why they are favoured for sauces. Shallots grow in small, tight clusters so that when you break one open there may be two or three bunched together at the root.

Their size makes them convenient for a recipe where only a little onion is required. Use shallots when only a small amount of onion is needed or when only a fine onion flavour is required. Shallots are a pleasant, if maybe extravagant, alternative to onions, but where recipes specify shallots (especially sauce recipes), they should be used if possible.

Although classic cookery frequently calls for particular ingredients, the art of improvisation should not be ignored. For instance, Coq au Vin is traditionally made with walnut-size white onions, but when substituted with shallots, the result is delightful.

History

Shallots are probably as ancient as onions. Roman commentators wrote eloquently about the excellence of shallots in sauces.

Varieties

Shallots are small slender onions with long necks and golden, copper-coloured skins. There are a number of varieties, although there is unlikely to be a choice in the supermarkets. In any case, differences are more in size and colour of skin than in flavour.

Buying and Storing

Like onions, shallots should be firm without any green shoots. They will keep well for several months in a cool dry place.

Preparing and Cooking

Skin shallots in the same way as onions, i.e. top and tail them and then peel off the outer skin. Pull apart the bulbs. Slice them carefully and thinly using a sharp knife – shallots are so small, it is easy to slip and cut yourself. When cooking them whole, fry over a very gentle heat without browning too much.

CHIVES

Chives: In culinary terms, chives are really classed as a herb, but as members of the onion family they are worth mentioning here. As anyone who has grown them knows, chives are tufts of aromatic grass with pretty pale lilac flowers, which are also edible.

Preparing and Serving

Chives are often snipped with scissors and added to egg dishes, or used as a garnish for salads and soups, adding a pleasant but faint onion flavour. Along with parsley, tarragon and chervil, they are an essential ingredient of *fines herbes*.

Chives are also a delicious addition to soft cheeses – far nicer than commercially bought cheeses, where the flavour of chives virtually disappears. Stir also into soft butter for an alternative to garlic butter. This can then be spread on to bread and baked like garlic bread.

If adding to cooked dishes, cook only very briefly, otherwise their flavour will be lost.

Chinese Chives: Chinese chives, sometimes called garlic chives, have a delicate garlic flavour, and if you see them for sale in your local Chinese supermarket, they are worth buying as they add a delicate onion flavour to stir-fries and other oriental dishes.

Preparing and Serving

Use them as you would chives - both the green and white parts are edible. They are also delicious served on their own as a vegetable accompaniment.

Buying and Storing

For both types of chives, look for plump, uniformly green specimens with no brown spots or signs of wilting. They can be stored for up to a week in the fridge. Unopened flowers on Chinese chives are an indication that the plant is young and therefore more tender than one with fully opened flowers.

Above left: Chives.
Above right: Chinese chives.

GARLIC

Garlic is an ingredient that almost anyone who does any cooking at all, and absolutely everyone who enjoys cooking, would not be without.

History

Garlic is known to have been first grown in around 3200 BC. Inscriptions and models of garlic found in the pyramids of ancient Egypt testify to the fact that garlic was not only an important foodstuff but that it had ceremonial significance as well. The Greeks and Romans likewise believed garlic to have magical qualities. Warriors would eat it for strength before going into battle, gods were appeased with gifts of garlic, and cloves of garlic were fastened round the necks of babies to ward off evil. Hence, vampire mythology has ancient precedents.

The Greeks and Romans also used garlic for its therapeutic qualities. Not only was it thought to be an aphrodisiac but also it was believed to be good for eczema, toothache and snake bites.

Although garlic found its way all over Europe – vats of butter, strongly flavoured with garlic, have been found by archaeologists working in Ireland which date back 200-300 years – fundamentally, its popularity today derives from our liking for Mediterranean, Indian and Asian food, in which garlic plays a very important part.

Nutrition

As is often the case, what was once dismissed as an old wives' tale is, after thorough scientific inquiry, found to be true. Garlic is a case in point; most authorities accept that it has many therapeutic properties. The most significant of these is that it lowers blood cholesterol, thus helping prevent heart disease. Also, raw garlic contains a powerful antibiotic and there is evidence that it has a beneficial effect against cancer and strokes, and increases the absorption of vitamins. Many garlic enthusiasts take their garlic in tablet form, but true devotees prefer to take it as it comes.

Right: A string of pink-skinned garlic.

Varieties

There are numerous varieties of garlic, from the large "elephant" garlic, to small tight bulbs. Their papery skin can be white, pink or purple. Colour makes no difference to taste but the particular attraction of the large purple bulbs is that they make a beautiful display in the kitchen.

As a general rule, the smaller the garlic bulb, the stronger it is likely to be. However, most garlic sold in shops is not classified in either shape or form (unless it is elephant garlic) and in practice you will simply pick up whatever you need, either loose, in bunches or on strings.

Garlic grown in a hot climate is likely to be the most pungent, and fresh new season's garlic has a subtle, mild flavour that is particularly good if it is to be used raw, for example, in salads and for dressings.

Above: Elephant garlic beside normal-size bulbs.

Buying and Storing

Garlic bulbs should be firm and round with clear, papery skins. Avoid any that are beginning to sprout. Garlic bulbs keep well stored in a cool, dry place; if the air is damp they will sprout and if it is too warm the cloves will eventually turn to grey powder.

Preparing and Cooking

First break the garlic bulb into cloves and then remove the papery skin. You can blanch this off with hot water but using a fingernail or knife is just as effective. When a garlic clove is split lengthways a shoot is revealed in the centre, which is occasionally green, and some people remove this whatever the colour. Cloves are the little segments which make up the bulb and most recipes call for one or more cloves of garlic. (Don't use a bulb when you just need a clove!)

Crush cloves either with the blade of a knife or use a garlic crusher. Crushed garlic cooks more evenly and distributes its flavour in food better than when it is used sliced or finely chopped (stir-fries are the exception). Prepare garlic according to the strength of flavour required: thinly sliced garlic is milder than chopped, which in turn is milder than crushed garlic and, of course, cooking mutes the pungency.

Garlic Breath

The taste and smell of garlic tends to linger on the breath and can be a problem to get rid of. Chewing parsley is a well-known remedy but is only moderately successful. Chewing the seeds of cardamom pods is also said to work but is rather unpleasant. The best suggestion is to eat garlic with your friends so that nobody notices!

LEEKS

Leeks are very versatile, having their own distinct, subtle flavour. They are excellent in pies and casseroles with other ingredients, braised in cream and served by themselves, or simmered in butter as an accompanying vegetable.

Leeks are also wonderful in soups and broths and have rightly earned the title, "king of the soup onions". Cock-a-leekie from Scotland and *Crème Vichyssoise*, invented by the chef of New York's Ritz-Carlton, are two classic leek soups, but many other soups call for leeks.

History

Leeks, like onions and garlic, have a long history. They were grown widely in ancient Egypt and were also eaten and enjoyed throughout the Greek and Roman period. In England, there is evidence that leeks were enjoyed during the Dark Ages. There is little mention of them during the Middle Ages, and history suggests that between the sixteenth and eighteenth centuries eating leeks was not considered fashionable.

However, while they may not have enjoyed a good reputation among the notoriously fickle aristocracy, the rural communities probably continued to eat leeks. They grow in all sorts of climates and are substantial enough to make a reasonable meal for a poor family. It was probably during this time that they were dubbed "poor man's asparagus" – a name which says more about people's snobbery about food than it does about leeks.

Many place names in England, such as Leckhampstead and Leighton Buzzard are derived from the word leek and, of course, the leek has been a national emblem of Wales for hundreds of years.

Varieties

There are many different varieties of leeks but among them there is little difference in flavour. Commercially grown leeks tend to be about 25cm/10in long and about 2cm/3⁄4in in diameter. Leeks nurtured in home gardens can be left to grow to an enormous size but these may develop a woody centre.

the first layer of white; then cut a slit from one end to the other through to the centre of the leek *(below)*. Wash under cold running water, pulling the sections apart so that the water rinses out any stubborn pieces of earth. If you slice the leeks – either slice thickly or thinly – place them in a colander and rinse thoroughly under cold water.

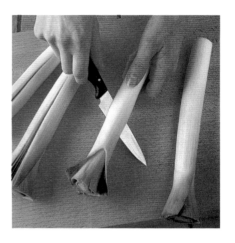

RAMP

Among the many wild onions and leeks, the Canadian ramp is perhaps the best known. Also called the wild leek, it looks a little like a spring onion, but has a stronger and more assertive garlic-onion flavour. Choose unblemished, clear white specimens with bright, fresh leaves and keep in a cool place, wrapped in a plastic bag to store.

Prepare and cook as you would spring onions, by trimming the root end and then slicing thinly. Use in cooking or in salads but remember the onion flavour is stronger, so use sparingly.

Buying and Storing

Buy leeks which look fresh and healthy. The white part should be firm and unblemished and the leaves green and lively. As leeks do not keep particularly well, it is best to buy them as and when you need them. If you need to store them, trim away the top of the leaves and keep them in the salad drawer of the fridge or in a cool place. After several days they will begin to shrivel.

Preparing

It is important to wash leeks thoroughly before cooking as earth and grit lodges itself between the white sections at the base. To prepare leeks, cut away the flags (leaves) and trim the base. Unless the leek is extremely fresh or home grown, you will probably have to remove

Cooking

Leeks can be steamed or boiled and then added to your recipe, or fry sliced leeks gently in butter for a minute or so and then cover with a lid to sweat so they cook without browning. Unlike onions, leeks shouldn't be allowed to brown; they become tough and unappetizing. They can be stir-fried, however, with a little garlic and ginger. If they begin to cook too fiercely, splash in a little stock and soy sauce and simmer until tender.

Left: Leeks.
Above: Ramp.

SHOOTS
AND
STEMS

ASPARAGUS

Asparagus is definitely a luxury vegetable. Its price, even in season, sets it apart from cabbages and cauliflowers, and it has a taste of luxury too. The spears, especially the thick, green spears, at their best in early summer, have an intense, rich flavour that is impossible to describe but easy to remember. If the gods eat, they will eat asparagus – served simply with a good Hollandaise!

History

The ancient Greeks enjoyed wild asparagus but it was not until the Roman period that we know it was cultivated. Even then asparagus was highly thought of; it is recorded that Julius Caesar liked to eat it with melted butter. There is little mention of asparagus being eaten in England until the seventeenth century. Mrs Beeton has 14 recipes for asparagus and from the prices quoted in her cookbook it is apparent that it was expensive even in Victorian times.

Nutrition

Asparagus provides vitamins A, B2 and C and is also a good source of potassium, iron and calcium. It is a well-known diuretic.

Varieties

There are many varieties of asparagus and many different ways of raising it too. Spanish and some Dutch asparagus is white with ivory tips; it is grown under mounds of soil and cut just as the tips begin to show. The purple variety is mostly grown in France, where the spears are cut once the tips are about 4cm/1 1/2 in above the ground. Consequently, the stalks are white and the tops tinged with green or purple. In contrast, English and American asparagus grows above the ground and the spears are entirely green. Arguments continue over which has the better flavour, most growers expressing a preference for their own asparagus!

Thin, short asparagus are called sprue and are excellent when briefly steamed or stir-fried and added to salads. In Italy, they are served by themselves, scattered with grated Parmesan cheese.

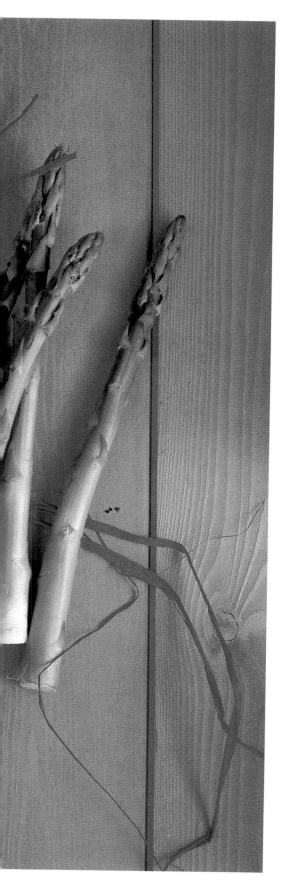

Preparing

Unless the asparagus comes straight from the garden, cut off the bottom of the stalk as it is usually hard and woody. If the bottom parts of the stem also feel hard, pare this away with a potato peeler *(below)*. However, if the asparagus is very fresh, this is not necessary, and sprue rarely needs trimming at all.

Buying and Storing

Asparagus has a relatively short growing season from late spring to early summer Nowadays, it is available in the shops almost all year round but outside the season it will have been imported. It is still good, but it is expensive and will not have the flavour of home-produced asparagus, since it starts to lose its flavour once it is cut.

When buying asparagus, the tips should be tightly furled and fresh looking, and the stalks fresh and straight. If the stalks are badly scarred or droopy, it indicates that they have been hanging around for too long and it is not worth buying. Asparagus will keep for several days if necessary. Untie the bundles and store in the salad drawer of the fridge.

Cooking

The problem with cooking asparagus is that the stalks take longer to cook than the tender tips, which need to be only briefly steamed. Ideally, use an asparagus kettle. Place the asparagus spears with the tips upwards in the wire basket and then lower into a little boiling salted water in the kettle. Cover and cook until the stems are tender.

Alternatively, if you don't have an asparagus kettle, place the bundle upright in a deep saucepan of boiling salted water. (The bundle can be wedged into place with potatoes.) Cover with a dome of foil and cook for 5-10 minutes or until the spears are tender. The cooking time depends largely on the thickness of the spears but take care not to overcook; the spears should still have a "bite" to them.

Asparagus can also be roasted in a little olive oil. This cooking method intensifies the flavour and is gratifyingly simple. Serve with just a sprinkling of sea salt – it's quite delicious! If steaming asparagus, serve simply with melted butter, which perfectly complements the luxury of the vegetable.

Left: Asparagus.
Above: White asparagus.

GLOBE ARTICHOKES

Globe artichokes have an exquisite flavour and are a very sociable food to eat. They grow in abundance in Brittany, and during July and August farmers can frequently be seen selling them by the roadside. The globes are huge hearty specimens and are extremely fresh, so they make a good buy.

History

It is not known for certain whether artichokes were eaten in antiquity. Although they are mentioned by writers, they could have been referring to the cardoon, which is the uncultivated form of artichoke. Cardoons grew wild in many southern European countries, and, as far as we know, cultivated artichokes first became a popular food in Italy. However, Goethe did not share the Italians' liking for the vegetable and remarks in his book, *Travels Through Italy*, that "the peasants eat thistles", something he didn't care for at all.

Nowadays, artichokes are grown all over southern Europe and in California. People in Italy, France and Spain eat artichokes while the vegetable is still young, before the choke has formed and the entire artichoke is edible. Unfortunately, such young delicacies are not exported but look out for them if you are in these countries.

Buying and Storing

It is only worth buying artichokes when they are in season, although they are available in supermarkets almost all year round. In winter, however, they are sad looking specimens, small and rather dry, and are really not worth the bother of cooking. At their best, artichokes should be lively looking with a good bloom on their leaves, the inner leaves wrapped tightly round the choke and heart inside. Artichokes will keep for 2-3 days in the salad drawer of the fridge but are best eaten as soon as possible.

Preparing and Cooking

First twist off the stalk which should also remove some of the fibres at the base and then cut the base flat and pull away any small base leaves. If the leaves are very spiky, trim them with a pair of scissors if liked *(above)*, then rinse under running water. Cook in boiling water, acidulated with the juice of half a lemon. Large artichokes need to be simmered for 30-40 minutes until tender. To test if they are done, pull off one of the outer leaves. It should come away easily and the base of the leaf should be tender.

prickly choke and discard, to reveal the heart. Eat the heart with a knife and fork, dipping it in the garlic butter or vinaigrette.

CARDOONS

This impressively large vegetable is closely related to the globe artichoke and has a superb flavour, a cross between artichokes and asparagus. Cultivated plants frequently grow to 2 metres/6 feet in height, and once mature, cardoons, like celery, are blanched as they grow. This process involves wrapping the stalks with newspaper and black bags for several weeks, so that when harvested, in late autumn, before the frosts, the stalks are a pale green.

The cardoon is a popular vegetable in southern Europe but less commonly available elsewhere. In Spain, for instance, it is much appreciated and often appears on the table, poached and served with chestnuts or walnuts. Only the inner ribs and heart are used.

Artichokes and Drink

Artichokes contain a chemical called cynarin, which in many people (although surprisingly not all) affects the taste buds by enhancing sweet flavours. Among other things, this will spoil the taste of wine. Consequently, don't waste good wine with artichokes but drink iced water instead, which should taste pleasantly sweet.

Eating Artichokes

Artichokes are fun to eat. They have to be eaten with fingers, which does away with any pomp and ceremony, always a handicap for a good dinner party.
Serve one artichoke between two, so that people can share the fun of pulling off the leaves and dipping them into garlic butter or vinaigrette. If you want to serve one each, serve them in succession. The dipping sauces are an essential part of eating artichokes; people can either spoon a little on to their plates or give everyone a little bowl each. After dipping, draw the leaf through your teeth, eating the fleshy part and piling the

remains of the leaf on your plate. When most of the leaves have been eaten, a few thin pointed leaves remain in the centre, which can be pulled off altogether. Then pull or cut away the fine

Far left: Globe artichokes.
Top: Baby globe artichokes.
Above: Cardoons.

CELERY

Some people say that the very act of eating celery has a slimming effect because chewing it uses up more calories than the vegetable itself contains. Although it may be insubstantial, celery nevertheless has a distinct and individual flavour, sharp and savoury, which makes it an excellent flavouring for soups and stuffings, as well as good on its own or in salads. The astringent flavour and crunchy texture of celery contrasts well with the other ingredients in salads such as Waldorf salad or walnut and avocado salad.

History

Celery has been eaten in this country for several hundred years, having been introduced from Italy where it was commonly eaten in salads.

Nutrition

Celery is very low in calories but contains potassium and calcium.

Varieties

Most greengrocers and supermarkets, depending on the time of year, sell both green and white celery and you would be excused for not knowing the difference. When celery is allowed to grow naturally, the stalks are green. However, by banking up earth against the shoots celery is blanched: the stalks are protected from sunlight and remain pale and white. Consequently, white celery is often "dirty" – covered loosely in soil – while green celery will always be clean. White celery, which is frost hardy, is only available in winter. It is more tender and less bitter than green celery and is generally considered superior. Celery is therefore thought of as a winter vegetable and is traditionally used at Christmas time, for stuffing and as a sauce to go with turkey or ham.

Buying and Storing

White celery is in season during the winter months. If possible, buy "dirty" celery which hasn't been washed. It has a better flavour than the pristine but rather bland supermarket variety. Look for celery with green fresh-looking leaves and straight stems. If the leaves or any

outer stalks are missing, it is likely to be rather old, so worth avoiding.

Celery will keep for several days in the salad drawer of the refrigerator. Limp celery can be revived by wrapping it in absorbent paper and standing it in a jar of water.

Preparing

Wash if necessary and pull the stalks apart, trimming the base with a sharp knife. Cut into thick or thin slices according to the recipe. When served raw and whole, the coarse outer "strings" should be removed from each stalk by pulling them up from the base.

Cooking and Serving

Serve celery raw and finely sliced in salads, mixed with cream cheese or soured cream. Braised celery is tasty, either whole or sliced. Celery has a distinctive, savoury, astringent flavour so is excellent in soups or stuffings.

CELERIAC

Strictly speaking, celeriac is a root vegetable, being the root of certain kinds of celery. It is knobbly with a patchy brown/white skin and has a similar but less pronounced flavour than celery. Grated and eaten raw, it has a crunchy texture, but when cooked it is more akin to potatoes. Thin slices of potato and celeriac cooked *au gratin* with cream is a popular way of serving this vegetable.

Buying and Preparing

If possible, buy smallish bulbs of celeriac. The flesh discolours when exposed to light, so as soon as you have peeled, sliced, diced or grated the celeriac, plunge it into a bowl of acidulated water (water with lemon juice added).

Cooking

Celeriac can be used in soups and broths, or can be diced and boiled and eaten in potato salads.

Left: Green celery.
Above: White celery.
Right: Celeriac.

FIDDLEHEAD FERN

Sometimes called the ostrich fern, these shoots are a rich green colour and are normally about 5cm/2in long. They have an unusual flavour, something like a cross between asparagus and okra, and have a slightly chewy texture, which makes them a popular choice for oriental dishes.

Preparing and Cooking

To prepare and cook, trim the ends and then steam or simmer in a little water or sauté in butter until tender. Use in salads or serve as a first course with a Hollandaise sauce.

Right: Fiddlehead ferns.
Below left: Alfalfa sprouts.
Below right: Mung bean sprouts.

ORIENTAL SHOOTS

BAMBOO SHOOTS

In the Far East, edible bamboo shoots are sold fresh in the market. The young shoots are stripped of their brown outer skins and the insides are then eaten. Although fresh bamboo shoots can occasionally be found in oriental stores, the most readily available variety is sold in cans. The flavour is undoubtedly spoilt. Fresh bamboo shoots have a mild but distinct taste, faintly reminiscent of globe artichokes, while canned ones really taste of nothing at all. However, the texture, which, in Chinese cuisines particularly, is as important as the flavour, is not so impaired, and bamboo shoots have a pleasantly crunchy bite.

Preparing and Cooking

Peel away the outer skin and then cook in boiling water for about half an hour. They should feel firm, but not "rock" hard. Once cooked, slice thinly and serve by themselves as a side dish, with garlic butter or a sauce, or add to stir-fries, spring rolls or any oriental dish

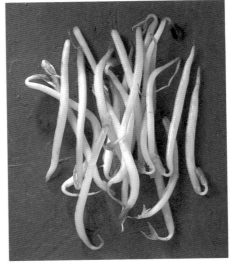

where you need a contrast of textures. Since canned bamboo shoots have been preserved in brine, always rinse well before using.

BEAN SPROUTS

Bean sprouts are rather a neglected vegetable, used almost carelessly for orien-

tal dishes but otherwise passed by as being insipid and not very interesting. It's a reputation they don't deserve: not only do they have a lovely fresh flavour, but they are also good for you.

All sorts of seeds can be sprouted, but the bean family are favourites among the sprouted vegetables. The bean sprouts most commonly available in the shops are sprouted mung beans, but aduki beans, alfalfa, lentils and soybeans can all be sprouted and taste delicious.

Nutrition

Beans sprouts contain a significant amount of protein, Vitamin C and many of the B vitamins. They have an excellent flavour too, best appreciated eaten raw in salads or sandwiches, and for slimmers they are an ideal food, low in calories, yet with sufficient substance to be filling, and with a flavour and texture that can be enjoyed without a dressing.

Buying and Storing

Bean sprouts should only be bought when absolutely fresh. They don't keep for long and they will taste sour if past their best. The sprouts should be firm, not limp, and and the tips should be green or yellow; avoid any that are beginning to turn brown.

Cooking

If stir-frying, add the bean sprouts at the last minute so they cook for the minimum period to keep plenty of crunch and retain their nutritional value. Most health food shops will have instructions on sprouting your own beans. Only buy seeds intended for sprouting.

PALM HEARTS

Fresh palm hearts are the buds of cabbage palm trees and are considered a delicacy in many parts of the world. They are available canned from oriental stores, but are most prized when fresh. These should be blanched before being cooked to eliminate any bitterness. They can be braised or sautéed and then served hot with a Hollandaise sauce, or cold with a simple vinaigrette.

WATER CHESTNUTS

Water chestnut is the common name for a number of aquatic herbs and their nut-like fruit, the best known and most popular variety being the Chinese water chestnut, sometimes known as the Chinese sedge. In China they are grown in exactly the same way as rice, the plants needing the same conditions of

high temperatures, shallow water and good soil. In spring, the corms are planted in paddy fields which are then flooded to a depth of some 10cm/4in. These are drained in autumn, and the corms are harvested and stored over the winter. Water chestnuts are much used in Chinese cooking and have a sweet crunchy flavour with nutty overtones.

They are edible cooked or raw and are excellent in all sorts of Chinese dishes.

Above: Sprouting mung beans.
Below: clockwise from top left: Canned Water chestnuts, canned Bamboo shoots, fresh Water chestnuts, canned Palm hearts.

FENNEL

The vegetable fennel is closely related to the herb and spice of the same name. It is called variously Florence fennel, sweet fennel, *finocchio dulce* or Italian fennel.

   Like the herb, Florence fennel has the distinct flavour of anise, a taste that seems to go particularly well with fish, so the vegetable is often served with fish dishes while the herb or spice is commonly used in fish stocks, sauces or soups. The leaves are edible and can be used in soups and stocks as well as for garnishing.

History

Florence fennel has only been popular in Britain for the last 20 years, although it has a long history of cultivation, having been eaten by the ancient Egyptians, Greeks and Romans. In Italy, fennel has been eaten for several centuries: many of the best fennel recipes come from Italy and other parts of the Mediterranean.

Buying and Storing

If possible, buy small tender bulbs. The bulbs should be clean and white with no bruises or blemishes and the feathery leaves should be green and lively. Fennel will keep for a day or two in the salad drawer of the fridge.

Preparing

Unless the bulbs are very young and tender, remove the first layer of skin, as it is likely to be tough (this can be used for a stock). Fennel can then be sliced into slivers by cutting downwards or into rings by cutting across the bulb. When used raw in salads, it needs to be cut into smaller pieces.

Cooking and Serving

Fennel can be served raw if it is thinly sliced and dressed with a light vinaigrette. In salads, its flavour contrasts well with apple, celery and other crunchy ingredients. Fennel is also excellent braised with onions, tomatoes and garlic.

Right: Florence fennel.
Far right: Marsh samphire.

SAMPHIRE

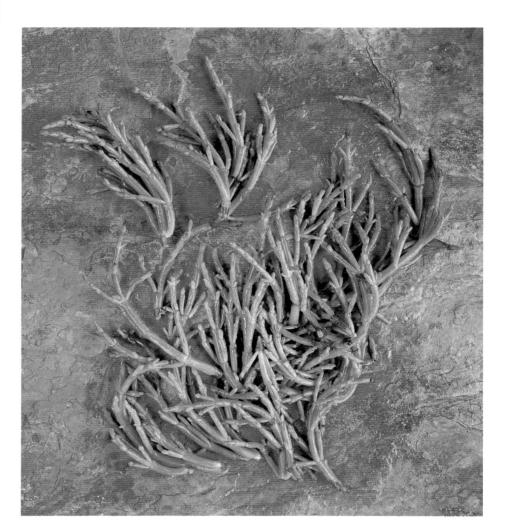

There are two types of samphire. Marsh samphire grows in estuaries and salt marshes while rock samphire, sometimes called sea fennel, grows on rocky shores. The two are understandably confused since they are both connected with the sea, yet they are completely different plants.

The type likely to be sold by a fishmonger is marsh samphire. It is also known as glasswort and is sometimes called sea asparagus, as its shoots are similar to sprue, small asparagus shoots.

Although marsh samphire grows easily and is commonly found all over Europe and North America, it is not cultivated and is only available for a short time while it is in season, normally in late summer and early autumn.

Samphire has a distinctly salty, iodine flavour and a pleasant crisp texture. The flavour is reminiscent of the sea and goes particularly well with fish and seafood. However, samphire can be enjoyed simply steamed and dipped into melted butter.

Buying and Storing

When in season, good fishmongers get regular stocks of marsh samphire, and it should look bright and fresh. Buy it as you need it, as it will not keep for long.

Preparing and Cooking

If necessary, wash marsh samphire under cold running water. It is best steamed over a pan of boiling water for no more than 3 minutes. Alternatively, blanch it in boiling water for 3-5 minutes and then drain. Samphire can be eaten raw but blanching it removes some of the saltiness.

To eat samphire, draw the shoots through the teeth to peel the succulent part from the thin central core.

THREE
ROOTS

Potatoes

Parsnips

Jerusalem Artichokes

Turnips and Swedes

Carrots

Horseradish

Beetroot

Salsify and Scorzonera

Exotic Roots

Ginger and Galangal

POTATOES

History

The potato originates from South America. Most people learned at school that Sir Walter Raleigh brought the tubers to England from Virginia, but this never convinced historians as the potato was completely unknown in North America until the eighteenth century. They now believe that Sir Francis Drake was responsible. In 1586, after battling against the Spaniards in the Caribbean, Drake stopped to pick up provisions from Cartegena in northern Colombia – and these included tobacco and potato tubers. En route home, he stopped off at Roanoke Island, off the coast of Virginia. The first group of English colonists had been sponsored to settle there by Sir Walter Raleigh but by this time they had had enough. Drake brought them back to England, along with some of Raleigh's men and, of course, the provisions – including the potato tubers.

Potatoes apparently fascinated Queen Elizabeth and intrigued horticulturists, but they were not an overnight success among the people. The wealthy frequently reviled them as being flavourless and the food of the poor. People distrusted the fact that they reached maturity underground, believing them to be the work of the devil. In Scotland, Presbyterian ministers darkly advised their congregations that there was no mention of potatoes in the Bible, and thus the eating of them was an ungodly act!

In spite of such a bad press, potatoes nevertheless were slowly recognized for their merit. By 1650 they were the staple food of Ireland, and elsewhere in Europe potatoes began to replace wheat as the most important crop, both for people and for livestock. In an early English cookery book, *Adam's Luxury and Eve's Cookery*, there are 20 different recipes for cooking and serving potatoes.

The first mention of potatoes in America is in 1719 in Londonderry, New Hampshire. They arrived not from the south, but via Irish settlers who brought their potatoes with them.

The current popularity of potatoes is probably thanks to a Frenchman called Antoine-Auguste Parmentier. A military pharmacist of the latter part of the eighteenth century, Parmentier recognized the virtues of the potato, both for its versatility and as an important food for the poor, and set out to improve its image. He persuaded Louis XVI to let him ostentatiously grow potatoes on royal land around the palace in Versailles to impress the fashion-conscious Parisians. He also produced a court dinner in which each course contained potatoes. Gradually, eating potatoes became chic, first among people in the French court and then in French Society. Today, if you see *Parmentier* in a recipe or on a menu, it means "with potato".

Nutrition

Potatoes are an important source of carbohydrate. Once thought to be fattening, we now know that, on the contrary, potatoes can be an excellent part of a calorie-controlled diet – provided, of course, they are not fried in oil or mashed with too much butter. Potatoes are also a very good source of vitamin C, and during the winter potatoes are often the main source of this vitamin. They also contain potassium, iron and vitamin B.

Varieties

There are more than 400 varieties of potato but unless you are a gardener, you will find only some 15 varieties generally available. Thanks to labelling laws, packaged potatoes carry their names which makes it easier to learn to differentiate between the varieties and find out which potato is good for what.

New Potatoes

Carlingford: Available as a new potato or as main crop, Carlingford has a close white flesh.

Jersey Royal: Often the first new potato of the season, Jersey Royals have been shipped from Jersey for over a hundred years and have acquired an enviable reputation among everyone who enjoys good food. Boiled or steamed and then served with butter and a sprinkling of parsley, they cannot be beaten.

Jersey Royals are kidney-shaped, with yellow firm flesh and a distinctive flavour. Don't confuse Jersey Royals with Jersey Whites, which are actually Maris Pipers, grown in Jersey.

Maris Bard: A regularly shaped, slightly waxy potato with white flesh.

Maris Peer: This variety has dry firm flesh and a waxy texture and doesn't disintegrate when cooked – consequently, it is good in salads.

Main Crop Potatoes

Desirée: A potato with a pink skin and yellow soft-textured flesh. It is good for baking, chipping, roasting and mashing.

Estima: A good all-rounder with yellow flesh and pale skin.

Golden Wonder: A russet-skinned potato that was the original favourite for making crisps. It has an excellent, distinctive flavour, and should you find them for sale, buy them at once for baked potatoes. They are also good boiled or roasted.

Kerr's Pink: A good cooking potato with pink skin and creamy flesh.

King Edward: Probably the best known of British potatoes, although not the best in flavour. King Edwards are creamy white in colour with a slightly floury texture.

Red King Edwards are virtually identical except for their red skin. Both are good roasted or baked. However, the flesh disintegrates when boiled, so while good for mashing do not use King Edwards if you want whole boiled potatoes.

Maris Piper: This is a widely grown variety of potato, popular with growers and cooks because it is good for all kinds of cooking methods – baking, chipping, roasting and mashing. It has a pale, smooth skin and creamy white flesh.

Left: Maris Bard potatoes.
Above: Kerr's pink (left) and Maris Piper (right) potatoes.
Right: Romano potatoes.

Pentland Dell: A long, oval-shaped potato with a floury texture that tends to disintegrate when boiled. For this reason, it is popular for roasting as the outside becomes soft when par-boiled and then crisps up with the fat during roasting.

Romano: The Romano has a distinctive red skin with creamy flesh and is a good all-rounder, similar to Desirée.

Wilja: Introduced from Holland, this is a pale, yellow-fleshed potato with a good, sweet flavour and waxy texture.

Other Varieties

Although most of these varieties are also main crop, they are less widely available than those listed above but are increasingly sold in supermarkets. They are recommended for salads but many are also excellent sautéed or simply boiled.

Cara: A large main crop potato, which is excellent baked or boiled but is a good all-rounder.

Finger Potatoes: Thumb-sized, long baby potatoes are sometimes called finger potatoes or fingerlings. Among the many varieties are the German Lady's Finger. Since they are new crop potatoes, they need simply be boiled and then served either in salads or with a little butter and a sprinkling of parsley.

La Ratte: A French potato with a smooth skin and waxy yellow flesh. It has a chestnut flavour and is good in salads.

Linzer Delikatess: These small, kidney-shaped potatoes look a little like Jersey Royals but have a pale smooth skin. They do not have much taste and are best in salads where their flavour can be enhanced with other ingredients.

Pink Fir Apple: This is an old English variety, with pink skin and a smooth yellow flesh. It is becoming increasingly popular and has a distinctive flavour.

Above left: Cara (left) and Estima (right) potatoes.
Left: Linzer Delikatess potatoes.
Above: Desirée (left) and King Edward (right) potatoes.
Right: Finger potatoes.

Purple Congo: If you want to startle your friends, serve some of these striking purple-blue potatoes. There are several varieties of blue potato, ranging from a pale lavender to a purple-black, but one of the most popular is the Purple Congo, which is a wonderful deep purple. They are best boiled and served simply with a little butter and do retain their colour when cooked.

Truffe de Chine: Another deep purple, almost black potato, of unknown origin but now grown in France. It has a nutty, slightly mealy flavour and is best served in a salad with a simple dressing. Like the Purple Congo, it retains its colour after being cooked.

Recommended Varieties for Cooking

Baking: As for roasting, use potatoes with a floury texture, such as Golden Wonder, Pentland Dell, King Edward and Maris Piper.

Boiling: Jersey Royal, Maris Bard and Maris Peer, or any of the Egyptian or Belgian new crop varieties. In addition, Pink Fir Apple, La Ratte and Linzer Delikatess are excellent.

Chipping: King Edward, Golden Wonder, Romano, Maris Piper and Desirée.

Mashing: Golden Wonder, Maris Piper, King Edward, Wilja, Romano and Pentland Dell.

Roasting: Pentland Dell, Golden Wonder, Maris Piper, King Edward, Desiree and Romano are among the best roasting potatoes. Ideally, use potatoes with a floury texture.

Salads: All the small, specialist potatoes, such as La Ratte, Pink Fir Apple and Linzer Delikatess as well as Finger Potatoes and small new potatoes.

Sautéing: Any waxy type of potato, such as Maris Bard, Maris Peer, any of the specialist potatoes, and Romano and Maris Piper.

Buying and Storing

Potatoes should always be stored in a dark, cool, dry place. If they are stored exposed to the light, green patches will develop which can be poisonous, and they will go mouldy if kept in the damp. When buying potatoes in bulk, it is best to buy them in paper sacks rather than plastic bags as humid conditions will cause them to go rotten. Similarly, if you buy potatoes in polythene bags, remove them when you get home and place them in a vegetable rack or in a paper bag, in a dark place.

Main crop potatoes will keep for several months in the right conditions but will gradually lose their nutritional value. New potatoes should be eaten within two or three days as they will go mouldy if stored for too long.

Preparing

Most of the minerals and vitamins contained in potatoes are contained in or just below the skin. It is therefore better to eat potatoes in their skins. New potatoes need only be washed under running water; older potatoes need to be scrubbed.

If you peel potatoes, use a peeler that removes only the very top surface (*below left*) or, alternatively, for salads and cold dishes, boil the potatoes in their skins and peel when cool.

Cooking

Baking: Cook baked potatoes in a low oven for well browned and crunchy skins and fluffy flesh. Baked potatoes can be cooked more quickly in a microwave oven; for a crunchy texture to the skin place them in a hot oven for 10 minutes.

Boiling: It is impossible to generalize on how long to boil as it depends so much on the variety of potato. Try to cut potatoes to an even size (new potatoes should not need to be cut), salt the water if liked, cover and cook over a moderate heat. Don't boil potatoes too fiercely; old ones especially may disintegrate and leave you with a pan of starchy water.

Chips: Home-cooked chips are a treat worth occasionally giving the family instead of the convenient but otherwise disappointing oven chips. However, they are fatty and therefore not good for you when eaten in great quantities or too often.

To make them, cut the potatoes into even-size chips and place in a bowl of cold water for about 10 minutes before frying. Drain and then dry in a piece of muslin or an old dish towel before frying. Fry only as many chips as will comfortably sit in the fat. Halfway through cooking, drain them and allow the oil to come back to temperature before plunging the chips back in. This browns the chips and they don't soak up excessive amounts of oil. Be warned; the cooking smells from making chips tend to linger!

Mashing: Boil the potatoes until tender, drain thoroughly and then tip them back into the pan; mash with a little milk and butter using a potato masher (*below right*), and season to taste with a little salt, if necessary, and pepper. Never use a food processor or liquidizer: the potatoes will turn into an inedible thick grey paste. You may lightly whisk potatoes with a fork after mashing to fluff them up but no more – for once modern machines have not improved on the basic utensil.

Roast Potatoes: The best roast potatoes are made using a floury textured potato such as Maris Piper or King Edward. Wash and cut them into even-size chunks and par-boil them in lightly salted water until they begin to go tender and the outside looks soft. Drain them through a sieve or colander and then tip them back into the saucepan, put the lid on and shake the pan two or three times. This roughens up the surface of the potato. Place the potatoes in a dish of hot oil or fat, or round a joint of meat, and turn them over so that they are evenly coated. Roast them in the oven for 40-50 minutes until golden. Serve as soon as possible once cooked as the outsides become leathery if they are kept in a warm oven for too long.

Sautéing: There are various ways to sauté potatoes and no one way is better than another. For sautéed sliced potatoes, par-boil whole potatoes for 5-10 minutes until they begin to soften. Drain them thoroughly and then slice into thick rounds. Using sunflower oil or a mixture of sunflower and olive oil (not butter as it will burn), fry them in a large frying pan. Turn the potatoes occasionally and cook until evenly browned. For sautéed, diced potatoes, cut the potatoes into small cubes, blanch for 2 minutes and then drain well. Either fry them on the stove or cook them in a little oil in the oven; turn them once or twice to brown evenly.

Steaming: New potatoes are excellent steamed. Place them on a bed of mint in a steamer or a colander over a pan of boiling water for 15-20 minutes.

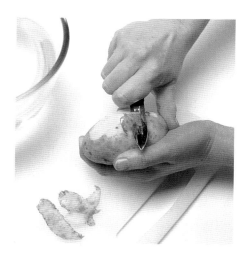

Above left: Purple Congo potatoes.
Below left: Truffe de Chine (left) and Pink Fir Apple (right) potatoes.

PARSNIPS

There's something very old-fashioned about parsnips. They conjure up images of cold winter evenings and warm comforting broths supped in front of a blazing wood fire. Nowadays parsnips are available all year through, but many people still feel they belong to winter, adding their characteristic flavour to soups and stews.

Parsnips are related to carrots, similarly sweet but with a distinct earthy flavour that blends well with other root vegetables and is also enhanced with spices and garlic.

History

Parsnips have a long history. The Romans grew and cooked them to make broths and stews. When they conquered Gaul and Britain, the Romans discovered that root vegetables grown in northerly areas had a better flavour than those grown in the south – they may have been the first to decree that parsnips should be eaten after the first frost!

Throughout the Dark Ages and early Middle Ages, parsnips were the main starchy vegetable for ordinary people (the potato had yet to be introduced). Parsnips were not only easy to grow but were a welcome food to eat during the lean winter months. They were also valued for their sugar content. Sweet parsnip dishes like jam and desserts became part of traditional English cookery, and they were also commonly used for making beer and wine. Parsnip wine is still one of the most popular of the country wines, with a beautiful golden colour and a rich sherry-like flavour.

Nutrition

Parsnips contain moderate amounts of vitamins A and C, along with some of the B vitamins. They are also a source of calcium, iron and potassium.

Buying and Storing

Parsnips are really a winter crop, although nowadays they are available all year round. Tradition has it that parsnips are best after the first frost, but many people like the very young tender parsnips available in the early summer. When buying parsnips, choose small or

medium-size specimens as the large ones tend to be rather fibrous. They should feel firm and be a pale ivory colour without any sprouting roots. Store parsnips in a cool place, ideally an airy larder or cool outhouse, where they will keep well for 8-10 days.

Preparing

Very small parsnips need little or no peeling; just trim the ends and cook according to your recipe. Medium-size and large parsnips need to be peeled. Larger parsnips also need to have the woody core removed; if it is cut out before cooking, the parsnips will cook more quickly and evenly.

Cooking

Roast parsnips are best par-boiled for a few minutes before adding to the roasting dish. Very young parsnips can be roasted whole but larger ones are best halved or quartered lengthways. Roast in butter or oil for about 40 minutes in an oven preheated to 200°C/400°F/Gas 6.

To boil parsnips, cut them into pieces about 5cm/2in long and boil for 15-20 minutes until tender. When boiled briefly like this, they keep their shape, but when added to a casserole or stew they eventually disintegrate. Don't worry if this happens; parsnips need plenty of cooking so that the flavour can blend with the other ingredients.

JERUSALEM ARTICHOKES

Jerusalem artichokes are related to the sunflower and have nothing to do with Jerusalem. One explanation for their name is that they were christened gira-sole, "Jerusalem", because their yellow flowers turned towards the sun. The Italian name for the Jerusalem artichoke is *girasole articocco*.

These small knobbly tubers have a lovely distinct flavour and are good in Palestine soup, a popular classic recipe. They are also delicious baked or braised.

History

Jerusalem artichokes are thought to have come from the central United States and Canada, where they were cultivated by the American-Indians as long ago as the fifteenth century. However, many writers have alluded to the fact that they cause "wind", which tempers their popularity.

Buying and Storing

Jerusalem artichokes are at their best during winter and early spring. They are invariably knobbly but if possible buy neat ones with the minimum of knobs to save waste. The skins should be pale brown without any dark or soft patches. If they are stored in a cool dark place they will keep well for up to 10 days.

Preparing

The white flesh of artichokes turns purplish brown when exposed to light, so when peeling or slicing them raw, place them in a bowl of acidulated water (water to which the juice of about half a lemon has been added). Because artichokes are so knobbly, it is often easier to boil them in acidulated water in their skins and peel them afterwards – the cooked skins should slip off easily.

Cooking

Jerusalem artichokes can be cooked in many of the ways in which you would cook potatoes or parsnips. They are excellent roasted, sautéed or dipped in batter and fried, but first par-boil them for 10-15 minutes until nearly tender. For creamed artichokes, mix with potatoes in equal amounts; this slightly blunts their flavour, making a tasty side dish which is not too overpowering.

Left: Parsnips.
Below: Jerusalem artichokes.

TURNIPS <u>AND</u> SWEDES

Turnips and swedes are both members of the cabbage family and are closely related to each other – so close that it is not surprising that their names are often confused. For instance, swedes are sometimes called Swedish turnips or swede-turnips and in Scotland, where they are thought of as turnips, they are called neeps.

Nowadays, the confusion is not so acute. Many greengrocers and super-markets sell early or baby turnips or, better still, French turnips – *navets*.

Both are small and white, tinged either with green or in the case of *navets*, with pink or purple. Consequently, people are learning to tell their swedes from their turnips and also discovering what a deli-cious vegetable the turnip is.

History

Turnips have been cultivated for cen-turies, principally as an important live-stock feed but also for humans. Although they were not considered the food of gourmets, they have been grown by

poorer families as a useful addition to the winter table.

Swedes were known as turnip-rooted cabbages until the 1780s, when Sweden began exporting the vegetable to Britain and the shorter name resulted.

Until recently, turnips and swedes have not enjoyed a very high reputation among cooks in many parts of the world. This is partly because they are perceived as cattle food and partly because few people have taken the trouble to find acceptable ways of cooking them. Schools and other

institutions tend to boil and then mash them to a watery pulp, and for many people this is the only way they have eaten either vegetable.

The French, in contrast, have had far more respect for the turnip, at least. For centuries they have devised recipes for their delicate *navets*, roasting them, caramelizing them in sugar and butter or simply steaming and serving with butter. Young, tender turnips have also been popular all over the Mediterranean region for many years, and there are many dishes using turnips with fish, poultry, or teamed with tomatoes, onions and spinach.

Nutrition

Both turnips and swedes are a good source of calcium and potassium.

Varieties

French *navets*, small round, squashed-shaped turnips tinged with pink or purple, are increasingly available in greengrocers and supermarkets in the spring. Less common, but even more prized by the French, are the long carrot-shaped turnips, called *vertus*. English turnips are generally larger and are mainly green and white.

Both have the characteristic peppery flavour, but this is less pronounced in *navets* which are generally sweeter.

Swedes generally have a more substantial, fuller-bodied flavour than turnips but at their best have a subtle, pleasant taste. The Marian is a yellow fleshed variety with a distinct "swede" flavour. White fleshed swedes, like Merrick, have a more watery, turnip-like flavour.

Buying and Storing

Turnips: If possible, buy French *navets* or failing that, the smallest and youngest turnips, available in the shops from spring. They should be firm, smooth and unblemished, ideally with fresh green tops. Store in a cool dry place.
Swedes: Unlike turnips, swedes generally seem to come large. However, if possible, choose small swedes with smooth and unblemished skins as large ones are likely to be tough and fibrous. Store as for turnips.

Preparing and Cooking

Turnips: Young turnips should not need peeling; simply trim, then simmer or steam until tender. They are delicious raw, thinly sliced or grated into salads.

Peel older turnips (*below*) and then slice or dice before cooking. Remember, turnips are members of the cabbage family and older specimens particularly can show signs of that unpleasant cabbage rankness if overcooked. To avoid this, blanch turnips if they are to be served as a vegetable dish, or add sparingly to soups and casseroles, so that the rank flavour is dispersed.

Swedes: Peel to remove the skin and then cut into chunks (*below*). Swedes will disintegrate if overcooked, and they are unpleasantly raw tasting if not cooked sufficiently. The only answer is to check frequently while they are cooking. Swedes are particularly good when teamed with other root vegetables in soups and casseroles, adding a pleasant, slightly nutty flavour.

*Far left: Navets and Turnips.
Below: Swedes.*

CARROTS

After potatoes, carrots are without doubt our best-known and best-loved root vegetable. In the days when vegetables were served merely as an accessory to meat, carrots always made an appearance – often overcooked but still eaten up because, we were told, they helped you to see in the dark.

Carrots have many different flavours, depending on how they are cooked. Young, new season carrots braised in butter and a splash of water are intensely flavoured and sweet; when steamed, they are tender and melting. Carrots grated into salads are fresh and clean tasting, while in casseroles they are savoury with the characteristic carrot flavour. In soups they are fragrant and mild, and in cakes their flavour can hardly be detected, yet their sweetness adds richness.

History

Until the Middle Ages, carrots were purple. The orange carrots came from Holland, from where they were exported in the seventeenth and eighteenth centuries. Although purple and white carrots continued to be eaten in France, nowadays they are something of a rarity.

Nutrition

Carrots contain large amounts of carotene and vitamin A, along with useful amounts of vitamins B3, C and E. When eaten raw, they also provide good quantities of potassium, calcium, iron and zinc, but these are reduced when carrots are boiled.

The idea that carrots are good for your night sight originated in the Second World War. Early radar stations were established along the south and east coasts of England in 1939 to detect aggressors in the air or at sea. The Germans attributed this sudden remarkable night vision to the British habit of eating carrots. Indeed, the vitamin A in carrots forms retinal, a lack of which brings on night blindness.

Buying and Storing

Home-grown carrots are so much nicer than shop bought ones. Almost all vegetables have a better flavour if grown organically, but this is particularly true of carrots.

When buying carrots, look out for the very young, pencil thin ones, which are beautifully tender either eaten raw or steamed for just a few minutes. Young carrots are commonly sold with their feathery tops intact, which should be fresh and green. Older carrots should be firm and unblemished. Avoid tired looking carrots as they will have little nutritional value.

Carrots should not be stored for too long. They will keep for several days if stored in a cool, airy place or in the salad drawer of the fridge.

Preparing

Preparation depends on the age of the carrots. The valuable nutrients lie either in or just beneath the skin, so if the carrots are young, simply wash them under cold running water. Medium-size carrots may need to be scraped and large carrots will need either scraping or peeling.

Cooking

Carrots are excellent cooked or raw. Children often like raw carrots as they have a very sweet flavour. They can be cut into julienne strips, with a dressing added, or grated into salads and coleslaw – their juices run and blend wonderfully with the dressing. Carrots can be cooked in almost any way you choose. As an accompaniment, cut them into julienne strips and braise in butter and cider, or cook in the minimum of stock and toss in butter and a sprinkling of caraway seeds.

Roasted carrots are delicious, with a melt-in-the-mouth sweetness. Par-boil large ones first, but younger carrots can be quickly blanched or added direct to the pan with a joint of meat.

HORSERADISH

Horseradish is grown for its pungent root, which is normally grated and mixed with cream or oil and vinegar and served with roast beef. Fresh horseradish is available in many supermarkets in the spring, and you can make your own horseradish sauce by simply peeling the root and then mixing 45ml/3 tbsp grated horseradish with 150m/^{1}/$_4$ pint/^{2}/$_3$ cup whipping cream and adding a little Dijon mustard, vinegar and sugar to taste. As well as being excellent with hot or cold beef, horseradish sauce is delicious with smoked trout or mackerel or spread thinly on sandwiches with a fine pâté.

Left: Carrots.
Right: Horseradish.

BEETROOT

Experience of vinegar-sodden beetroot has doubtlessly put many people off beetroot. Those who love it know to buy their beetroot fresh, so that they can cook it themselves. It can be served in a number of different ways: baked and served with soured cream, braised in a creamy sauce, grated in a salad or used for the classic soup *borscht*.

History

Beetroot is closely related to sugar beet and mangelwurzels. As the demand for sugar increased over the centuries, when sugar could successfully be extracted from beet, sugar production became a big industry in Britain and Europe.

Mangelwurzels were eaten in parts of Europe and in England in times of famine, although they were primarily grown as cattle fodder.

Beetroot, however, has probably been eaten since Roman times. By the mid-nineteenth century it was clearly a popular vegetable, and Mrs Beeton in her famous cookbook has 11 recipes for it, including a beetroot and carrot jam and beetroot fritters.

Nutrition

Beetroot is an excellent provider of potassium. The leaves, which have the flavour of spinach, are high in vitamin A, iron and calcium.

Buying and Storing

If possible, buy small beetroots which have their whiskers intact and have at least 5cm/2in of stalk at the top; if they are too closely cropped they will bleed during cooking. Beetroots will keep for several weeks if stored in a cool place.

Preparing

To cook beetroot whole, first rinse under cold running water. Cut the stalks to about 2.5cm/1in above the beetroot and don't cut away the root or peel it – or the glorious deep red colour will bleed away. When serving cold in salads, or where the recipe calls for chopped or grated beetroot, peel away the skin with a potato peeler or sharp knife.

Cooking

To bake in the oven, place the cleaned beetroot in a dish with a tight-fitting lid, and add 60-75ml/4-5 tbsp of water. Lay a double layer of foil over the dish before covering with the lid, then bake in a low oven for 2-3 hours or until the beetroot is tender. Check occasionally to ensure the pan doesn't dry out and to see whether the beetroot is cooked. It is ready when the skin begins to wrinkle and can be easily rubbed away with your fingers. Alternatively, simply wrap the beetroot in a double layer of foil and cook as above. To boil beetroot, prepare as above and simmer for about 1½ hours.

BEET GREENS

The tops of several root vegetables are not only edible, but are also extremely nutritious. Beet greens are particularly good, being very high in vitamins A and C, and indeed have more iron and calcium than spinach itself. They are delicious, but not easily available unless you grow your own. If you are lucky enough to get some, boil the greens for a few minutes, then drain well and serve with butter or olive oil.

Left: Beetroot.
Above right: Scorzonera.
Below right: Salsify.

SALSIFY AND SCORZONERA

These two root vegetables are closely related to each other as well as to members of the same family as dandelion and lettuce. All have long tapering roots.

Salsify has a white or pale brownish skin and scorzonera, sometimes called black salsify, has a black skin. They both have a pale creamy flesh and a fairly similar flavour reminiscent of artichokes and asparagus. Salsify is said to have the superior flavour and has been likened to oysters (it is sometimes referred to as the oyster plant), although many people fail to detect this.

Both salsify and scorzonera make an unusual and pleasant accompaniment, either creamed or fried in butter. They can also be also used in soups.

History

Salsify is native to the Mediterranean but now grows in most areas of Europe and North America. Scorzonera is a southern European plant.

Both roots are classified as herbs and, like many wild plants and herbs, their history is bound up with their use in medicines. The roots, together with their leaves and flowers, were used for the treatment of heartburn, loss of appetite and various liver diseases.

Buying and Storing

Choose specimens that are firm and smooth and, if possible, still with their tops on, which should look fresh and lively. Salsify will keep for several days stored in a cool dark place.

Preparing

Salsify and scorzonera are difficult to clean and peel. Either scrub the root under cold running water and then peel after cooking, or peel with a sharp stainless steel knife (*below*). As the flesh discolours quickly, place the trimmed pieces into acidulated water (water to which lemon juice has been added).

Cooking

Cut into short lengths and simmer for 20–30 minutes until tender. Drain well and sauté in butter, or serve with lemon juice, melted butter or chopped parsley.

Alternatively, they can be puréed for soups or mashed. Cooked and cooled salsify and scorzonera can be served in a mustard or garlic vinaigrette with a simple salad.

EXOTIC ROOTS

Throughout the tropical regions of the world all sorts of tubers are grown and used for a fabulous variety of dishes. Yams, sweet potatoes, cassava and taro, to name but a few, are for many people a staple food, not only cooked whole as a vegetable accompaniment, but ground or pounded for bread and cakes. There is an enormous variety of these tropical and subtropical tubers, and while they cannot be cultivated in a moderate climate, the more common tubers are now widely available in specialist shops and in most supermarkets.

SWEET POTATOES

Sweet potatoes are another one of those vegetables that once tasted, are never forgotten. They are, as the name suggests, sweet, but they also have a slightly spicy taste. It's this distinct sweet and savoury flavour which makes them such an excellent foil to many savoury dishes

and they are fittingly paired with meat dishes that need a touch of sweetness, like turkey or pork.

History

Sweet potatoes are native to tropical America, but today they are grown all over the tropical world. They have been grown in South America from before the Inca civilizations and were introduced into Spain before the ordinary potato. They also have a long history of cultivation in Asia spreading from Polynesia to New Zealand in the fourteenth century.

They are an important staple food in the Caribbean and southern United States, and many famous recipes feature these vegetables. Candied sweet potatoes, for instance, are traditionally served with ham or turkey at Thanksgiving all over the United States, while Jamaica and the West Indies abound with sweet potato dishes, from the simple baked

potato to Caribbean pudding, a typically sweet and spicy dish with sweet potatoes, coconut, limes and cinnamon.

Sweet potatoes appear to have been introduced to England even earlier than regular potatoes. Henry VIII was said to have been very partial to them baked in a pie, believing they would improve his love life! If Henry VIII was eating sweet potatoes in the early/mid-sixteenth century, then it's likely he received them via the Spanish, who, thanks to Christopher Columbus, were busy conquering the New World, thus experiencing a whole range of tropical vegetables and fruit.

Varieties

The skin colour ranges from white to pink to reddish brown. The red-skinned variety, which has a whitish flesh, is the one most commonly used in African and Caribbean cooking.

Buying and Storing

Choose small or medium-size ones if possible as larger specimens tend to be rather fibrous. They should be firm and evenly shaped; avoid those that seem withered, have damp patches or are sprouting. They will keep for several days in a cool place.

Preparing and Cooking

If baking, scrub the potatoes well and cook exactly as you would for ordinary potatoes. To boil, either cook in their skins and remove these after cooking, or peel and place in acidulated water (water to which lemon juice has been added). This prevents them turning brown and it's worth boiling them in lightly acidulated water for the same reason. Sweet potatoes can be cooked in any of the ways you would cook ordinary potatoes - roast, boiled, mashed or baked. However, avoid using them in creamy or gratin-type dishes. They are both too sweet and too spicy for that.

It is preferable to roast or sauté them with onions and other savoury ingredients to bring out their flavour, or mash them and serve them American-style over chunks of chicken for a crusted chicken pie.

YAMS

Yams have been a staple food for many cultures for thousands of years. There are today almost countless varieties, of different shapes, sizes and colours and called different names by different people. Most varieties are thought to have been native to China, although they found their way to Africa at a very early period and became a basic food, being easy to grow in tropical and subtropical conditions, and containing the essential carbohydrate of all staple foods.

Although cush-cush or Indian yam was indigenous to America, most yams were introduced to the New World as a result of the slave trade in the sixteenth century. Today with such a huge variety of this popular vegetable available, there are innumerable recipes for yam, many probably not printed and published, but handed down by word of mouth from mother to daughter and making their appearance at mealtimes all over the hot regions of the world.

Varieties

The greater yam, as the name suggests, can grow to a huge size. A weight of 62kg/150lb has been recorded. The varieties you are likely to find in shops will be about the size of a small marrow, although smaller yams are also available such as the sweet yam, which looks like a large potato and is normally covered with whiskery roots. All sizes have a coarse brown skin and can be white or red-fleshed.

In Chinese stores, you may find the Chinese yam, which is a more elongated, club-like shape and is covered with fine whiskers.

Buying

Look out for firm specimens with unbroken skins. The flesh inside should be creamy and moist and if you buy from a grocery, the shopkeeper may well cut open a yam so you can check that it is fresh. They can be stored for several weeks in a cool, dark place.

Preparing

Peel away the skin thickly to remove the outer skin and the layer underneath that contains the poison dioscorine. This in fact is destroyed during cooking, but discard the peel carefully. Place the peeled yam in salted water as it discolours easily.

Cooking

Yams, like potatoes, are used as the main starchy element in a meal, boiled and mashed, fried, sautéed or roasted. They tend to have an affinity with spicy sauces and are delicious cut into discs, fried and sprinkled with a little salt and cayenne pepper. African cooks frequently pound boiled yam to make a dough which is then served with spicy stews and soups.

TARO/EDDO

Like yams, taro is another hugely important tuber in tropical areas, and for thousands of years it has been a staple food for many people. It goes under many different names; in South-east Asia, South and Central America, all over Africa and in the Caribbean it is called variously eddo and dasheen.

There are two basic varieties of taro - a large barrel-shaped tuber and a smaller variety, which is often called eddo or dasheen. They are all a dark mahogany brown with a rather shaggy skin, looking like a cross between a beetroot and a swede.

Although they look very similar, taro belongs to a completely different family from yam and in flavour and texture is noticeably different. Boiled, it has a completely unique flavour, something like a floury water chestnut.

Buying and Storing

Try to buy small specimens; the really small smooth bulbs are tiny attachments to the larger taro and are either called eddoes, or rather sweetly, "sons of taro". Stored in a cool, dark place, they should keep for several weeks.

Left: Sweet potatoes.
Above: Yams.

Preparing

Taros, like yams, contain a poison just under the skin which produces an allergic reaction. Consequently, either peel taros thickly, wearing rubber gloves, or cook in their skins. The toxins are completely eliminated by boiling, and the skins peel off easily.

Cooking

Taros soak up large quantities of liquid during cooking, and this can be turned to advantage by cooking in well flavoured stock or with tomatoes and other vegetables. For this reason, they are excellent in soups and casseroles, adding bulk and flavour in a similar way to potatoes. They can also be steamed or boiled, deep-fried or puréed for fritters but must be served hot as they become sticky if allowed to cool.

CALLALOO

Callaloo are the leaves of the taro plant, poisonous if eaten raw, but used widely in Asian and Caribbean recipes. They are cooked thoroughly, then used for wrapping meat and vegetables. Callaloo can also be shredded and cooked together with pork, bacon, crab, prawns, okra, chilli, onions and garlic, together with lime and coconut milk to make one of the Caribbean's most famous dishes, named after the leaves themselves, Callaloo.

JICAMA

Also known as the Mexican potato, this large root vegetable is a native of central America. It has a thin brown skin and white, crunchy flesh which has a sweet, nutty taste. It can be eaten cooked in the same way as potatoes or sliced and added raw to salads.

Buy specimens that are firm to the touch. Jicama in good condition will keep for about two weeks if stored in a plastic bag in the refrigerator.

Top: Taros (Eddoes).
Above: Jicama.
Left: Callaloo.

CASSAVA

This is another very popular West Indian root, used in numerous Caribbean dishes. It is native to Brazil, and found its way to the West Indies surprisingly via Africa, where it also became a popular vegetable. Known as cassava in the West Indies, it is called manioc or mandioc in Brazil, and juca or yucca is used in other parts of South America.

Cassava is used to make tapioca, and in South America a sauce and an intoxicating beverage are prepared from the juice. However, in Africa and the West Indies it is eaten as a vegetable either boiled, baked or fried, or cooked and pounded to a dough to make *fufu,* a traditional savoury African pudding.

Right: Cassava.
Below left: Ginger.
Below right: Galangal.

GINGER AND GALANGAL

GINGER

This is probably the world's most important and popular spice and is associated with a number of different cuisines - Chinese, Indian and Caribbean, to name but a few. It was known in Europe during the Roman period, but was still fairly rare until the spice routes opened up trade in the sixteenth and seventeenth centuries. Like many spices, ginger has the quality of enhancing and complementing both sweet and savoury food, adding a fragrant spiciness to all sorts of dishes. However, while ground ginger is best in recipes which will be baked, and stem ginger, where the ginger is preserved in syrup, tastes wonderful in desserts, for savoury dishes, always use fresh root ginger.

Nowadays, the pale, knobbly roots of fresh ginger are widely available in supermarkets and whenever possible, buy just a small quantity, as you will not need a great deal and fresh ginger will not keep indefinitely.

To prepare, simply peel away the skin with a sharp knife and grate or thinly slice according to the recipe.

GREATER GALANGAL

Galangal looks similar to ginger except that the rhizome is thinner and the young shoots are bright pink. The roots should be prepared in the same way as ginger and can be used in curries and satay sauces.

FOUR
GREENS

SPINACH

For many people, spinach is inextricably linked with Popeye, the cartoon character who used to eat huge amounts of spinach. It is a wonderfully versatile vegetable, popular worldwide, with nearly every cuisine featuring spinach somewhere in its repertoire. The Italians are particularly partial to spinach and have hundreds of dishes using the vegetable. The words *à la florentine* mean the dish contains spinach.

As well as being delicious on its own, chopped or puréed spinach can be mixed with a range of other ingredients with superb results. It has a particular affinity with dairy products and in the Middle East, feta or helim cheese is used to make boreks or other spinach pies. The Italians mix spinach with ricotta or Parmesan cheese for a huge range of recipes, and the English use eggs and sometimes Cheddar for a spinach soufflé.

History

Spinach was first cultivated in Persia several thousands of years ago. It came to Europe via the Arab world; the Moors introduced it to Spain, and Arabs in the Middle East took it to Greece. It first appeared in England in the fourteenth century, probably via Spain. It is mentioned in the first known English cookery book, where it is referred to as *spynoches*; which echoes the Spanish word for spinach which is *espinacas*. It quickly became a popular vegetable, probably because it is quick and easy to grow and similarly easy and quick to cook.

Nutrition

Spinach is an excellent source of vitamin C if eaten raw, as well as vitamins A and B, calcium, potassium and iron. Spinach was originally thought to provide far more iron than it actually does, but the iron is "bound" up by oxalic acid in cooked spinach, which prevents the body absorbing anything but the smallest amounts. Even so, it is still an extremely healthy vegetable whether eaten cooked or raw.

Buying and Storing

Spinach grows all year round, so you should have no difficulty in buying it fresh. Frozen spinach is a poor substitute, mainly because it has so little flavour, so it is worth the effort to use the fresh product.

Spinach leaves should be green and lively; if they look tired and the stalks are floppy, shop round until you find something in better condition. Spinach reduces significantly when cooked; about 450g/1lb will serve two people. Store it in the salad drawer of the fridge, where it will keep for 1-2 days.

Preparing

Wash well in a bowl of cold water and remove any tough or large stalks.

Cooking

Throw the leaves into a large pan with just the water that clings to the leaves and place over a low heat with a sprinkling of salt. Cover the pan so the spinach steams in its own liquid and shake the pan occasionally to prevent the spinach sticking to the bottom. It cooks in 4-6 minutes, wilting down to about an eighth of its former volume. Drain and press out the remaining liquid with the back of a spoon.

Spinach can be used in a variety of ways. It can be chopped and served with lots of butter, or similarly served with other spring vegetables such as

BRUSSELS SPROUTS

baby carrots or young broad beans. For frittatas, chop the spinach finely, stir in a little Parmesan cheese, a good sprinkling of salt and pepper and a dash of cream, if liked, and stir into the omelette before cooking. Alternatively, purée it for sauces or blend it for soups. Spinach is also delicious raw, served with chopped bacon or croûtons. A fresh spinach salad is delicious as the leaves have just the right balance of flavour – sharp but not overpowering.

Below: Spinach.
Below right: Brussels sprouts.

Brussels sprouts have a pronounced and sweet nutty flavour, quite unlike cabbage, although the two are closely related. They are traditionally served at Christmas with chestnuts and indeed have a definite affinity for certain nuts – particularly the sweet flavoured nuts, e.g. almonds pair well rather than hazelnuts or walnuts.

History

Brussels sprouts were cultivated in Flanders (now Belgium) during the Middle Ages. They are basically miniature cabbages which grow in a knobbly row on a long tough stalk. The Germans call sprouts *rosenkohl* – rose cabbage – a pretty and descriptive name as they look like small rosebuds.

Buying and Storing

Buy Brussels sprouts as fresh as possible as older ones are more likely to have that strong unpleasant "cabbage" flavour. They should be small and hard with tightly wrapped leaves. Avoid any

that are turning yellow or brown or have loose leaves.

Brussels sprouts will keep for several days in a cool place such as a larder or salad drawer of a fridge, but it is far better to buy them as you need them.

Preparing

Cut away the bottom of the stalk and remove the outer leaves. Some people cut a cross through the bottom of the stalk although this is not really necessary. If you haven't been able to avoid buying big Brussels sprouts, cut them in half or into quarters, or slice them thinly for stir-frying.

Cooking

As with cabbage, either cook Brussels sprouts very briefly or braise slowly in the oven. Cook in small amounts of fast boiling water for about 3 minutes until just tender. To stir-fry Brussel sprouts, slice into three or four pieces and then fry in a little oil and butter – they taste great with onions and ginger.

CAULIFLOWER

Cauliflower is a member of the cabbage family, *Brassica oleracea*. Like all cabbages, cauliflower suffers terribly from overcooking. A properly cooked cauliflower has a pleasant fresh flavour but when overcooked it turns grey and becomes unpalatably soft, taking on a nasty rank flavour with an unpleasant aftertaste. Children often like raw cauliflower even though they may not connect it with the same vegetable served up at school.

History

Cauliflower is thought to have come originally from China and thence to the Middle East. The Moors introduced it to Spain in the twelfth century and from there it found its way to England via established trading routes. The early cauliflower was the size of a tennis ball but they have gradually been cultivated to the enormous sizes we see today. Ironically, baby cauliflowers are now fashionable.

Varieties

Green and occasionally purple cauliflowers are available in the shops. The purple variety was originally grown in Sardinia and Italy but is increasingly grown by other market gardeners. They look pretty and unusual but are otherwise similar to white cauliflowers.

Dwarf varieties of cauliflowers are now commonly available in shops, as well as baby white cauliflowers.

Romanescoes: These pretty green or white vegetables look like a cross between broccoli and cauliflower, but are more closely related to cauliflowers. They taste very much like cauliflowers, but since they are quite small, they are less likely to be overcooked and consequently retain their excellent flavour.

Broccoflower: A cross between broccoli and cauliflower, this looks like a pale green cauliflower. It has a mild flavour and should be cooked in the same way as you would cauliflower.

Right: Baby cauliflowers.
Far right top: Romanescoes.
Far right bottom: Green cauliflowers.

Nutrition

Cauliflower contains potassium, iron and zinc, although cooking reduces the amounts. It is also a good source of vitamins A and C.

Buying and Storing

In top condition, a cauliflower is a creamy white colour with the outer leaves curled round the flower. The head should be unblemished with no black or discoloured areas and the outer leaves should look fresh and crisp. Keep cauliflower in a cool place for no longer than 1-2 days; after that it will deteriorate and valuable nutrients will be lost.

Preparing

To cook a cauliflower whole, first trim away the coarse bottom leaves (leave the inner ones on, if liked). Very large cauliflowers are best halved or broken into florets, as the outside will overcook before the inside is tender. Some people trim away the stalk, but others like this part and only trim off the very thick stalk at the bottom of the plant.

Cooking

Cauliflowers are excellent steamed, either whole or in florets. Place in a steamer or colander over a pan of boiling water, cover and steam until just tender and immediately remove from the heat. The florets can then be fried in olive oil or butter for a few minutes to give a lightly browned finish.

When cooking a cauliflower whole, start testing it after 10 minutes; it should feel tender but still have plenty of "bite" left in it. Cauliflower is a popular vegetable accompaniment, either served with just a little butter, or with a tomato or cheese sauce. It is also good stir-fried with onions and garlic together with a few tomatoes and capers.

Cauliflower is excellent in salads or used for crudités. Either use it raw or blanch it in boiling water for 1-2 minutes, then refresh under cold running water.

Small cauliflowers and romanescoes are intended to be cooked whole, and can be steamed or boiled, covered with a lid, in the minimum of water for 4-5 minutes until just tender.

SPROUTING BROCCOLI AND CALABRESE

Varieties

Calabrese Broccoli: This is the vegetable we today commonly call broccoli, with large beautiful, blue-green heads on succulent stalks. It is named after the Italian province of Calabria where this variety was first developed.

Purple Sprouting Broccoli: The original variety – has long thin stalks with small flowerheads that are normally purple but can be white or green. Heads, stalks and tender leaves are all edible. The purple heads turn green when cooked but the others keep their colour. Purple sprouting broccoli is more seasonal than the easily available calabrese; it is usually available from late winter onwards.

Buying and Storing

If possible, buy loose broccoli rather than the pre-wrapped bundles, because it is easier to check that it is fresh and also because wrapped vegetables tend to deteriorate more quickly.

Purple sprouting broccoli can also be sold loose or prepacked. Check that the stalk, flowerhead and leaves all look fresh and that the flowerlets are tightly closed and bright green. Neither type will keep for long.

Broccoli or calabrese is a relatively modern vegetable and is one of the most popular. It is quick and easy to prepare with little or no waste and similarly easy to cook. It is attractive, whether served raw or cooked, and you can buy it in the quantity you require, unlike cauliflower or cabbage.

History

Before calabrese came into our shops, people bought and ate purple sprouting broccoli. This is basically an "untidy" version of calabrese, with long shoots and clusters of flowerheads at the end – the broccoli we know today has neat tidy heads. The stalks of purple sprouting broccoli have a faint asparagus flavour.

The Romans cooked purple sprouting broccoli in wine or served it with sauces and it is still a popular vegetable today in Italy, cooked in the oven with anchovies and onions or served with pasta in a garlic and tomato sauce.

Preparing and Cooking

Trim the ends and remove any discoloured leaves.

Calabrese Broccoli: Break into even-size pieces, dividing the stem and flowerhead lengthways if they are thick. Cook in a little boiling water for 4-5 minutes until just tender and then drain. Do not steam this variety of broccoli as its vibrant green colour tends to turn grey.

Purple Sprouting Broccoli: Either steam in long, even-size lengths in a steamer or, if you have an asparagus kettle, cook as you would asparagus. Alternatively, tie the stems loosely together and stand in a little water – if necessary, wedge it in with a potato or rolled up piece of foil. Cover with a dome of foil and steam for 4-5 minutes until tender.

Serving

Serve both varieties simply with butter and lemon juice or with a Hollandaise or Béarnaise sauce as an accompaniment. They are also excellent stir-fried.

Above far left: Purple cauliflower.
Below far left: Purple sprouting broccoli.
Above: Calabrese broccoli.
Below: Turnip tops.

TURNIP TOPS

Turnip tops, like beet greens, are both delicious and nutritious. They are not widely available but if you are able to buy some or if you grow your own, slice them *(below)* and boil or steam for a few minutes, then drain and serve with butter.

CABBAGE

Cabbage, sliced and cooked, can be one of two things: deliciously crisp, with a mild pleasant flavour – or overcooked and horrible! Cabbage and other brassicas contain the chemical hydrogen sulphide, which is activated during cooking at about the point the vegetable starts to soften. It eventually disappears, but during the in-between time, cabbage acquires its characteristic rank smell and flavour. So, either cook cabbage briefly, or cook it long and slow, preferably with other ingredients so that flavours can mingle.

History

Cabbage has a long and varied history. However, because there are many varieties of cabbage under the general heading of "brassica", it is difficult to be sure whether the variety the Greeks and Romans enjoyed is the same as today's round cabbage, or something more akin to kale or even Chinese cabbage.

The round cabbages we know today were an important food during the Dark Ages, and by the Middle Ages they were in abundance, as you will see if you study the paintings of that period. These commonly show kitchen tables or baskets at market positively groaning with fruit and vegetables, and cabbages in all their shapes and sizes were often featured.

Medieval recipes suggest cooking cabbages with leeks, onions and herbs. In the days when all except the very wealthy cooked everything in one pot, it is fair to assume that cabbages were cooked long and slow until fairly recently.

Varieties

Savoy Cabbage: This is a variety of green cabbage with crimped or curly leaves. It has a mild flavour and is particularly tender, thus needing less cooking than other varieties.

Spring Greens: These have fresh loose heads with a pale yellow-green heart. They are available in spring and are delicious simply sliced, steamed and served with butter.

Right: Savoy cabbage.
Above far right: Green cabbage.
Below far right: Spring greens.

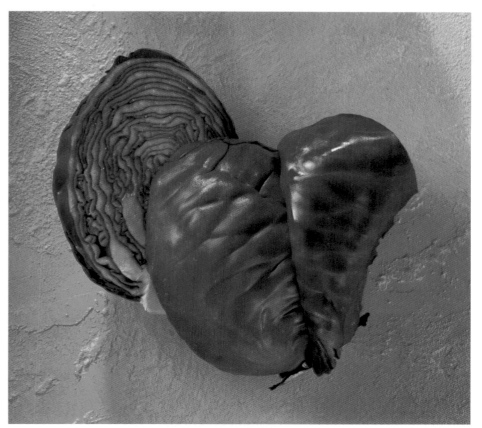

Green Cabbage: The early green, or spring, cabbages are dark green, loose leafed and have a slightly pointed head. They have little or no heart as they are picked before this has had time to develop. Nevertheless, they are a very good cabbage and all but the very outside leaves should be tender. As the season progresses, larger, firmer and more pale green cabbages are available. These are a little tougher than the spring cabbages and need longer cooking.

Red Cabbage: A beautifully coloured cabbage with smooth firm leaves. The colour fades during cooking unless a little vinegar is added to the water. Red cabbage can be pickled or stewed with spices and flavourings.

White Cabbage: Sometimes called Dutch cabbages, white cabbages have smooth firm pale green leaves. They are available throughout the winter. They are good cooked or raw. To cook, slice them thinly, then boil or steam and serve with butter. To serve raw, slice thinly and use in a coleslaw.

Buying and Storing

Cabbages should be fresh looking and unblemished. When buying, avoid any with wilted leaves or those that look or feel puffy. Savoys and spring greens will keep in a cool place for several days; firmer cabbages will keep happily for much longer.

Preparing

Remove the outer leaves, if necessary, and then cut into quarters. Remove the stalk and then slice or shred according to your recipe or to taste.

Cooking

For green or white cabbages, place the shredded leaves in a pan with a knob of butter and a couple of tablespoons of water to prevent burning. Cover and cook over a medium heat until the leaves are tender, occasionally shaking the pan or stirring.

Red cabbage is cooked quite differently and is commonly sautéed in oil or butter and then braised in a low oven for up to 1½ hours with apples, currants, onions, vinegar, wine, sugar and spices.

KALE <u>AND</u> CURLY KALE

Kale is the name used for a variety of green leafed vegetable of the brassica family. Most kales have thick stems and robust leaves that do not form a head. Many kales have curly leaves, which are the variety most commonly eaten. Large coarse-leafed kales are grown for cattle and sheep feeds.

History

Kale is thought to be one of the first cultivated brassicas. Colewort, the wild ancestor, still grows along the coasts of western Europe.

Varieties

Collards: Collards, or collard greens, are a popular green vegetable in the southern United States. They are grown in summer and autumn for harvesting in the spring and are a good source of vitamin A.
Curly Kale: With its crimped, curly leaves, this is the most commonly available kale, although even this can be quite hard to come by. If you are a big fan and don't grow your own, try farm shops in early spring.
Purple or Silver Kale: This is an ornamental variety, and is grown almost exclusively for display.

Preparing and Cooking

Kale is probably the strongest tasting of the brassicas and is best cooked simply, paired with a bland-flavoured vegetable, such as potatoes. To prepare, break the leaves from the stalk and then cut out any thick stalk from the leaf. This can then be rolled and sliced or cooked whole. Boil the leaves in a little salted water for 3-5 minutes until tender. Owing to its robust nature, kale is frequently teamed with fairly hot spices and is consequently popular in many Indian dishes.

Above far left: White cabbage.
Below far left: Red cabbage.
Left: Collards.
Above: Curly kale.

GARDEN AND WILD LEAVES

VINE LEAVES

All leaves from vines that produce grapes can be eaten when young. They make an ideal wrapping for various meats and vegetables as they are surprisingly strong and of course edible. Most countries that produce wine will have dishes where vine leaves appear. *Dolmades*, commonly eaten in Greece and the Middle East is perhaps the best known dish, but in France, Spain and Italy, there are recipes using vine leaves to wrap small birds, like quail or snipe.

Vine leaves have a faintly lemon/cabbage flavour which can be detected at its best in a good *dolmades*. The leaves need to be cooked briefly before using, so that they are pliable and don't crack or break as you wrap the food. Bring to the boil and simmer for about 1 minute. The leaves should then be drained and separated until you are ready to use them.

DANDELION

Any child who has picked dandelions for his or her rabbits or guinea pigs and has watched them gobble them up greedily will know that this weed, though hated by the gardener in the family, has something going for it. Some gardeners, of course, are very partial to dandelion and raise the plant carefully so that the leaves are fresh and tender for the salad, and in France dandelions can often be seen for sale at market.

Look in any book of herbal remedies,

and dandelions will feature prominently. They are a well-known diuretic, their French name - *pissenlits* (piss-a-bed) – attesting to this in no uncertain terms.

Although it's gratifying to pick your own vegetables for free, it is generally recommended, if you like dandelions, that you buy the domestic seeds and grow your own. These are likely to be the juiciest

and least bitter plants. If you do pick your own, do so well away from the roadside and wash the leaves carefully.

Dandelion leaves can be added to salads or used in *pissenlits au lard*, where whole young dandelion plants are dressed in vinaigrette and then covered in finely chopped pieces of salt pork or bacon and bacon fat.

SORREL

Sorrel is not always available commercially, although it is in France and is greatly prized. However, it grows wild in cool soils or you could grow it in your own garden. Young leaves are delicious in salads, or, later in the year, use it in soups or sauces to accompany fish. It has a sharp, distinct, lemon flavour and is commonly teamed with eggs and cream.

ORACHE

Although not related to spinach, this beautiful red or golden-leaved plant is called mountain spinach and its large leaves can be treated like spinach.

GOOD KING HENRY AND FAT HEN

These are both members of the goose-foot family and were popular vegetables in Tudor times. Today, Good King Henry has all but disappeared, and Fat Hen only grows wild as a weed. Both were superseded by spinach, which they are said to resemble in taste, although Fat Hen is milder.

NETTLES

Wild food enthusiasts get very excited about nettles as food, perhaps because they are plentiful and free and maybe because they take pleasure in eating something that everyone else avoids. Of course, once cooked, the sting completely disappears. They should be picked when they are very young and are good used in soups.

Above far left: Fresh vine leaves.
Above left: Dandelion.
Below left: Sorrel.
Right: Good King Henry.
Below: Nettles.

CHINESE GREENS

CHINESE CABBAGE/LEAVES (PE-TSAI)

Chinese cabbage, also called Napa cabbage, has pale green, crinkly leaves with long, wide, white ribs. Its shape is a little like a very fat head of celery, which gives rise to another of its alternative names, celery cabbage. It is pleasantly crunchy with a faint cabbage flavour and since it is available all year round, it makes a useful winter salad component. Chinese cabbage is also very good stir-fried with a tasty sauce. It is an essential ingredient of many oriental recipes.

Buying and Storing

For some reason, Chinese leaves almost always look fresh and perky when on sale in the supermarket, which probably indicates that they travel well and are transported quickly. Avoid any with discoloured or damaged stems. The leaves should be pale green and straight without blemishes or bruises. They will keep for up to six days in the salad drawer of the fridge.

Preparing

Remove the outside leaves and slice as much as you need.

Many Chinese greens are members of the brassica family. If you go into a popular and reasonably large Chinese supermarket, you'll be astonished at the varieties of green vegetables for sale. Discovering the name of these vegetables, on the other hand, can be a bit of a hit-and-miss undertaking, as the shop keepers, although always well intentioned, rarely know the English name, if indeed there is one.

CHINESE MUSTARD GREENS

Mustard greens are worth buying if you can, as they are very good to eat. The plant is a member of the cabbage family, but is grown in Europe solely for its mustard seed. In India and Asia it has long been grown for its oil seed, but the Chinese developed the plant for its leaves as well. These are deep green and slightly puckered-looking and have a definite mustard flavour, which can be quite fiery.

If you grow your own, then you'll be able to enjoy the young leaves which can be added to lettuce to spice up salads. Older leaves are best stir-fried and then dressed with a light Chinese sauce. They are also good cooked with onion and garlic and served as a side dish to accompany pork or bacon.

Preparing and Cooking

Break apart the stalks, rinse, then cut both stalks and leaves into thick or thin slices. These can then be stir-fried with garlic and onions, or cooked and served as you would Swiss chard. It has a pleasant flavour, milder than mustard greens, yet with more bite than the bland Chinese cabbage.

CHINESE BROCCOLI

This is another leafy vegetable, but with slender heads of flowers that look a little like our own broccoli, except that the flowers are usually white or yellow. Once the thicker stalks are trimmed, the greens can be sliced and cooked and served in the same way as Chinese mustard greens.

Far left: Chinese mustard greens.
Left: Chinese leaves.
Above: Chinese broccoli.
Below: Pak-Choi.

Cooking and Serving

If adding to salads, combine Chinese cabbage with something fairly forceful, like endive or rocket, and add a well-flavoured dressing. If adding to a stir-fry, cook with garlic, ginger and other fairly strong flavours. While the faint cabbage flavour will be lost, you will still get the pleasant crunchy "bite" of the stalk, and the leaves will carry the sauce.

PAK-CHOI

If you frequent your local Chinese super-market, you will almost certainly have come across *Pak-Choi*. In English it should correctly be called Chinese celery cabbage and its thick stalks, joined at the end in a small root, are vaguely celery-like. Its leaves, on the other hand, are generally large and spoon shaped. There are many different species of this vegetable, and smaller specimens look more like the tops of radish and have small slim stalks. Consequently the vegetable can be known by all sorts of picturesque names, like "horse's ear" and "horse's tail". There is no rule for discovering exactly what you are buying, but the important thing is to choose a fresh plant whatever its size: look for fresh green leaves and crisp stalks.

KOHLRABI

Kohlrabi looks like a cross between a cabbage and a turnip and is often classified as a root vegetable, even though it grows above ground. It is a member of the brassica family, but, unlike cabbages, it is the bulbous stalk that is edible rather than the flowering heads.

There are two varieties of kohlrabi: one is purple and the other pale green. They both have the same mild and fresh tasting flavour, not dissimilar to water chestnuts. It is neither as peppery as turnip nor as distinctive as cabbage, but it is easy to see why people think it a little like both. It can be served as an alternative to carrots or turnips.

History

Although kohlrabi is not a very popular vegetable in Britain, it is commonly eaten in other parts of Europe, as well as in China, India and Asia. In Kashmir, where it is grown extensively, there are many recipes – the bulbs are often finely sliced and eaten in salads and the greens are cooked in mustard oil with garlic and chillies.

Buying and Storing

Kohlrabi is best when small and young, since larger specimens tend to be coarse and fibrous. It keeps well for 7-10 days if stored in a cool place.

Preparing

Peel the skin with a knife and then cook whole or slice.

Cooking

Very small kohlrabi are tender and can be cooked whole. However, if they are any bigger than 5cm/2in in diameter, they can be stuffed. To do this, hollow out a little before cooking and then stuff with fried onions and tomatoes for instance. For sliced kohlrabi, cook until just tender and serve with butter or a creamy sauce. They can also be cooked long and slow in gratin dishes, with, for instance, potatoes as a variation of *Gratin Dauphinois*. Alternatively, par-boil them and bake in the oven covered with a cheese sauce.

SWISS CHARD

Swiss chard is one of those vegetables that needs plenty of water when growing, which explains why it is a popular garden vegetable in many places which have a high rainfall. Gardeners are very fond of Swiss chard, not only because it is delicious to eat but also because it is so very striking.

Swiss chard is often likened to spinach. The leaves have similarities, although they are not related and chard is on an altogether larger scale. Swiss chard leaves are large and fleshy with distinctive white ribs, and the flavour is stronger and more robust than spinach. It is popular in France where it is baked with rice, eggs and milk in *tians*, and cooked in a celebrated pastry from Nice – *tourte de blettes* – which is a sweet tart filled with raisins, pinenuts, apples and Swiss chard bound together with eggs. It is also often combined with eggs in frittatas and tortillas.

Swiss chard is a member of the beet family and is called by several names on this theme, including seakale beet and spinach beet.

Ruby or rhubarb chard has striking red ribs and leaf beet is often cultivated as a decorative plant, but they both have the same flavour, and unlike sugar beet and beetroot, they are cultivated only for their leaves.

Buying and Storing

Heads of chard should be fresh and bright green; avoid those with withered leaves or flabby stems. It keeps better than spinach but should be eaten within a couple of days.

Preparing

Some people buy or grow chard for the white stems alone and discard the leaves (or give them to pet guinea pigs), but this is a waste of a delicious vegetable. The leaf needs to be separated from the ribs, and this can be done roughly with a sharp knife (*right*) or more precisely using scissors. The ribs can then be sliced. Either shred the leaves, or blanch them and use them to wrap little parcels of fragrant rice or other food. If the chard is young and small, the ribs do not need to be removed.

Cooking

For pies, frittatas and gratins, the leaves and ribs can be cooked together. Gently sauté the ribs in butter and oil and then add the leaves a minute or so later. Alternatively, the ribs can be simmered in a little water until tender and the leaves added a few minutes later or steamed over the top.

Above far left: Kohlrabi.
Left: Purple kohlrabi.
Above: Swiss chard.

FIVE

BEANS, PEAS AND SEEDS

Broad Beans

Runner Beans

Peas

Green Beans

Sweetcorn

Okra

Dried Beans and Peas

BROAD BEANS

One of the delights of having a garden is discovering how truly delicious some vegetables are when garden fresh. This seems particularly true of broad beans, which have a superb sweet flavour that sadly can never be reproduced in the frozen product. If you're lucky enough to grow or be given fresh broad beans, don't worry about recipes; just cook them until tender and serve with butter. It will be a revelation! However, if you're not one of those lucky few, don't dismiss broad beans, as they are still a wonderfully versatile vegetable. They can be used in soups or casseroles, and, since they have a mealy texture, they also purée well.

History

People have been eating broad beans almost since time began. A variety of wild broad bean grew all over southern Europe, North Africa and Asia, and they would have been a useful food for early man. There is archaeological evidence that by Neolithic times broad beans were being farmed, making them one of the first foods to be cultivated.

Broad beans will grow in most climates and most soils. They were a staple food for people throughout the Dark Ages and the Middle Ages, grown for feeding people and livestock until being replaced by the potato in the seventeenth and eighteenth centuries. Broad beans were an important source of protein for the poor, and because they dry well, they would have provided nourishing meals for families until the next growing season.

Nutrition

Beans are high in protein and carbohydrates and are also a good source of vitamins A, B1 and B2. They also provide potassium and iron as well as several other minerals.

Buying and Storing

Buy beans as fresh as possible. The pods should preferably be small and tender. Use as soon as possible.

Preparing

Very young beans in tender pods, no more than 7.5cm/3in in length, can be eaten pod and all; top and tail, and then slice roughly. Usually, however, you will need to shell the beans. Elderly beans are often better skinned after they are cooked to rid them of the strong, bitter flavour that puts many people off this vegetable.

Cooking

Plunge shelled beans (or in their pods if very young) into rapidly boiling water and cook until just tender. They can also be par-boiled and then finished off braised in butter. For a simple broad bean purée, blend the cooked beans with garlic cooked in butter, cream and a pinch of fresh herbs, such as savory or thyme.

LIMA BEANS

These are popular in the US, named after the capital of Peru, and are sold mainly shelled. They are an essential ingredient in the American-Indian dish *succotash*. Lima beans should be cooked in a little boiling water until tender. Elderly beans need skinning after they are cooked. The dried bean, also known as the butter bean, can be large or small. These large beans tend to become mushy when cooked so are best used in soups or purées.

Above: Broad beans.
Below: Lima beans.
Right: Runner beans.

RUNNER BEANS

The runner bean is native to South America, where it has been cultivated for more than 2000 years and there is archaeological evidence of its existence much earlier than that.

It is a popular vegetable to grow. Most home vegetable gardeners have a patch of runner beans – they are easy to grow and, like all legumes, their roots contain bacteria that help renew nitrogen supplies in the soil.

They have a more robust flavour and texture than French beans and are distinct from green beans in several ways: they are generally much larger with long, flattened pods; their skin is rough textured, although in young beans this softens during cooking; and they contain purple beans within the pods, unlike green beans whose beans are mostly white or pale green. Nevertheless, runner beans belong to the same family as all the green beans.

Buying and Storing

Always buy young beans as the pods of larger beans are likely to be tough. The pods should feel firm and fresh; if you can see the outline of the bean inside the pod it is likely to be fibrous – although you could leave the beans to dry out and use the dried beans later in the season. Ideally, the beans inside should be no larger than your small fingernail.

Use as soon as possible after buying; they do not store well.

Preparing

Runner beans need to be topped and tailed and may also need stringing. Carefully put your knife through the top of the bean without cutting right through, and then pull downwards; if a thick thread comes away, the beans need stringing, so do the same on the other side. The beans can then be sliced either using a sharp knife or a slicer.

Slice through lengthways, not diagonally, so that you will be able to serve the beans with just a little skin and lots of succulent flesh.

Cooking

Plunge the beans into boiling salted water and cook until *al dente*.

PEAS

Fresh peas are wonderful – try tasting them raw, straight from the pod. Unfortunately, the season for garden peas is short, and frozen peas, which are the next best thing, never quite come up to the mark. If you grow your own peas, for three or four weeks in early summer, you can eat like kings; otherwise you can buy them from a good greengrocer, who may be able to keep you supplied all through the early summer.

History

Peas have an even longer history than broad beans, with archaeological evidence showing they were cultivated as long ago as 5700 BC. High in protein and carbohydrate, they would have been another important staple food and were eaten fresh or dried for soups or potage.

Pease porridge is mentioned in a Greek play written in 5 BC. Pease pudding, probably something similar, made with split peas with onion and herbs, is an old-fashioned but still very popular dish, especially in the north of England, traditionally eaten with ham and pork.

One of the first recipes for peas, however, comes from *Le Cuisinier Français*, which was translated into English in the middle of the seventeenth century and gives a recipe for *petits pois à la française* (peas cooked with small hearted lettuces) – still a popular recipe today.

Varieties

Mangetouts: These are eaten whole and have a delicate flavour, providing they are not overcooked. Unfortunately, they are easy to overcook and the texture then becomes rather slippery. Alternatively, blanch or stir-fry them. They are also good served raw in salads.

Petits Pois: These are not, as you might expect, immature peas but are a dwarf variety. Gardeners grow their own, but they are not available fresh in the shops as they are mainly grown commercially for canning or freezing.

Snow Peas, Sugar Peas, Sugar Snaps: These have the distinct fresh flavour of raw peas and are more plump and have more snap than mangetouts.

Buying and Storing

Only buy fresh peas; if they are old they are bound to be disappointing and you would be better off buying them frozen. In top condition, the pods are bright green and lively looking; the more withered the pod, the longer they have been hanging around. It is possible to surreptitiously sample peas on occasions, to check if they are fresh (greengrocers don't seem to mind if you buy some). Use fresh peas as soon as possible.

Preparing

Shelling peas can be very relaxing. Press open the pods and use your thumb to push out the peas *(below)*. Mangetouts and sugar snaps just need to be topped and tailed *(below right)*.

Left: Peas.
Above right: Sugar snap peas.

Cooking

Cook peas with a sprig of mint in a pan of rapidly boiling water or in a covered steamer until tender. Alternatively, melt butter in a flameproof casserole, add the peas and then cover and sweat over a gentle heat for 4–5 minutes. Cook mangetouts and sugar snaps in any of the same ways but for a shorter time.

GREEN BEANS

slim in shape. They should be eaten when very young, no more than 6-7.5cm/2½-3in in length.

Thai Beans: These long beans are similar to French beans and can be prepared and cooked in the same way.

Yellow Wax Beans: This is also a French bean and has a mild, slightly buttery taste.

Buying and Storing

Whatever variety, beans should be bright and crisp. Avoid wilted ones, or those with overly mature pods which feel spongy when lightly squeezed. They do not keep well, so use as soon as possible after buying or picking.

Preparing

To top and tail the beans: gather them together in one hand and then slice away the top 5mm/¼in *(below)*, then do the same at the other end. If necessary, pull off any stringy bits.

Cooking

Plunge beans into rapidly boiling salted water and cook until *al dente*. When overcooked, beans have a flabby texture and also lose much of their flavour. Drain and toss them in butter or serve in a sauce with shallots and bacon. For salads, cook until just tender and then refresh under cold water. They are excellent with a garlicky vinaigrette. Serve with carrots or other root vegetables and savour the contrast in flavours.

Whether you call beans French beans, wax beans, haricots or green beans, they all belong to a large and varied family.

History

The bean is a New World vegetable that had been cultivated for thousands of years by native people in both the north and south of the continent, which accounts for its wide diversity.

Varieties

One variety or another is available all year round and so they are one of the most convenient fresh green vegetables.

French Beans: This name encompasses a range of green beans, including the snap bean and bobby bean. They are mostly fat and fleshy and when fresh, should be firm so that they break in half with a satisfying snapping sound.

Haricots Verts: These are considered the best French beans and are delicate and

Above left: Bobby beans.
Below left: Yellow wax beans.
Right: Haricots verts.
Below: Thai beans.

SWEETCORN

Fresh sweetcorn, eaten on the cob with salt and a little butter is deliciously sweet. Some gardeners who grow it have a pan ready on the boil, so that when they cut the corn it goes into the pan in only the time it takes to race up from the garden to the kitchen. Buying it from the supermarket is inevitably a bit hit-or-miss, although if purchased in season, sweetcorn can be very good indeed.

History

In 1492, as Christopher Columbus disembarked on the island now called Cuba, he was met by American Indians offering two gifts of hospitality – one was tobacco and the other something the Indians called *maïs*. The English word for staple food was then corn, so that when Columbus and his crew saw that maize was the staple food for the Indians, it was dubbed "Indian corn".

Corn originated in South America and had enormous significance to the native Indians of the whole continent, who were said to have lived and died by corn. They referred to it as their "first mother and father, the source of life". By far their most important food, corn was used in many other ways as well. They used the plant for their shelters and for fences, and they wore it and decorated their bodies with it.

The Aztecs had corn planting cere-monies that included human sacrifices, and other tribes had similar customs to appease the god "corn". Countless myths and legends have been woven around corn, each tribe telling a slightly different story, but each on the same theme of planting and harvesting corn. For anthropologists and historians, they make compelling study.

Nutrition

Corn is a good carbohydrate food and is rich in vitamins A, B and C. It contains protein, but less so than most other cere-als. It is also a good source of potassium, magnesium, phosphorus and iron.

Varieties

There are five main varieties of corn – popcorn, sweet corn, dent corn, flint corn and flour corn. Dent corn is the most commonly grown worldwide, for

animal feeds and oil, and the corn we eat on the cobs is sweet corn. Baby sweetcorn cobs are picked when immature and are cooked and eaten whole.

Buying and Storing

As soon as corn is picked, its sugar begins to turn to starch and therefore the sooner it goes into the pot, the better. Wherever possible, buy locally grown sweetcorn.

Look for husks that are clean and green and tassels which are golden, with no sign of matting. The corn itself should look plump and yellow. Avoid cobs with pale or white kernels or those with older shrivelled kernels which will undoubtedly be disappointing.

Preparing

Strip away the husks. To use the kernels for recipes, cut downwards using a sharp knife from top to bottom *(left)*.

Cooking

Cook corn-on-the-cob in plenty of boiling salted water until tender. Timing depends on the size of the cobs but 10-15 minutes will normally be enough. Serve them with sea salt and butter, but if the cobs are really sweet, leave out the butter. Stir-fry baby sweetcorn cobs briefly and serve in oriental dishes.

Far left: Sweetcorn cobs.
Below: Baby sweetcorn cobs.

OKRA

History

Okra originated in Africa. In the sixteenth century, when African people were enslaved by the Spanish and shipped to the New World, they took with them the few things they could, including the plants and seeds from home – dried peas, yams, ackee – and okra. This lantern-shaped pod containing rows of seeds oozes a sticky mucilaginous liquid when cooked, and it was popular not only for its subtle flavour but also for thickening soups and stews.

The plant thrived in the tropical climate and by the early nineteenth century, when the slave trade was finally abolished, okra was an important part of the cuisine of the Caribbean and the southern United States. In and around New Orleans, the Creoles, the American-born descendants of European-born settlers, adopted a popular native American-Indian dish called *gumbo*. An essential quality of this famous dish was its thick gluey consistency. The Indians used filé powder (the dry pounded leaves of the sassafras tree), but okra was welcomed as a more satisfactory alternative.

Gumbos are now the hallmark of Creole cooking, and in some parts of America, the word "gumbo" is an alternative word for okra itself.

Buying and Storing

Choose young, small pods as older ones are likely to be fibrous. They should be bright green, firm and slightly springy when squeezed. Avoid any that are shrivelled or bruised. They will keep for a few days in the salad drawer of the fridge.

Preparing

When cooking whole, trim the top but don't expose the seeds inside or the viscous liquid will ooze into the rest of the dish. If, however, this is what you want, slice thickly or thinly according to the recipe *(right)*. If you want to eliminate some of this liquid, first soak the whole pods in acidulated water (water to which lemon juice has been added) for about an hour.

Cooking

The pods can be steamed, boiled or lightly fried, and then added to or used with other ingredients. If cooked whole, okra is not mucilaginous but is pleasantly tender. Whether cooked whole or sliced, use garlic, ginger or chilli to perk up the flavour, or cook Indian-style, with onions, tomatoes and spices.

Above: Okra.

DRIED BEANS AND PEAS

Dried beans feature in traditional cuisines all over the world, from Mexican re-fried beans to Italy's *pasta e fagioli*. They are nutritious, providing a good source of protein when combined with rice, and are a marvellous store cupboard standby.

Black-eyed Beans: Sometimes called black-eyed peas, these small cream-coloured beans have a black spot or eye. When cooked, they have a tender, creamy texture and a mildly smoky flavour. Black-eyed beans are widely used in Indian cooking.

Chana Dhal: Chana dhal is very similar to yellow split peas but smaller in size and with a slightly sweeter taste. It is used in a variety of vegetable dishes.

Chick-peas: These round beige-coloured pulses have a strong, nutty flavour when cooked. As well as being used for curries, chick-peas are also ground into a flour which is widely used in many Indian dishes such as *pakoras* and *bhajees*.

Flageolet Beans: Small oval beans which are either white or pale green in colour. They have a very mild, refreshing flavour and feature in classic French dishes.

Green Lentils: Also known as continental lentils, these have quite a strong flavour and retain their shape during cooking. They are very versatile and are used in a number of dishes.

Haricot Beans: Small, white oval beans which come in different varieties. Haricot beans are ideal for Indian cooking because not only do they retain their shape but they also absorb the flavours of the spices.

Kidney Beans: Kidney beans are one of the most popular pulses. They are dark red/brown, kidney-shaped beans with a strong flavour.

Mung Beans: These are small, round green beans with a slightly sweet flavour and creamy texture. When sprouted they produce the familiar beansprouts.

Red Split Lentils: A readily available lentil that can be used for making dhal. Use instead of toovar dhal.

Toovar Dhal: A dull orange-coloured split pea with a very distinctive earthy flavour. Toovar dhal is available plain and in an oily variety.

Soaking and Cooking Tips

Most dried pulses, except lentils, need to be soaked overnight before cooking. Wash the beans thoroughly and remove any small stones and damaged beans. Put into a large bowl and cover with plenty of cold water. When cooking, allow double the volume of water to beans and boil for 10 minutes. This initial boiling period is essential to remove any harmful toxins. Drain, rinse and cook in fresh water. The cooking time for all pulses varies depending on the type and their freshness. Pulses can be cooked in a pressure cooker to save time. Lentils, on the whole, do not need soaking. They should be washed in several changes of cold water before being cooked.

Left: Clockwise from bottom right: Mung beans, Flageolet beans, Chick-peas, Haricot beans, Black-eyed beans, Kidney beans.
Above: Clockwise from top: Red split lentils, Green lentils, Toovar dhal, Chana dhal.

SQUASHES

COURGETTES

Courgettes are the best loved of all the squashes as they are so versatile. They are quick and easy to cook and are succulent and tender with a delicate, unassuming flavour. Unlike other squashes, they are available all year round.

Vegetables taste best when eaten immediately after they have been picked, and this particularly applies to courgettes. They have a long season and are good to grow since the more you cut, the more the plants produce. Left unchecked, they turn into marrows.

Varieties

The courgette is classified as a summer squash, *cucurbita pepo*, along with marrows and pattypan squashes.

Courgettes: Sometimes called zucchini, courgettes are basically immature marrows. The word is a diminutive of the French *courge*, meaning marrow, and similarly zucchini means miniature *zucca*, Italian for gourd. Courgettes have a deep green skin, with firm pale flesh. The seeds and pith found in marrows have yet to form but are visible in more mature courgettes. Conversely, the prized baby courgettes have no suggestion of seeds or pith and the flesh is completely firm.

Yellow Courgettes: These are bright yellow and somewhat straighter than green courgettes. They have a slightly firmer flesh than green courgettes but are otherwise similar.

Pattypan Squashes: These little squashes look like tiny custard squashes. They can be pale green, yellow or white and have a slightly firmer texture than courgettes, but a similar flavour. They can be sliced and grilled in the same way as courgettes but, to make the most of their size and shape, steam them whole until tender.

Summer Crooknecks: Pale yellow with curves at the neck and a bumpy skin, crooknecks are prepared and cooked in the same way as courgettes.

Italian Courgettes: These very long, thin courgettes are grown in Italy. They are treated liked ordinary courgettes but are strictly a bottle gourd.

Buying and Storing

Courgettes should be firm with a glossy, healthy looking skin. Avoid any that feel squashy or generally look limp, as they will be dry and not worth using. Choose small courgettes whenever possible and buy in small quantities as needed.

Preparing

The tiny young courgettes need no preparation at all, and if they still have their flowers, so much the better. Other courgettes should be topped and tailed and then prepared according to the recipe, either sliced or slit for stuffing.

Cooking

Baby courgettes need little or no cooking. Steam them whole or just blanch them. Sliced larger courgettes can be steamed or boiled but take care that they do not overcook as they go soggy very quickly. Alternatively, grill, roast or fry them. Try dipping slices in a light batter and then shallow frying in a blend of olive and sunflower oil. To roast, place them in a ovenproof dish, scatter with crushed garlic and a few torn basil leaves and sprinkle with olive oil; then bake in a very hot oven until tender, turning the slices occasionally.

Top left: Pattypan squashes.
Far left: Baby courgettes.
Left: Yellow courgettes.
Right: Italian courgettes beside white and green courgettes.

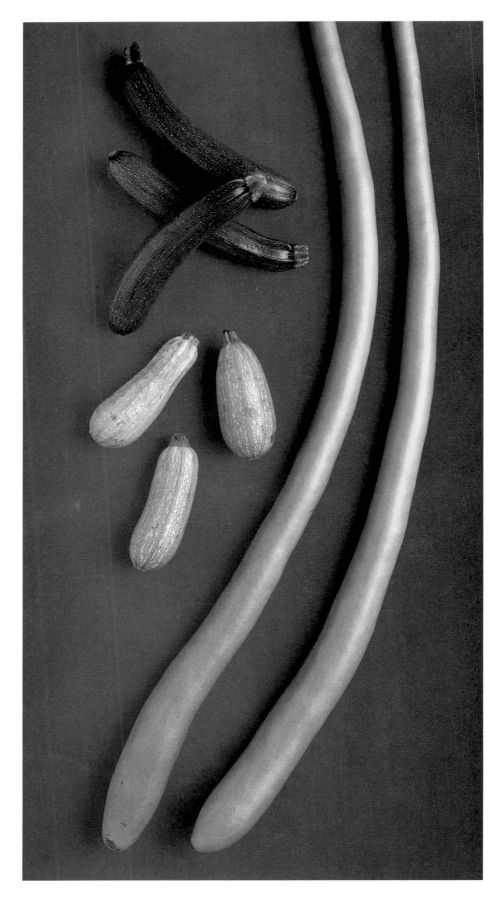

MARROWS AND SUMMER SQUASHES

Vegetable marrow is classified as a summer squash yet it is rather the poor relation of squashes. Most of the edible flesh is water and at best it is a rather bland vegetable, with a slightly sweet flavour. At worst, it is insipid and if cooked to a mush (which isn't unheard of), it is completely tasteless.

Marrows can be stuffed, although it involves a lot of energy expended for very little reward; but marrow cooked over a low heat in butter with no added water (so that it steams in its own juice) brings out the best in it.

History

Marrows, like all the summer and winter squashes, are native to America. Squashes were eaten by native American-Indians, traditionally with corn and beans, and in an Iroquois myth the three vegetables are represented as three inseparable sisters. Although the early explorers would almost certainly have come into contact with them, they were not brought back home, and vegetable marrow was not known in England until the nineteenth century. Once introduced, however, it quickly became very popular. Mrs Beeton gives eight recipes for vegetable marrow and observes that "it is now extensively used". No mention at all is made of courgettes, which of course are simply immature marrows, as any gardener will know.

Varieties

The word "marrow" as a general term tends to refer to the summer squashes. At the end of summer and in the early autumn a good variety of the large summer squashes is available.

Vegetable Marrows: This is the proper name for the large prize marrows, beloved of harvest festivals and country fairs. Buy small specimens whenever possible.

Spaghetti Squashes/Marrows: Long and pale yellow, like all marrows these squashes can grow to enormous sizes, but buy small specimens for convenience as well as flavour. They earned their name from the resemblance of the cooked flesh to spaghetti.

To boil a spaghetti squash, first pierce the end, so that the heat can reach the middle, then cook for about 25 minutes or until the skin feels tender. Cut the squash in half lengthways, remove the seeds, and then fork the strands of flesh out on to a plate. It has a fragrant, almost honey and lemon flavour and tastes good with garlic butter or pesto.

Custard Marrows: These are pretty, pale green squashes with scalloped edges and a similar flavour to courgettes. If possible, buy small specimens, about 10cm/4in across. Boil these whole until tender, then cut a slice off the tops, scoop out the seeds and serve with a knob of butter.

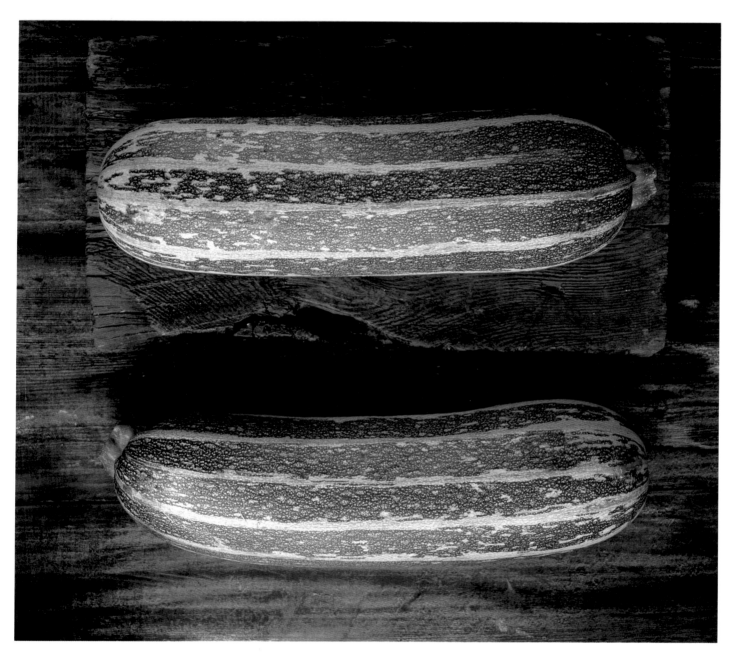

Buying and Storing

Buy vegetables that have clear, unblemished flesh and avoid any with soft or brown patches. Vegetable marrows and spaghetti squashes will keep for several months provided they are kept in a cool, dark place. Custard marrows will keep up to a week.

Preparing

Wash the skin. For sautéing or steaming, or if the skin is tough, peel it away. For braised marrow, cut into chunks and discard the seeds and pith (*right*). For stuff-

ing, cut into thick slices or cut lengthways and discard the seeds and pith.

Cooking

Place chunks of marrow in a heavy-based pan with a little butter, cover and cook until tender. It can then be livened up with garlic, herbs or tomatoes. For stuffed marrow, blanch first, stuff, then cover or wrap it in foil to cook.

Above: Vegetable marrows.
Far left: Spaghetti squashes.
Left: Custard marrows.

PUMPKINS AND WINTER SQUASHES

Pumpkins are the most famous of the winter squashes; aesthetically they are one of nature's most pleasing vegetables for their huge size, their colour and the smoothness of their skin. They originally came from America and, from a culinary point of view, they have their home there.

The name squash comes from America and as well as pumpkins, the family includes acorn, butternut and turban squashes to name but a few. There are simply hundreds of different squashes, including Sweet Dumpling, Queensland Blue (from Australia), Calabaza, Cushaw and Golden Nugget.

History

The tradition of eating pumpkin at Thanksgiving came from when the Pilgrim Fathers, who had settled in New England, proclaimed a day of thanksgiving and prayer for the harvest. The early tradition was to serve the pumpkin with its head and seeds removed, the cavity filled with milk, honey and spices, and baked until tender. The custom of eating pumpkin at Thanksgiving has remained but it is now served in a different way: puréed pumpkin, either fresh or tinned, is used to make golden tarts.

Varieties

There are a huge number of varieties of winter squashes and, confusingly, many are known by several different names. However, from a cooking point of view, most are interchangeable although it is best to taste dishes as you cook them, as seasoning may differ from one to the other. In general, they all have a floury and slightly fibrous flesh and a mild, almost bland flavour tinged with sweetness. Because of this blandness, they harmonize well with other ingredients.

Acorn Squashes: These are small and heart-shaped with a beautiful deep green or orange skin, or a mixture of the two. Peel, then use as for pumpkins or bake whole, then split and serve with butter.

Butternut Squashes: Perfectly pear-shaped, these are a buttery colour. Use in soups or in any pumpkin recipe.

Delicata Squashes: This pretty pale yellow squash has a succulent yellow flesh,

tasting like a cross between sweet potato and butternut squash.

English Pumpkins: These have a softer flesh than the American variety and are good for soups or, if puréed, combined with potatoes or other root vegetables.

Hubbard Squashes: These large winter squashes have a thick, bumpy, hard shell which can range in colour from bright orange to dark green. If they are exceptionally large, they are sometimes sold in halves or large wedges. They have a grainy texture and are best mashed with butter and seasoning.

Kabocha Squashes: Attractive bright green squashes with a pale orange flesh.

They are similar in flavour and texture to acorn squashes and can be prepared and cooked in the same way.

Onion Squashes: Round, yellow or pale orange, onion squashes have a mild flavour, less sweet than pumpkin but still with a slightly fruity or honey taste. They are good in risottos or in most pumpkin recipes, but taste for flavour – you may need to add extra seasoning or sugar.

Above: Clockwise from right: A pumpkin hybrid, Kabocha squash, Acorn squash.
Right: Clockwise from top right: Hybrid squash, two golden acorn squashes, two small and one large pumpkin.

Pumpkins: Large, bright yellow or orange squashes, with a deep orange flesh. They have a sweet, slightly honeyed, flavour and are very much a taste North Americans and Australians grow up with. However, they are not to everyone's liking and some people find them rather cloying. Pumpkin soup, pumpkin bread and pumpkin pie are part of the American tradition, as are faces carved from the shell at Hallowe'en.

Buying and Storing

All winter squashes may be stored for long periods. Buy firm, unblemished vegetables with clear smooth skins.

Preparing

For larger squashes, or for those being used for soups or purées, peel and cut into pieces, removing the seeds (*left*).

Cooking

Boil in a little water for about 20 minutes until tender, then mash and serve with butter and plenty of salt and pepper. Smaller squashes can be baked whole in their skins, then halved, seeded and served with butter and maple syrup. Pumpkin and other squashes can also be lightly sautéed in butter before adding stock, cream or chopped tomatoes.

EXOTIC GOURDS

While the squashes are native to America, most gourds originated in the Old World – from Africa, India and the Far East. However, over the millennia, seeds crossed water and, over the centuries, people crossed continents so that squashes and gourds are now common all over the world. Both belong to the family *Cucurbitacea,* and both are characterised by their rapid-growing vines.

Bottle Gourds: Bottle gourds are still a familiar sight in Africa, where they are principally grown not for their fruit, but for their dried shells. The gourds can grow to enormous sizes and the shells are used for water bottles, cups and musical instruments. The young fruit can be eaten, but it is extremely bitter and is normally only added to highly flavoured stews, like curries.

Chayotes: The chayote (pronounced chow-chow) is a popular gourd in all sorts of regions of the world and can be found in just about any ethnic supermarket, be it Chinese, African, Indian or Caribbean. In each it is known by a different name, christophine being the Caribbean term, but choko, shu-shu and chinchayote are among its many other names used elsewhere. Unlike most gourds, it originated in Mexico but was widely grown throughout the tropics after the invasions of the Spanish.

It is a pear-shaped fruit with a large central stone and has a cream-coloured or green skin. It has a bland flavour, similar to marrow, and a slightly firmer texture something like pumpkin. It is commonly used in Caribbean cooking, primarily as a side dish or in soufflés. Alternatively it can be used raw in salads.

Chinese Bitter Melons: These are a common vegetable in all parts of Asia and go by a myriad of names – bitter gourd and bitter cucumber to name but two. They are popular throughout Asia, eaten when very young, but are extremely bitter and rarely eaten in the West. They are easily recognized as they have warty, spiny skins, looking like a toy dinosaur. The skins are white when young but will probably have ripened to a dark green by the time they appear in the shops.

Most recipes from China suggest halving the gourd, removing the pulp and then slicing before boiling for several minutes to remove their bitterness. They can then be added to stir-fries or other oriental dishes.

Far left: Sweet dumpling.
Left: Pumpkin.
Top right: Chinese bitter melon.
Middle right: Loofah.
Below: Chayote.

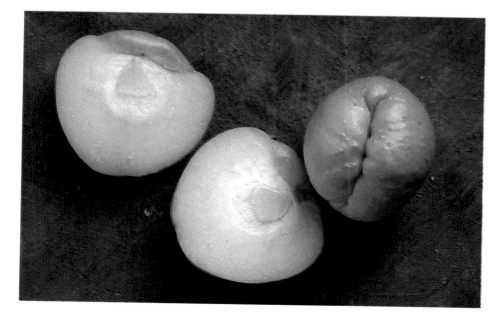

Smooth and Angled Loofahs: The smooth loofah must be one of the strangest plants. When young it can be eaten, although it is not much valued. However, the plant is grown almost exclusively for sponges, used everywhere as a back rub in the bath. The ripe loofahs are picked and, once the skin has been stripped off and the seeds shaken out, allowed to dry. The plant then gradually dries to a fibrous skeleton and thence to bathrooms everywhere - so now you know!

Angled or ribbed loofahs are more commonly eaten but again are only edible when young as they become unpleasantly bitter when mature. They taste something like courgettes and are best cooked in a similar way, either fried in butter or cooked with tomatoes, garlic and oil.

CUCUMBERS

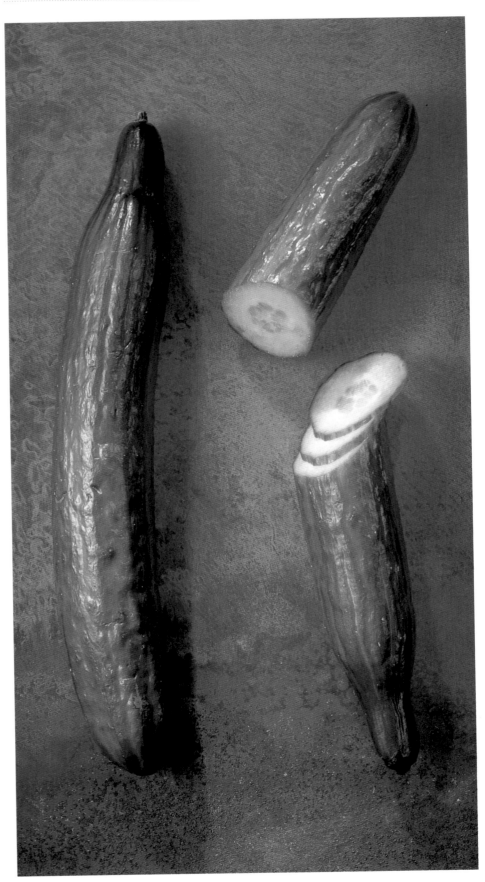

The Chinese say food should be enjoyed for its texture as well as flavour; cucumbers have a unique texture and refreshing cool taste. An afternoon tea with cucumber sandwiches, thinly sliced cucumber between wafer thin brown buttered bread, provides a delight of contrasts – the soft bread, the smooth butter and the cool crisp cucumber.

Varieties

English Cucumbers: These are the cucumbers the English are most familiar with. They have fewer seeds and thinner skin than the ridged cucumber.

Gherkins: These are tiny cucumbers with bumpy, almost warty skins and are mostly pickled in vinegar and eaten with cold meats or chopped into mayonnaise.

Kirbys: Small cucumbers, available in the United States and used for pickling.

Ridged Cucumbers: These are smaller than most cucumbers with more seeds and a thick, bumpy skin. You can buy them all over France but otherwise they tend to be available only in specialist greengrocers. The waxed ones need to be peeled before eating but most ridged cucumbers on the Continent are unwaxed and good without peeling.

Buying and Storing

Cucumbers should be firm from top to bottom. They are often sold pre-wrapped in plastic and can be stored in the salad drawer of the fridge for up to a week. Remove the plastic packaging once you've "started" a cucumber. Discard once it begins to go soggy.

Preparing

Whether you peel a cucumber or not is a matter of personal preference, but wash it if you don't intend to peel it. Some producers use wax coatings to give a glossy finish and these cucumbers must be peeled. If you are in doubt, buy organic cucumbers. Special citrus peelers can remove strips of peel to give an attractive striped effect when sliced.

Left: Cucumbers.
Above right: Ridged cucumbers.
Above far right: Baby cucumbers.
Below right: Kirbys.

Serving

Thinly sliced cucumber is most frequently served with a light dressing or soured cream. In Greece cucumber is an essential part of a Greek country salad, *horiatiki salata*, cut into thick chunks and served with tomatoes, peppers and feta and dressed simply with olive oil and a little wine vinegar.

Iced cucumber soup is delicious, and cucumber can also be puréed with yogurt, garlic and herbs and served with soured cream stirred in.

Cooking

Cucumbers are normally served raw, but are surprisingly good cooked. Cut the cucumber into wedges, remove the seeds and then simmer for a few minutes until tender. Once drained, return the cucumber to the pan and stir in a little cream and seasoning.

FRUIT

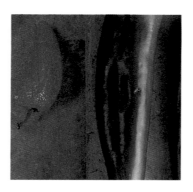

TOMATOES

Next to onions, tomatoes are one of the most important fresh ingredients in the kitchen. In Mediterranean cooking, they are fundamental. Along with garlic and olive oil, they form the basis of so many Italian, Spanish and Provençal recipes that it is hard to find many dishes in which they are not included.

History

Tomatoes are related to potatoes, aubergines and sweet and chilli peppers, and all are members of the nightshade family. Some very poisonous members of this family may well have deterred our ancestors from taking to tomatoes. Indeed, the leaves of tomatoes are toxic and can result in very bad stomach aches.

Tomatoes are native to western South America. By the time of the Spanish invasions in the sixteenth century, they were widely cultivated throughout the whole of South America and Mexico. Hernán Cortés, conqueror of the Aztecs, sent the first tomato plants, a yellow variety, to Spain (no doubt along with the plundered Aztec gold).

However, people did not instinctively take to this "golden apple". English horticulturists mostly grew them as ornamental plants to adorn their gardens and had little positive to say about them as food. Spain is recorded as the first country to use tomatoes in cooking, stewing them with oil and seasoning. Italy followed suit, but elsewhere they were treated with suspicion.

The first red tomatoes arrived in Europe in the eighteenth century, brought to Italy by two Jesuit priests. They were gradually accepted in northern Europe where, by the mid-nineteenth century, they were grown extensively, eaten raw, cooked or used for pickles.

Above right: Red and yellow cherry tomatoes.
Below right: Yellow pear tomatoes.
Opposite above: Round or salad tomatoes on the vine.
Opposite below: Beefsteak tomatoes.

Varieties

There are countless varieties of tomatoes, ranging from the huge beef tomatoes that measure 10cm/4in across, to tiny cherry tomatoes, not much bigger than a thumb nail. They come in all shapes too - elongated, plum-shaped or slightly squarish and even pear-shaped.

Beefsteak Tomatoes: Large, ridged and deep red or orange in colour, these have a good flavour so are good in salads.

Canned Tomatoes: Keep a store of canned tomatoes, especially in the winter when fresh ones tend to taste insipid. Tomatoes are one of the few vegetables that take well to canning, but steer clear of any that are flavoured with garlic or herbs. It is far better to add flavouring yourself.

Cherry Tomatoes: These small, dainty tomatoes were once the prized treasures of gardeners but are now widely available. Although more expensive than round tomatoes, they have a delightful sweet flavour and are worth paying the

extra money for serving in salads or for cooking whole.

Plum Tomatoes: Richly flavoured with fewer seeds than regular tomatoes, these Italian-grown tomatoes are usually recommended for cooking, although they can be used in salads.

Round or Salad Tomatoes: These are the common tomatoes found in greengrocers and supermarkets. They vary in size according to the exact type and season. Sun-ripened tomatoes have the best flavour, however, for year-round availability the fruit is often picked and ripened off the plant. These tomatoes are versatile in everyday cooking. Adding a pinch of sugar and taking care to season the dish well helps to overcome any weakness in flavour.

Sun-Dried Tomatoes: This is one of the fashionable foods of the late Eighties and early Nineties. They add an evocative flavour to many Mediterranean dishes but don't use them too indiscriminately.

Tomato Purée: This is good for adding an intense tomato flavour, but use carefully or the flavour will be overpowering. Tubes have screw tops and are better than cans as, once they are opened,

they can be kept for up to 4-6 weeks in the fridge.

Yellow Tomatoes: These are exactly like red tomatoes - they may be round, plum or cherry-sixed - except they are yellow.

Buying and Storing

Ideally, tomatoes should be allowed to ripen slowly on the plant so that their flavour can develop. Consequently, home-grown tomatoes are best, followed by those grown and sold locally. When buying from a supermarket or green-grocer, look at the leafy green tops; the fresher they look the better. Buy locally grown beefsteak or cherry tomatoes for salads and plum tomatoes for rich sauces. Paler tomatoes or those tinged with green will redden if kept in a brown paper bag or the salad drawer of the fridge, but if you intend to use tomatoes straight away, buy bright red speci-mens. Overripe tomatoes, where the skin has split and they seem to be bursting with juice, are excellent in soups. However, check for any sign of mould or decay, as this will spoil all your good efforts.

Preparing

Slice tomatoes across rather than down-wards for salads and pizza toppings. For wedges, cut downwards; halve or quarter and cut into two or three depending on the size of the tomato.

Cooking

Among the many classic tomato dishes is tomato soup, cooked to a delicate orange colour with stock or milk, or simmered with vegetables, garlic and basil. Recipes *à la provençale* indicate that tomatoes are in the dish; in Provençal cooking and Italian dishes, tomatoes are used with fish, meat and vegetables, in sauces and stuffings, with pasta and in superb salads. The Italian *tri colore salata* is a combination of large tomatoes, mozzarella and basil (the three colours of the Italian flag). The natural astringency of tomatoes means that, in salads, they need only be sprinkled with a fruity olive oil.

Chopped Tomatoes

Chopped tomatoes add a depth of flavour to all sorts of meat and vegetarian dishes. Ideally, even in fairly rustic meals, the tomatoes should be peeled, since the skin can be irritating to eat once cooked. Some sauces also recom-mend seeding tomatoes, in which case cut the tomato into halves and scoop out the seeds before chopping (above).

Skinning Tomatoes

Cut a cross in the tops of the tomatoes, then place in a bowl and pour over boiling water. Leave for a minute (*above*), then use a sharp knife to peel away the skin, which should come away easily. Do a few at a time (five at most) otherwise they will begin to cook while soaking; boil the kettle for the next batch when you have finished peeling. The water must be boiling.

AUBERGINES

Many varieties of aubergines are cultivated and cooked all over the world. In Europe, Asia or America, they feature in a multitude of different dishes.

History

Although aubergines are a member of the nightshade family and thus related to potatoes, tomatoes and peppers, they were not discovered in the New World. The first mention of their cultivation is in China in 5 BC, and they are thought to have been eaten in India long before that. The Moors introduced the aubergine to Spain some 1200 years ago and it was grown in Andalucia. It is likely that they also introduced it to Italy, and possibly from there to other southern and eastern parts of Europe.

In spite of their popularity in Europe, aubergines did not become popular in Britain or the United States until very recently; although previous generations of food writers knew about them, they gave only the occasional recipe for cooking with them.

Above left: Plum tomatoes.
Top: Aubergines.
Above: Baby aubergines.
Left: Japanese aubergines.

Meanwhile, in the southern and eastern parts of Europe, aubergines had become extremely well liked, and today they are one of the most popular vegetables in the Mediterranean. Indeed, Italy, Greece and Turkey claim to have 100 ways of cooking them. In the Middle East, aubergines are also a central part of their cuisine.

Varieties

There are many different varieties of aubergines, differing in colour, size and shape according to their country of origin. Small ivory-white and plump aubergines look like large eggs (hence their name in the States: eggplant). Pretty striped aubergines may be either purple or pink and flecked with white irregular stripes. The Japanese or Asian aubergine is straight and very narrow, ranging in colour from a pretty variegated purple and white to a solid purple. It has a tender, slightly sweet flesh. Most aubergines, however, are either glossy purple or almost black and can be long and slim or fat like zeppelins. All aubergines have a similar flavour and

texture; they taste bland yet slightly smoky when cooked, and the flesh is spongy to touch when raw, but soft after cooking.

Buying and Storing

Aubergines should feel heavy and firm to the touch, with glossy, unblemished skins. They will keep well in the salad drawer of the fridge for up to two weeks.

Preparing

When frying aubergines for any dish where they need slicing (e.g. ratatouille), it is a good idea to salt the slices first in order to draw out some of their moisture, otherwise, they absorb enormous quantities of oil during cooking (they absorb copious amounts anyway, but salting

reduces this slightly). Salting also used to be advised to reduce their bitterness but today's varieties are rarely bitter. To salt aubergines, cut into slices, about 1cm/½ in thick for fried slices, *(top right)* or segments *(above right)* and sprinkle generously with salt. Leave them to drain in a colander for about one hour, then rinse well and gently squeeze out the moisture from each slice or carefully pat dry with a piece of muslin.

Cooking

Aubergine slices can be fried in olive oil, as they are or first coated in batter – both popular Italian and Greek starters.

For moussaka, *parmigiana* and other dishes where aubergines are layered with other ingredients, fry the slices

briefly in olive oil. This gives them a tasty crust, while the inside stays soft.

To make a purée, such as for Poor Man's Caviar, first prick the aubergine all over with a fork and then roast in a moderately hot oven for about 30 minutes until tender. Scoop out the flesh and mix with spring onions, lemon juice and olive oil. One of the most famous aubergine dishes is *Imam Bayaldi* – "the Iman fainted" – fried aubergines stuffed with onions, garlic, tomato, spices and lots of olive oil.

Above left: White aubergines.
Below left: Striped aubergines.
Above: Thai aubergines, including white, yellow and Pea aubergines.

PEPPERS

In spite of their name, peppers have nothing to do with the spice pepper used as a seasoning, although early explorers may have been mistaken in thinking the fruit of the shrubby plant looked like the spice they were seeking. It is thanks to this 400-year-old mistake that the name "pepper" has stuck.

History

The journeys Christopher Columbus and the conquistadors made were partly to find the spices Marco Polo had found a hundred years earlier in the Far East. Instead of the Orient, however, Columbus discovered the Americas, and instead of spices, he found maize, potatoes and tomatoes. He would have noted, though, that the American-Indians flavoured their food with ground peppers, and since it was hot, like pepper, perhaps wishful thinking coloured his objectivity. In any case, he returned with the new vegetables, describing them as peppers and advertising them as more pungent than those from Caucasus.

Varieties

Peppers and chillies are both members of the capsicum family. To distinguish between them, peppers are called sweet peppers, bell peppers and even bullnose peppers and come in a variety of colours – red, green, yellow, white, orange and a dark purple-black.

The colour of the pepper tells you something about its flavour. Green peppers are the least mature and have a fresh "raw" flavour. Red peppers are ripened green peppers and are distinctly sweeter. Yellow/orange peppers taste more or less like red peppers, although perhaps slightly less sweet and if you have a fine palate you may be able to detect a difference. Black peppers have a similar flavour to green peppers but when cooked are a bit disappointing as they turn green; so if you buy them for their dramatic colour, they are best used in salads.

In Greece and other parts of southern Europe, longer, slimmer peppers are often available which have a more pronounced sweet and pungent flavour than the bell-shaped peppers – although this may be because they are locally picked and therefore absolutely fresh. Whichever is the case, they are quite delicious.

Buying and Storing

Peppers should look glossy and sprightly and feel hard and crisp; avoid any that look wrinkled or have damp soft patches. They will keep for a few days at the bottom of the fridge.

Preparing

To prepare stuffed peppers, cut off the top and then cut away the inner core and pith, and shake out the seeds. The seeds and core are easily removed when halving, quartering or slicing.

Cooking

There are countless ways of cooking peppers. Sliced, they can be fried with onions and garlic in olive oil and then braised with tomatoes and herbs. This is the basic ratatouille; other vegetables, such as courgettes and aubergines, can of course be added.

Peppers can be roasted, either with ratatouille ingredients or with only onions and garlic. Cut into large pieces, place in a roasting pan and sprinkle with olive oil, torn basil and seasoning. Roast in a very hot oven (220°C/425°F/Gas 7) for about 30 minutes, turning occasionally. Grilled peppers are another superb dish. Once grilled they can be skinned to reveal a soft, luxurious texture and added to salads.

Above far left: Red, green and orange peppers.
Below far left: Yellow peppers.
Above left: White peppers.
Above right: Purple peppers.

Skinning Peppers

Cut the pepper into quarters lengthways and grill, skin side up (*above*), until the skin is charred and evenly blistered. Place the pieces immediately into a plastic bag (you will need tongs or a fork as they will be hot) and close the top of the bag with a tie or a loose knot. Leave for a few minutes and then remove from the bag and the skin will peel off easily.

CHILLIES

Some people apparently become so addicted to the taste of hot food that they carry little jars of chopped dried chillies around with them and scatter them over every meal. Although this is a bit extreme, it is chillies more than any other ingredient that spice up our mealtimes.

Varieties

Chillies are the most important seasoning in the world after salt. Unlike peppers, to which they are closely related, the different varieties of chilli can have widely different heat values – from the "just about bearable" to the "knock your head off" variety.

Anaheim Chilli: A long, thin chilli with a blunt end, named after the Californian city. It can be red or green and has a mild, sweet taste.

Ancho Chilli/Pepper: These look like tiny peppers. They are mild enough to taste their underlying sweetness.

Birdseye or Bird Chilli: These small red chillies are fiery hot. Also known as pequin chillies.

Cayenne Pepper: This is made from the dried, ground seeds and pods of chillies. The name comes from the capital of French Guiana, north of Brazil, although the cayenne chilli does not grow there any longer and the pepper is made from chillies grown all over the world.

Early Jalapeño: A popular American chilli, which starts dark green and gradually turns to red.

Habanero: Often called Scotch Bonnet, this is the hottest of all chillies and is small and can be green, red or yellow. Colour is no real guide to its heat properties, so don't be fooled into thinking that green ones are mild. They are all *very* hot. The habanero comes from Mexico and is frequently used in Mexican and Caribbean dishes.

Hot Gold Spike: A large, pale, yellow-green fruit grown in the south-western United States: it is very hot.

Above: Birdseye chillies.
Left: Habanero chillies (in and below bowl) and Yellow wax peppers.

Preparing

The capsaicin in chillies is most concentrated in the pith inside the pod and this, together with the seeds, should be cut away (*below*) unless you want maximum heat. Capsaicin irritates the skin and especially the eyes, so take care when preparing chillies. Either wear gloves or wash your hands thoroughly after handling chillies.If you rub your eyes, even if you have washed your hands carefully, it will be painful.

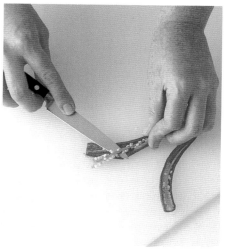

Poblano: A small, dark green chilli, served whole in Spain either roasted or grilled. They are mostly mild but you can get the rogue fiery one, so beware if eating them whole.

Red Chilli: These are long, rather wrinkled chillies which are green at first and then gradually ripen to red. They are of variable hotness and, because they are so long and thin, are rather fiddly to prepare.

Serrano Chilli: A long, red and extremely hot chilli.

Tabasco: A sauce made with chillies, salt and vinegar and first made in New Orleans. It is a fiery sauce, popular in Creole, Caribbean and Mexican cookery – or indeed in any dish requiring last minute heat.

Yellow Wax Pepper: Pale yellow to green, these can vary from mild to hot.

Buying and Storing

Some fresh chillies look wrinkled even in their prime and therefore this is not a good guide to their freshness. They should, however, be unblemished, and avoid any which are soft or bruised.

The substance which makes the chilli hot is a volatile oil called capsaicin. This differs not only from one type to another but also from plant to plant, depending on growing conditions; the more the plant has to struggle to survive in terms of light, water, soil, etc, the more capsaicin will be produced. It is therefore impossible to tell how hot a chilli will be before tasting, although some types are naturally hotter than others. The belief that green chillies are milder than red ones does not necessarily follow; generally red chillies will have ripened for longer in the sun with the result that they will only be sweeter for all that sunshine.

Chillies can be stored in a plastic bag in the fridge for a few days.

Cooking

In Mexican cooking chillies play a vital, and almost central role. It is difficult to think of any savoury Mexican dish that does not contain either fresh chillies or some form of processed chilli, whether canned, dried or ground. Other cuisines, however, are equally enthusiastic about chillies. They are essential in curries and similar dishes from India and the Far East, and in Caribbean and Creole food they are also used extensively.

If you have developed a tolerance for really hot food, then there is no reason why you shouldn't add as many as you wish. In general, however, use chillies discreetly, if for no better reason than you can't take the heat away if you make a mistake.

Above: Ancho chillies (left) and Anaheim chillies (on board).

PLANTAINS AND GREEN BANANAS

While bananas are well and truly fruit, eaten almost exclusively as a dessert or by themselves as fruit, plantains can reasonably be considered among the vegetable fraternity as they have a definite savoury flavour, are normally eaten as a first or main course and can only be eaten once cooked.

Varieties

Plantains: Also known as cooking bananas, these have a coarser flesh and more savoury flavour than sweet bananas. While superficially they look exactly like our own bananas, they are, on closer inspection, altogether larger and heavier looking. They can vary in colour from the unripe fruit, which is green, through yellow to a mottled black colour, which is when the fruit is completely ripe.

Green Bananas: Only certain types of green bananas are used in African and Caribbean cooking, and the "greenish" bananas you find in most western supermarkets are normally eating bananas, just waiting to ripen. If you need green bananas for a recipe, look out for them in West Indian or African cookshops.

Preparing

Plantains: These are inedible raw and must be cooked before eating. Unless very ripe, the skin can be tricky to remove. With yellow and green plantains, cut the fruit into short lengths, then slit the skin along the natural ridge of each piece of plantain. Gently ease the skin away from the flesh and pull the skin until it peels off completely (*below*).

Once peeled, plantains can be sliced horizontally or into lengths and then roasted or fried. Like bananas, plantains will discolour if exposed to the air so, if not using immediately, sprinkle with lemon juice or place in a bowl of salted water.

Green Bananas: These should be prepared in a similar way. As with plantains, green bananas should not be eaten raw and are usually boiled, either in their skins or not, according to the recipe.

If making green banana crisps, use a potato peeler to produce the thinnest slices (*left*).

If cooking plantains or green bananas in their skins, slit the skin lengthways along the sides and place in a saucepan of salted water. Bring to the boil, simmer gently for about 20 minutes until tender and then cool. The peel can then easily be removed before slicing.

Cooking and Serving

Plantains and green bananas both have an excellent flavour. In many African and Caribbean recipes they are roasted or fried and then served simply with salt. However, if boiled, they can be sliced and served in a simple salad with a few sliced onions, or added to something far more elaborate like a gado gado salad, with mango, avocado, lettuce and prawns.

Plantains also make a delicious soup, where they are often teamed with sweetcorn. After frying an onion and a little garlic, add two sliced and peeled plantains, together with tomatoes, if liked. Fry gently for a few minutes and then add vegetable stock to cover and one or two sliced chillies, together with about 175g/6oz sweetcorn. Simmer gently together until the plantain is tender.

Above left: Plantains.
Below left: Green bananas.
Below: Canned ackee.

ACKEE

Ackee is a tropical fruit which is used in a variety of savoury dishes, mainly of Caribbean origin, where the fruit is very popular. The fruit itself is bright red and, when ripe, bursts open to reveal three large black seeds and a soft, creamy flesh resembling scrambled eggs. It has a slightly lemony flavour and is traditionally served with saltfish to make one of Jamaica's national dishes. Only buy ripe fruit as, when under-ripe, certain parts of the fruit are toxic.

However, unless you are visiting the Caribbean you are probably only likely to find ackee in cans, and indeed most recipes call for canned ackee which is a good substitute for the fresh fruit.

Jamaican cooks also use ackee to add a subtle flavour to a variety of vegetable and bean dishes. The canned ackee needs very little cooking, and should be added to dishes in the last few minutes of cooking. Take care when stirring into a dish as it breaks up very easily.

AVOCADOS

The avocado has been known by many names – butter pear and alligator pear to name but two. It earned the title butter pear clearly because of its consistency, but alligator pear was the original Spanish name. Although you would be forgiven for thinking this was due to its knobbly skin (among some varieties anyway), the name in fact derives from the Spanish which was based on the Aztec word, the basically unpronounceable *ahuacatl*. From this to the easily-said alligator and thence to avocado was but one short step.

History

The avocado is a New World fruit, native to Mexico, but while it would have been "discovered" by the Old World explorers, it didn't become a popular food in Europe until the middle of this century, when modern transport meant that growers in California, who started farming avocados in the middle of the nineteenth century, could market this fruit worldwide. Avocados are now also exported by South Africa and Australia.

Nutrition

The avocado is high in protein and carbohydrate. It is one of the few fruits that contains fat, and it is also rich in potassium, Vitamin C, some B vitamins and Vitamin E. Its rich oils, particularly its Vitamin E content, mean that it is not only useful as food, but for skin and hair care too, something the Aztecs and Incas were aware of a thousand years ago. The cosmetic industry may have been in its infancy, but it still knew a good thing when it saw it.

Because of their valuable protein and vitamin content, avocados are a popular food for babies. They are easily blended, and small children generally enjoy their creamy texture and pleasant flavour.

Varieties

There are four varieties: Hass, the purple-black small knobbly avocado, the Ettinger and Fuerte, which are pear-shaped and have smooth green skin, and the Nabal, which is rounder in shape. The black-coloured Hass has golden-yellow flesh, while green avocados have pale green to yellow flesh.

Buying and Storing

The big problem in buying avocados is that they're never ripe when you want them to be. How often do you see shoppers standing by the avocado shelves, feeling around for that rare creature, the perfectly ripe avocado? Most times they all feel as hard as rocks; that or else they're hopelessly soft and squashy and clearly past their best. The proper and sensible thing to do is buy fruit a few days before you need it. An unripe avocado will ripen in between 4-7 days at room temperature. Once it is ripe, it will keep well in the fridge for a few days, but you still need to plan well in advance if you want to be sure of the perfect avocado.

The alternative is to hope for the best and keep feeling around until you find a ripe fruit. A perfect avocado should have a clean, unblemished skin without any brown or black patches. If ripe, it should "give" slightly if squeezed gently in the hand, but not so much that it actually feels soft. Over-ripe avocados are really not worth bothering with, however persuasive and generous the offer from the man on the market. The flesh will be unattractively brown and stringy and the bits of good flesh you do manage to salvage will be soft and pulpy. Good for a dip, but nothing much else.

Preparing

Although they are simple fruits, avocados can be the devil to prepare. Once peeled, you are left with a slippery object which is then almost impossible to remove from the stone.

If you intend to eat the avocados in halves, it's fairly simple to just prise out the stone once halved. If you want to slice the fruit, use this tip I learnt from my chef friend. The only thing you need is a very sharp knife. Cut the avocado in half, remove the stone and then, with the skin still on, cut through the flesh and the skin to make slices. It is then relatively simple to strip off the peel.

Remember to sprinkle the slices with lemon juice as the flesh discolours once exposed to the air.

Cooking

Most popular raw, avocados can also be baked, grilled or used in sautéed and sauced dishes.

Serving Ideas

As well as prawns or vinaigrette, a half avocado can hold a mixture of chopped tomatoes and cucumber, a mild garlic cheese dip or a soured cream potato salad. Slices of avocado are delicious served with sliced tomatoes and mozzarella, sprinkled simply with olive oil, lemon juice and plenty of black pepper. Avocado can be chopped and added to a salad, or puréed for a rich dressing.

In Mexico, where avocados grow in abundance, there are countless avocado recipes. Guacamole is perhaps the best known, but they are also eaten in soups and stews and commonly used to garnish tacos and enchiladas.

BREADFRUIT

Breadfruit is the name for a tropical tree that grows on the islands of the South Pacific ocean. The fruit of the tree is about the size of a small melon with a rough rind and a pale, mealy flesh.

Preparing

The fruit should be peeled and the core removed.

Cooking

Breadfruit can be treated like potatoes: the flesh may be boiled, baked or fried. It is a staple food for the people of the Pacific islands who bake the flesh, or dry and grind it for biscuits, bread and puddings. It has a sweet flavour and soft texture when ripe.

Left: Clockwise from the right: Fuerte, Hass and Nabal avocados.
Right: Breadfruit.

EIGHT

SALAD
VEGETABLES

Lettuce

Rocket

Chicory and Radicchio

Radishes

Watercress

Mustard and Cress

LETTUCE

One aspect of lettuce that sets it apart from any other vegetable is that you can only buy it in one form – fresh.

History

Lettuce has been cultivated for thousands of years. In Egyptian times it was sacred to the god Min, and tubs of lettuce were ceremoniously carried before this fertility god. It was then considered a powerful aphrodisiac, yet for the Greeks and the Romans lettuce was thought to have quite the opposite effect, making one sleepy and generally soporific. Chemists today confirm that lettuce contains a hypnotic similar to opium, and in herbal remedies lettuce is recommended for insomniacs.

Varieties

There are hundreds of different varieties of lettuce. Today, an increasing variety is available in the shops so that the salad bowl can be wealth of colour and texture.

Round Lettuces

Sometimes called head or cabbage lettuces, round lettuces have cabbage-like heads and include:
Butterheads: These are the classic lettuces seen in kitchen gardens. They have a pale heart and floppy, loosely packed leaves. They have a pleasant flavour as long as they are fresh.
Crispheads: Crisp lettuces, such as Iceberg, have an excellent crunchy texture and will keep their vitality long after butterheads have faded and died.
Looseheads: These are non-hearting lettuce with loose leaves and include *lollo rosso* and *lollo biondo*, oakleaf lettuce and Red Salad Bowl. Although they are not particularly remarkable for their flavour, they look superb.

Cos Lettuces

The cos is the only lettuce that would have been known in antiquity. It is known

Above: Butterhead lettuce.
Right: Lollo rosso lettuce.
Above far right: Cos lettuce.
Below far right: Lamb's lettuce.
Below extreme right: Little Gem.

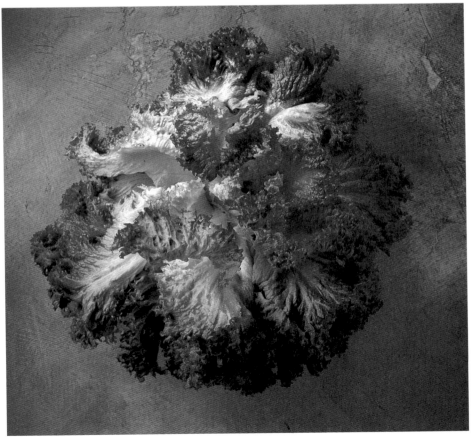

by two names: cos, derived from the Greek island where it was found by the Romans; and romaine, the name used by the French after it was introduced to France from Rome. There are two cos lettuces, both with long, erect heads.

Cos: Considered the most delicious lettuce, this has a firm texture and a faintly nutty texture. It is the correct lettuce for Caesar Salad, one of the classic salads.

Little Gems: In appearance Little Gems look like something between a baby cos and a tightly furled butterhead. They have firm hearts and are enjoyed for their distinct flavour. Like other lettuce hearts, they cope well with being cooked.

LAMB'S LETTUCE OR CORN SALAD

This popular winter leaf does not actually belong to the lettuce family (it is related to Fuller's teasel), but as it makes a lovely addition to salads, this seems a good place to include it. Called *mâche* in France, it has spoon-shaped leaves and an excellent nutty flavour.

Nutrition

As well as containing vitamins A, C and E, lettuce provides potassium, iron and calcium and traces of other minerals.

Buying and Storing

The best lettuce is that fresh from the garden. The next best thing is to buy lettuce from a farm shop or pick-your-own (although if fertilizers and pesticides are used, their flavour will be disappointing compared to the organic product). Nowadays, lettuce is frequently sold ready shredded and packed with herbs etc, an acceptable and convenient form of buying lettuce. Whether you buy lettuce prepacked or from the shelf, it must be fresh. Soil and bugs can be washed off but those with limp or yellow leaves are of no use. Eat lettuce as soon as possible after purchasing; in the meantime keep it in a cool dark place, such as the salad drawer of the fridge.

Making Salads

Salads can be made using only one lettuce or a mixture of many. There are no rules but, when mixing salads, choose leaves to give contrast in texture and colour as well as flavour. Fresh herbs, such as parsley, coriander and basil also add an interesting dimension.

Tear rather than cut the leaves of loose-leafed lettuce; icebergs and other large lettuces are commonly sliced or shredded. Eat as soon as possible after preparing.

Dressings should be well-flavoured with a hint of sharpness but never too astringent. Make them in a blender, a screw top jar or in a large bowl so that the ingredients can be thoroughly blended. Always use the best possible oils and vinegars, in roughly the proportion of five oil to one vinegar or lemon juice. Use half good olive oil and half sunflower oil, or for a more fragrant dressing, a combination of walnut oil and sunflower oil. A pinch of salt and pepper is essential, French mustard is optional and the addition of a little sugar will blunt the flavour.

Add the dressing to the salad when you're ready to serve – never before.

ROCKET

Rocket has a wonderful peppery flavour and is excellent in a mixed green salad. It has small, bright green dandelion-shaped leaves. The Greeks and Romans commonly ate rocket (*arugula*) in mixed salads, apparently to help counterbalance the dampening effect lettuce had on the libido – rocket's aphrodisiac properties in antiquity are well catalogued. It used to be sown around the statues of Priapus, the mythological Greek god of fertility and protector of gardens and herbs and son of Aphrodite and Dionysus.

Buying and Storing

Rocket is to be found either among the salads or fresh herbs in supermarkets. Buy fresh green leaves and use soon after purchasing. If necessary, the leaves can be kept immersed in cold water.

Preparing and Serving

Discard any discoloured leaves. Add rocket to plain green salads, or grind with garlic, pine nuts and olive oil for a dressing for pasta.

Since it has such a striking flavour, a little rocket goes a long way, making it an excellent leaf for garnishing. It tastes superb contrasted with grilled goat's cheese, or one or two leaves can be added to sandwiches, or loosely packed into pitta bread pockets along with tomatoes, avocado, peanuts and bean sprouts.

Above left: Oak Leaf lettuce.
Below left: Curly endive.
Below: Rocket.

CHICORY AND RADICCHIO

Chicory, radicchio, endive and escarole are all related to each other and when they are tasted together you can easily detect their family resemblance. Their names are occasionally interchanged: chicory is often referred to as Belgian or French endive and French and Belgian *chicorée* is the English curly endive.

CHICORY

During the late eighteenth century, chicory was grown in Europe for its root, which was added to coffee. A Belgian, M. Brezier, discovered that the white leaves could be eaten, a fact he kept secret during his lifetime; but after his death chicory became a popular vegetable, first in Belgium and later elsewhere in Europe. Its Flemish name is *witloof*, meaning "white leaf", and its characteristic pale leaf is due to its being grown in darkness; the paler it is, the less bitter its flavour.

Chicory can be eaten raw but is commonly cooked, either baked, stir-fried or poached. To eat raw, separate the leaves and serve with fruit, such as oranges or grapefruit, which counteract chicory's slight bitterness.

RADICCHIO

This is one of many varieties developed from wild chicory. It looks like a small lettuce with deep wine-red leaves and striking cream ribs and owes its splendid foliage to careful shading. If it is grown completely in the dark the leaves are marbled pink, and those that have been exposed to some light can be patched with a green or copper colour. Its flavour tends to be bitter but contrasts well with green salads. Radicchio can be stir-fried or poached, although the leaves turn dark green when cooked.

CURLY ENDIVE AND ESCAROLE

These are robust salad ingredients in both flavour and texture. The curly-leaved endive looks like a green frizzy mop and the escarole is broad-leaved, but both have a distinct bitter flavour. Serve mixed with each other and a well-flavoured dressing. This dampens down the bitter flavour but gives the salad a pleasant "bite".

Preparing

To prepare chicory, take out the core at the base with a sharp knife (*left*) and discard any wilted or damaged leaves. Rinse thoroughly, then dry the leaves.

Above: Chicory.
Above right: Radicchio.
Below right: Escarole.

Preparing Salad Leaves

Pull the leaves away from the stalk, discarding any wilted or damaged leaves.

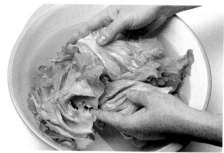

Wash the leaves in plenty of cold water, swirling gently to make sure all dirt and any insects are washed away.

Place the washed leaves in a soft tea towel and then gently pat dry.

Place in a dry dish towel in a large plastic bag. Chill in the fridge for about 1 hour.

RADISHES

Radishes have a peppery flavour that can almost be felt in the nostrils as you bite into one. Their pungency depends not only on the varieties but also on the soil in which they are grown. Freshly harvested radishes have the most pronounced flavour and crisp texture.

Varieties

Radishes were loved throughout antiquity and consequently there are many varieties worldwide. Both the small red types and the large white radishes are internationally popular.

Red Radishes: These small red orbs have many pretty names, such as Cherry Belle and Scarlet Globe, but are mostly sold simply as radish. They are available all year round, have a deep pink skin, sometimes paler or white at the roots and a firm white flesh. Their peppery flavour is milder in the spring and they are almost always eaten raw. Finely sliced and sandwiched in bread and butter, they make an interesting *hors d'oeuvre*.

French Breakfast Radishes: These are red and white and slightly more elongated than the red radish. They tend to be milder than English radishes and are popular in France either eaten on their own or served with other raw vegetables as *crudités*.

Mooli or Daikon Radishes: Sometimes known as the oriental radish, the mooli is a smooth-skinned, long, white radish. Those bought in the shops have a mild flavour, less peppery than the red radish – perhaps because they lose their flavour after long storage (moolis straight from the garden are hot and peppery). They can be eaten raw or pickled, or added to stir-fries.

Buying and Storing

Buy red radishes that are firm with crisp leaves. If at all possible, buy moolis or daikons which still have their leaves; this is a good indication of their freshness as they wilt quickly. The leaves should be green and lively and the skins clear with no bruises or blemishes. They can be stored in the fridge for a few days.

Preparing and Serving

Red radishes need only to be washed. They can then be sliced or eaten whole by themselves or in salads. You can make a feature of them by slicing into a salad of, say, oranges and walnuts, perhaps with a scattering of rocket and dressed with a walnut oil vinaigrette. To use moolis in a stir-fry, cut into slices and add to the dish for the last few minutes of cooking. They add not only flavour but also a wonderfully juicy and crunchy texture.

Left: Red radishes.
Above: French breakfast radishes.
Right: Mooli or daikon radishes.

WATERCRESS

Watercress is perhaps the most robustly flavoured of all the salad ingredients and a handful of watercress is all you need to perk up a rather dull green salad. It has a distinctive "raw" flavour, both peppery and slightly pungent and this, together with its bright green leaves, make it a popular garnish.

Watercress, as the name suggests, grows in water. It needs fast flowing clean water to thrive and is really only successful around freshwater springs on chalk hills. The first watercress beds were cultivated in Europe but the vegetable is now grown worldwide.

WINTER CRESS

Winter cress or land cress is often grown as an alternative to watercress, when flowing water is not available. It looks like a robust form of watercress and indeed has a similar if even more assertive flavour, with a distinct peppery taste. Use as you would use watercress, either in salads or in soups.

Nutrition

Watercress is extremely rich in vitamins A, B2, C, D and E. It is also rich in calcium, potassium and iron and provides significant quantities of sulphur and chloride.

Buying and Storing

Only buy fresh looking watercress – the darker and larger the leaves the better. Avoid any with wilted or yellow leaves. It will keep for several days in the fridge or better still, submerged in a bowl, or arranged in a jar of cold water, and kept in a cool place.

Preparing and Cooking

Discard any yellow leaves and remove thick stalks which will be too coarse for salads or soups. Small sprigs can be added to salads.

For soups and purées, either blend watercress raw or cook briefly in stock, milk or water. Cooking inevitably destroys some of the nutrients but cooked watercress has a less harsh flavour, while still retaining its characteristic peppery taste.

MUSTARD AND CRESS

Mustard and cress are often grown together, to provide spicy greenery as a garnish or for salads. They are available all year round.

Mustard seedlings germinate 3-4 days sooner than the cress, so if you buy mustard and cress from the supermarket, or grow your own on the windowsill, initially the punnets will only show mustard seedlings.

History

Cress has been grown for thousands of years, known first to the Persians. There is a story that the Persians would always eat cress before they baked bread, and there are other references in antiquity to people eating cress with bread.

Serving

Today mustard and cress are often enjoyed in sandwiches, either served simply on buttered bread, or with avocado or cucumber added. Cress probably wouldn't be substantial enough as a salad in itself, but, with its faint spicy flavour, it can perk up a plain green salad, and it is also excellent in a tomato salad, dressed simply with olive oil and tarragon vinegar.

Above left: Watercress.
Below left: Winter cress.
Above right: Mustard seedlings.
Right: Cress seedlings.

MUSHROOMS

BUTTON MUSHROOMS

There is nothing like fried mushrooms on toast for breakfast, served sizzling hot straight from the pan. Once cooked, mushrooms, especially fried ones, go soft and flabby quite quickly; they still taste OK but the pleasure is not so great.

History

In the past, mushrooms have had a firm association with the supernatural and even today their connection with the mysterious side of life hasn't completely disappeared. Fairy rings – circles of mushrooms – inexplicably appear overnight in woods and fields and thunder is still thought to bring forth fresh crops of mushrooms.

Many types of mushrooms and fungi are either poisonous or hallucinogenic, and in the past their poisons have been distilled for various murderous reasons.

The use of the term mushroom to mean edible species, and toadstool to mean those considered poisonous, has no scientific basis, and there is no simple rule for distinguishing between the two. Picking wild mushrooms is not safe unless you are confident about identifying edible types. In France, during the autumn, people take the wild mushrooms they have gathered to the local pharmacy for identification.

Varieties

Button/White Mushrooms: Cultivated mushrooms are widely available in shops and are sold when very young and tiny as button mushrooms. The slightly larger ones are known as closed cap, while larger ones still are open capped or open cup mushrooms. They have ivory or white caps with pinky/beige gills which darken as they mature. All have a pleasant unassuming flavour.

Chestnut Mushrooms: These have a thicker stem and a darker, pale brown cap. They have a more pronounced "mushroomy" flavour and a meatier texture than white mushrooms.

Buying and Storing

It is easy to see whether or not button mushrooms are fresh – their caps will be clean and white, without bruises or blemishes. The longer they stay on the shelves, the darker and more discoloured the caps become, while the gills underneath turn from pink to brown.

If possible, use the paper bags provided in many supermarkets nowadays when buying mushrooms. Mushrooms in plastic bags sweat in their own heat, eventually turning slippery and unappetizing. If you have no choice or you buy mushrooms in cellophane-wrapped

cartons, transfer loose to the bottom of the fridge as soon as possible. They will keep only for a day or two.

Preparing

Mushrooms should not be washed but wiped with a damp cloth or a piece of kitchen paper (*below*). This is partly because you don't want to increase their water content, and also because they should be fried as dry as possible.

Unless the skins are very discoloured, it should not be necessary to peel them, although you probably will need to trim the very base of the stem.

Cooking

Mushrooms are largely composed of water and shrink noticeably during cooking. They also take up a lot of fat as they cook so it is best to use butter or a good olive oil for frying. Fry mushrooms briskly over a moderately high heat so that as they shrink the water evaporates and they don't stew in their own juice. For the same reason do not fry too many mushrooms at once in the same pan.

Most of the recipes in this book use fried mushrooms as their base and they are completely interchangeable – so if you can't get wild mushrooms or chestnut mushrooms for instance, button mushrooms can be used instead.

Above: Button mushrooms.
Top right: Flat mushrooms.
Far right: Field mushrooms.
Right: Chestnut mushrooms (top) and open capped or cup mushrooms.

FIELD MUSHROOMS

Field mushrooms are the wild relatives of the cultivated mushroom and when cooked have a wonderful aroma. Flat mushrooms, although indistinguishable from field mushrooms in appearance, have probably been cultivated and are also excellent. Connoisseurs say that only wild mushrooms have any flavour but many would argue against this. However, if you know where to find field mushrooms, keep the secret to yourself (most mushroom devotees seem to know this) and count yourself lucky!

Buying and Storing

Field mushrooms are sometimes available during the autumn in farm shops. Since they are likely to have been picked recently, they should be fresh unless obviously wilting. Unless you intend to stuff them, don't worry if they are broken in places as you will be slicing them anyway. Use as soon as possible after purchase.

Preparing

Trim the stalk bottoms if necessary and wipe the caps with a damp cloth. Slice according to the recipe.

Cooking

For true field mushrooms, you need do nothing more complicated than simply fry them in butter or olive oil with a suggestion of garlic if liked. However, like flat mushrooms, field mushrooms can be used for stuffing, in soups or indeed any mushroom recipe. They are darker than button mushrooms and will colour soups and sauces brown, but the flavour will be extremely good.

When stuffing mushrooms, gently fry the caps on both sides for a few minutes. The stalks can be chopped and added to the stuffing or can be used for soups or stocks.

WOODLAND MUSHROOMS

Varieties

Ceps: Popular in France, where they are known as *cèpes* and in Italy where they are called *porcini*, these meaty, bun-shaped mushrooms have a fine almost suede-like texture and a good flavour. Instead of gills they have a spongy texture beneath the cap and unless they are very young it is best to scrape this away as it goes soggy when cooked. Ceps are excellent fried in oil or butter over a brisk heat to evaporate the liquid and then added to omelettes.

Alternatively, an Italian way of cooking is to remove the stalk and the spongy tubes, and brush the tops with olive oil. Grill for about 10 minutes under a moderate grill and then turn them over and pour olive oil and a sprinkling of garlic into the centre. Grill for a further 5 minutes and then serve sprinkled with seasoning and parsley.

Chanterelles: Frilly, trumpet-shaped chanterelles are delicate mushrooms which range in colour from cream to a vivid yellow. Later, winter chanterelles, have greyish-lilac gills on the underside of their dark caps. Chanterelles have a delicate, slightly fruity flavour and a firm, almost rubbery texture. They are difficult to clean as their tiny gills tend to trap grit and earth. Rinse them gently under cold running water and then shake dry. Fry in butter over a gentle heat to start with so they exude their liquid and then increase the heat to boil it off. They are delicious with scrambled eggs, or served by themselves with finely cut toast.

Horn of Plenty/Black Trumpets: Taking its name from its shape, this mushroom ranges in colour from mid-brown to black. As it is hollow, it will need to be brushed well to clean or, if a large specimen, sliced in half. It is very versatile, but goes particularly well with fish.

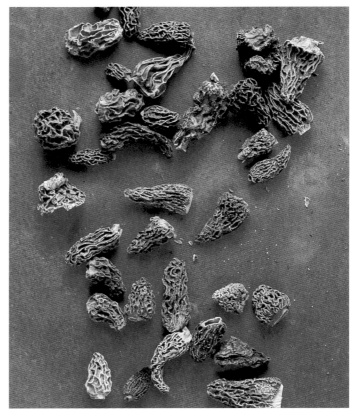

Morels: These are the first mushroom of the year, appearing not in autumn but in spring. In Scandinavia they are called the "truffles of the north" and are considered among the great edible fungi. They are cone-shaped with a crinkled spongy cap but are hollow inside. You will need to wash them well under running water as insects tend to creep into their dark crevices. Morels need longer cooking than most mushrooms: sauté them in butter, add a squeeze of lemon and then cover and simmer for up to an hour until tender. The juices can then be thickened with cream or egg yolks.

Dried Mushrooms: Most wild mushrooms are available dried. To reconstitute, soak in warm water for about 20-30 minutes; in the case of morels when they are added to stews, soak for about 10 minutes. Dried wild mushrooms, particularly ceps, have an intense flavour.

Left: Clockwise from the top: Ceps, Horn of Plenty, Chanterelles.
Above right: Dried mushrooms.
Above left: Morels.
Right: Winter chanterelles.

WILD MUSHROOMS AND OTHER FUNGI

Mushroom gathering, a seasonal event throughout Eastern Europe, Italy and France, is increasingly popular in Britain. The French are particularly enthusiastic: in autumn whole families drive to secret locations to comb the ground for prizes like shaggy ink caps or ceps. Wild mushrooms are sold in supermarkets.

OYSTER MUSHROOMS

These ear-shaped fungi grow on rotting wood. Cap, gills and stem are all the same colour, which can be greyish brown, pink or yellow. They are now widely cultivated, although they are generally thought of as wild mushrooms. Delicious both in flavour and texture, they are softer than the button mushroom when cooked but seem more substantial, having more "bite" to them.

Buying and Storing

Fresh specimens are erect and lively looking with clear gills and smooth caps. They are often sold packed in plastic

boxes under cellophane wrappings and will wilt and go soggy if left on the shelf for too long. Once purchased, remove them from the plastic packaging and use as soon as possible.

Preparing

Oyster mushrooms rarely need trimming at all but if they are large, tear rather than cut them into pieces. In very large specimens the stems can be tough and should be discarded.

Cooking

Fry in butter until tender – they take less time to cook than white mushrooms. Do not overcook oyster mushrooms as the flavour will be lost and the soft texture will become more rubbery.

Left: Pink and yellow oyster mushrooms.
Above: Grey oyster mushrooms.

ENOKITAKI MUSHROOMS

This is another Japanese mushroom. The wild variety is orangy-brown with shiny caps but outside Japan, you will probably only be able to find the cultivated variety, which are similarly fine, with pin-sized heads, but are pale coloured with snowy white caps. They have a fine, sweet and almost fruity flavour. In Japanese cookery they are added to salads or used as a garnish for soups or hot dishes. Since they become tough if overcooked, add enokitaki mushrooms at the very last minute of cooking.

SHIITAKE MUSHROOMS

These Japanese fungi are now commonly available in supermarkets. They are among a variety of tree mushrooms (called *take* in Japan, the *shii* being the hardwood tree from which they are harvested). They have a meaty, slightly acid flavour and a distinct slippery texture. Shiitake mushrooms, though once only available in oriental stores, are now widely available in most supermarkets. Unlike button mushrooms that can be flash-fried, shiitake need to be cooked through, although even this only takes 3-5 minutes. Add them to stir-fries for a delicious flavour and texture. Alternatively, fry them in oil until tender. Sprinkle with sesame oil and then serve with a little soy sauce.

Above: Enokitaki mushrooms.
Right: Shiitake mushrooms.

SOUPS
AND
STARTERS

*Tantalizing appetizers, ranging from traditional
favourites to more exotic dishes for elegant dinners.
There are soups that could be served either before a
main course or on their own as light, wholesome meals,
in addition to dozens of ideas for tempting starters.*

VEGETABLE STOCK

*USE THIS VERSATILE STOCK AS
THE BASIS FOR ALL GOOD SOUPS
AND SAUCES. IF YOU'VE AN
EXTRA LARGE SAUCEPAN OR A
PRESERVING PAN, WHY NOT MAKE
DOUBLE THE QUANTITY AND
FREEZE SEVERAL BATCHES?*

MAKES 2.25 LITRES/4 PINTS/10 CUPS

INGREDIENTS

   2 leeks, roughly chopped
   3 sticks celery, roughly chopped
   1 large onion, with skin, chopped
   2 pieces fresh root ginger, chopped
   3 garlic cloves, unpeeled
   1 yellow pepper, seeded and chopped
   1 parsnip, chopped
   mushroom stalks
   tomato peelings
   45ml/3 tbsp light soy sauce
   3 bay leaves
   bundle of parsley stalks

   3 sprigs of fresh thyme
   1 sprig of fresh rosemary
   10ml/2 tsp salt
   ground black pepper
   3.5 litres/6 pints/15 cups cold water

1 Put all the ingredients into a very large saucepan or a small preserving pan.

2 Bring slowly to the boil, then lower the heat and simmer for 30 minutes, stirring from time to time.

3 Allow the liquid and vegetables to cool. Strain, discard the vegetables and the stock is ready to use. Alternatively, chill or freeze the stock and keep it to use as required.

CRISP CROÛTONS

Easy to make and simple to store, these croûtons add a delightful touch to fresh homemade soups. They are also an ideal way of using up stale bread. Speciality bread such as ciabatta or baguettes can be thinly sliced to make the nicest, crunchiest croûtons, but everyday sliced loaves can be cut into interesting shapes for fun entertaining. Use a good quality, flavourless oil such as sunflower or groundnut, or for a fuller flavour brush with extra virgin olive oil. Alternatively, you could use a flavoured oil such as one with garlic and herbs or chilli.

   Preheat the oven to 200°C/400°F/Gas 6. Place the croûtons on a baking sheet, brush with your chosen oil, then bake for about 15 minutes until golden and crisp. They crisp up further as they cool. Store them in an airtight container for up to a week. Reheat in a warm oven if liked, before serving.

SWEETCORN AND POTATO CHOWDER

THIS HEARTY AMERICAN SOUP IS HIGH IN BOTH FIBRE AND FLAVOUR. IT'S WONDERFUL SERVED WITH CRUSTY BREAD AND TOPPED WITH MELTED CHEDDAR CHEESE.

SERVES FOUR

INGREDIENTS

 1 onion, chopped
 1 garlic clove, crushed
 1 medium size potato, chopped
 2 sticks celery, sliced
 1 small green pepper, seeded, halved
  and sliced
 30ml/2 tbsp sunflower oil
 25g/1oz/2 tbsp butter
 600ml/1 pint/2½ cups stock or water
 salt and ground black pepper
 300ml/½ pint/1¼ cups milk
 1 × 200g/7oz can butter beans
 1 × 300g/11oz can sweetcorn kernels
 good pinch dried sage

1 Put the onion, garlic, potato, celery and green pepper into a large saucepan with the oil and butter.

2 Heat the ingredients until sizzling, then turn the heat down to low. Cover and sweat the vegetables gently for 10 minutes, shaking the pan occasionally.

3 Pour in the stock or water, season to taste and bring to the boil. Turn down the heat, cover and simmer gently for about 15 minutes.

4 Add the milk, beans and sweetcorn – including their liquors – and the sage. Simmer again for 5 minutes. Check the seasoning and serve hot.

ALL THE REDS SOUP

VIBRANT IN COLOUR AND TASTE,
THIS SOUP IS QUICKLY MADE.
USE A RED ONION – IF YOU CAN
FIND ONE – IT WILL ENHANCE
THE FINAL APPEARANCE.

SERVES FOUR TO SIX

INGREDIENTS
  1 red pepper, seeded and chopped
  1 onion, chopped
  1 garlic clove, crushed
  30ml/2 tbsp olive oil
  1 × 400g/14oz can chopped tomatoes
  1 litre/1 pints/4 cups stock
  30ml/2 tbsp long grain rice
  30ml/2 tbsp Worcestershire sauce
  1 × 200g/7oz can red kidney beans
  5ml/1 tsp dried oregano
  5ml/1 tsp sugar
  salt and ground black pepper
  fresh parsley, chopped, and Cheddar
    cheese, grated, to garnish

1 Put the pepper, onion, garlic and oil into a large saucepan. Heat until sizzling then turn down to low. Cover and cook gently for 5 minutes.

2 Add the rest of the ingredients, except the garnishes, and bring to a boil. Stir well, then simmer – covered – for 15 minutes. Check the seasoning, garnish and serve hot. Omit the Cheddar cheese for a vegan soup.

CHINESE TOFU AND LETTUCE SOUP

THIS LIGHT, CLEAR SOUP IS
BRIMMING FULL OF NOURISHING
TASTY PIECES. IDEALLY, MAKE
THIS IN A WOK WITH THE HOME-
MADE VEGETABLE STOCK.

SERVES FOUR

INGREDIENTS
  30ml/2 tbsp groundnut or sunflower oil
  200g/7oz smoked or marinated tofu,
    cubed
  3 spring onions, sliced diagonally
  2 garlic cloves, cut in thin strips
  1 carrot, thinly sliced in rounds
  1.2 litre/2 pints/5 cups stock
  30ml/2 tbsp soy sauce
  15ml/1 tbsp dry sherry or vermouth
  5ml/1 tsp sugar
  115g/4oz cos or romaine lettuce,
    shredded
  salt and ground black pepper

1 Heat the oil in a wok, then stir-fry the tofu cubes until browned. Drain and set aside on kitchen paper towel.

2 In the same oil, stir-fry the onions, garlic and carrot for 2 minutes. Pour in the stock, soy sauce and sherry or vermouth and sugar.

3 Bring to a boil and cook for 1 minute or so. Stir in the lettuce until it just wilts. Add the tofu, season to taste and serve the soup immediately.

SPICED INDIAN CAULIFLOWER SOUP

THIS MILDLY SPICY SOUP MAKES A WARMING FIRST COURSE, AN APPETIZING QUICK MEAL AND A DELICIOUS SUMMER STARTER.

SERVES FOUR TO SIX

INGREDIENTS

1 large potato, peeled and diced
1 small cauliflower, chopped
1 onion, chopped
15ml/1 tbsp sunflower oil
1 garlic clove, crushed
15ml/1 tbsp fresh ginger, grated
10ml/2 tsp ground turmeric
5ml/1 tsp cumin seeds
5ml/1 tsp black mustard seeds
10ml/2 tsp ground coriander
1 litre/1¾ pints/4 cups vegetable stock
300ml/½ pint/1¼ cups natural yogurt
salt and ground black pepper
fresh coriander or parsley, to garnish

1 Put the potato, cauliflower and onion into a large saucepan with the oil and 45ml/3 tbsp water. Heat until hot and bubbling, then cover and turn the heat down. Continue cooking the mixture for about 10 minutes.

2 Add the garlic, ginger and spices. Stir well and cook for another 2 minutes, stirring occasionally. Pour in the stock and season well. Bring to a boil, then cover and simmer for about 20 minutes. Stir in the yogurt, season well and garnish with coriander or parsley.

WINTER WARMER SOUP

SIMMER A SELECTION OF POPULAR WINTER ROOT VEGETABLES TOGETHER FOR A WARMING AND SATISFYING SOUP.

SERVES SIX

INGREDIENTS

3 medium carrots, chopped
1 large potato, chopped
1 large parsnip, chopped
1 large turnip or small swede, chopped
1 onion, chopped
30ml/2 tbsp sunflower oil
25g/1oz/2 tbsp butter
1.5 litres/2½ pints/6 cups water
salt and ground black pepper
1 piece fresh root ginger, grated
300ml/½ pint/1¼ cups milk
45ml/3 tbsp crème fraîche, fromage
  frais or natural yogurt
30ml/2 tbsp fresh dill, chopped
fresh lemon juice

1 Put the carrots, potato, parsnip, turnip or swede and onion into a large saucepan with the oil and butter. Fry lightly, then cover and sweat the vegetables on a very low heat for 15 minutes, shaking the pan occasionally.

2 Pour in the water, bring to a boil and season well. Cover and simmer for 20 minutes until the vegetables are soft.

3 Strain the vegetables, reserving the stock, add the ginger and purée in a food processor or blender until smooth.

4 Return the purée and stock to the pan. Add the milk and stir while the soup gently reheats.

5 Remove from the heat, stir in the crème fraîche, fromage frais or yogurt plus the dill, lemon juice and extra seasoning, if necessary. Reheat the soup, if you wish, but do not allow it to boil as you do so, or it may curdle.

EGG FLOWER SOUP

FOR THE VERY BEST FLAVOUR,
YOU DO NEED TO USE A HOME-
MADE STOCK FOR THIS SOUP. THE
EGG SETS INTO PRETTY STRANDS
GIVING THE SOUP A FLOWERY
LOOK, HENCE THE NAME.

SERVES SIX

INGREDIENTS
1 litre/1¾ pints/4 cups stock
45ml/3 tbsp light soy sauce
30ml/2 tbsp dry sherry or vermouth
3 spring onions, diagonally sliced
small piece fresh root ginger, shredded
4 large lettuce leaves, shredded
5ml/1 tsp sesame seed oil
2 eggs, beaten
salt and ground black pepper
sesame seeds, to garnish

1 Pour the stock into a large saucepan. Add all the ingredients except the eggs and seeds. Bring to the boil and then cook for about 2 minutes.

2 Very carefully, pour the eggs in a thin, steady stream into the centre of the boiling liquid.

3 Count to three, then quickly stir the soup. The egg will begin to cook and form long threads. Season to taste, ladle the soup into warm bowls and serve immediately sprinkled with sesame seeds.

BROCCOLI AND BLUE BRIE SOUP

A POPULAR VEGETABLE,
BROCCOLI MAKES A DELICIOUS
SOUP WITH AN APPETIZING DEEP
GREEN COLOUR. FOR A TASTY
TANG, STIR IN SOME CUBES OF
BLUE BRIE CHEESE JUST BEFORE
SERVING.

SERVES SIX

INGREDIENTS
1 onion, chopped
450g/1lb broccoli spears, chopped
1 large courgette, chopped
1 large carrot, chopped
1 medium potato, chopped
25g/1oz/2 tbsp butter
30ml/2 tbsp sunflower oil
2 litres/3½ pints/8 cups stock or water
about 75g/3oz blue Brie (or Dolcellate)
  cheese, cubed
salt and ground black pepper
almond flakes, to garnish (optional)

1 Put all the vegetables into a large saucepan, together with the butter and oil, plus about 45ml/3 tbsp stock or water.

2 Heat the ingredients until sizzling and stir well. Cover and cook gently for 15 minutes, shaking the pan occasionally, until all the vegetables soften.

3 Add the rest of the stock or water, season and bring to a boil, then cover and simmer gently for about 25–30 minutes.

4 Strain the vegetables and reserve the liquid. Purée the vegetables in a food processor or blender then return them to the pan with the reserved liquid.

5 Bring the soup back to a gentle boil and stir in the cheese until it melts. (Don't let the soup boil too hard or the cheese will become stringy.) Season to taste and garnish with a scattering of almond flakes.

HOME-MADE MUSHROOM SOUP

HOME-MADE MUSHROOM SOUP IS QUITE, QUITE DIFFERENT FROM CANNED OR PACKET SOUPS.

SERVES FOUR TO SIX

INGREDIENTS

    450g/1lb open cup mushrooms, sliced
    115g/4oz shiitake mushrooms, sliced
    45ml/3 tbsp sunflower oil
    1 onion, chopped
    1 stick celery, chopped
    1.2 litres/2 pints/5 cups stock or water
    30ml/2 tbsp soy sauce
    50g/2oz/¼ cup long grain rice
    salt and ground black pepper
    300ml/½ pint/1¼ cups milk
    fresh parsley, chopped, and almond
        flakes, to garnish

1 Put all the mushrooms into a large saucepan with the oil, onion and celery. Heat until sizzling, then cover and simmer for about 10 minutes, shaking the pan occasionally.

2 Add the stock or water, soy sauce, rice and seasoning. Bring to the boil, then cover and simmer gently for 20 minutes until the vegetables and rice are soft.

3 Strain the vegetables, reserving the stock, and purée in a food processor or blender until smooth. Return the vegetables and reserved stock to the pan.

4 Stir in the milk, reheat until boiling and taste for seasoning. Serve hot, sprinkled with a little chopped parsley and a few almond flakes.

CLASSIC MINESTRONE

THE HOME-MADE VERSION OF THIS FAMOUS SOUP IS A DELICIOUS REVELATION AND MOUTH-WATERINGLY HEALTHY.

SERVES FOUR

INGREDIENTS

1 large leek, thinly sliced
2 carrots, chopped
1 courgette, thinly sliced
115g/4oz whole green beans, halved
2 sticks celery, thinly sliced
45ml/3 tbsp olive oil
1.5 litres/2½ pints/6¼ cups stock
1 × 400g/14oz can chopped tomatoes
15ml/1 tbsp fresh basil, chopped
5ml/1 tsp fresh thyme leaves, chopped
    or 2.5ml/½ tsp dried thyme
salt and ground black pepper
1 × 400g/14oz can cannellini or kidney
    beans
50g/2oz/⅓ cup small pasta shapes or
    macaroni
fresh Parmesan cheese, finely grated
    (optional) and fresh parsley, chopped,
    to garnish

1 Put all the fresh vegetables into a large saucepan with the olive oil. Heat until sizzling, then cover, lower the heat and sweat the vegetables for 15 minutes, shaking the pan occasionally.

2 Add the stock (use water if you wish), tomatoes, herbs and seasoning. Bring to the boil, replace the lid and simmer gently for about 30 minutes.

3 Add the beans and their liquor together with the pasta, and simmer for a further 10 minutes. Check the seasoning and serve hot sprinkled with the Parmesan cheese (if used) and parsley.

COOK'S TIP
Minestrone is also delicious served cold on a hot summer's day. In fact, the flavour improves if it is made a day or two ahead and stored in the refrigerator. It can also be frozen and reheated.

CLASSIC FRENCH ONION SOUP

WHEN FRENCH ONION SOUP IS MADE SLOWLY AND CAREFULLY, THE ONIONS ALMOST CARAMELIZE TO A DEEP MAHOGANY COLOUR. IT HAS A SUPERB FLAVOUR AND IS A PERFECT WINTER SUPPER DISH.

SERVES FOUR

INGREDIENTS
    4 large onions
    30ml/2 tbsp sunflower or olive oil, or
        15ml/1 tbsp of each
    25g/1oz butter
    900ml/1½ pints/3¾ cups stock
    salt and freshly ground black pepper
    4 slices French bread
    40–50g/1½–2oz Gruyère or Cheddar
        cheese, grated

1 Peel and quarter the onions and slice or chop them into 5mm/¼in pieces. Heat the oil and butter together in a deep heavy-based saucepan, preferably with a medium-size base so that the onions form a thick layer.

3 When the onions are a rich mahogany brown, add the stock and a little seasoning. Simmer, partially covered, for 30 minutes, then taste and adjust the seasoning according to taste.

4 Preheat the grill and toast the French bread. Spoon the soup into four oven-proof serving dishes and place a piece of bread in each. Sprinkle with the cheese and grill for a few minutes until golden.

2 Fry the onions briskly for a few minutes, stirring constantly, and then reduce the heat and cook gently for 45–60 minutes. At first, the onions need to be stirred only occasionally but as they begin to colour, stir frequently. The colour of the onions gradually turns golden and then more rapidly to brown, so take care to stir constantly at this stage so they do not burn on the base.

ASPARAGUS SOUP

HOME-MADE ASPARAGUS SOUP HAS A DELICATE FLAVOUR, QUITE UNLIKE THAT FROM A CAN. THIS SOUP IS BEST MADE WITH YOUNG ASPARAGUS, WHICH IS TENDER AND BLENDS WELL. SERVE IT WITH WAFER THIN SLICES OF BREAD.

SERVES FOUR

INGREDIENTS
    450g/1lb young asparagus
    40g/1½ oz butter
    6 shallots, sliced
    15g/½ oz plain flour
    600ml/1 pint/2½ cups vegetable
      stock or water
    15ml/1 tbsp lemon juice
    salt and freshly ground black pepper
    250ml/8fl oz/1 cup milk
    120ml/4fl oz/½ cup single cream
    10ml/2 tsp chopped fresh chervil

1 Trim the stalks of the asparagus if necessary. Cut 4cm/1½in off the tops of half the asparagus and set aside for a garnish. Slice the remaining asparagus.

2 Melt 25g/1oz of the butter in a large saucepan and gently fry the sliced shallots for 2–3 minutes until soft but not brown, stirring occasionally.

3 Add the sliced asparagus and fry over a gentle heat for about 1 minute. Stir in the flour, cook for 1 minute. Stir in the stock or water, lemon juice and season to taste. Bring to the boil and then simmer, partially covered, for 15–20 minutes until the asparagus is very tender.

4 Cool slightly and then process the soup in a food processor or blender until smooth. Then press the puréed asparagus through a sieve placed over a clean saucepan. Add the milk by pouring and stirring it through the sieve with the asparagus so as to extract the maximum amount of asparagus purée.

5 Melt the remaining butter and fry the reserved asparagus tips gently for about 3–4 minutes to soften.

6 Heat the soup gently for 3–4 minutes. Stir in the cream and the asparagus tips. Heat gently and serve sprinkled with the chopped fresh chervil.

CARROT AND CORIANDER SOUP

NEARLY ALL ROOT VEGETABLES MAKE EXCELLENT SOUPS AS THEY PURÉE WELL AND HAVE AN EARTHY FLAVOUR WHICH COMPLEMENTS THE SHARPER FLAVOURS OF HERBS AND SPICES. CARROTS ARE PARTICULARLY VERSATILE, AND THIS SIMPLE SOUP IS ELEGANT IN BOTH FLAVOUR AND APPEARANCE.

SERVES FOUR TO SIX

INGREDIENTS

450g/1lb carrots, preferably young
  and tender
15ml/1 tbsp sunflower oil
40g/1½oz butter
1 onion, chopped
1 celery stalk, sliced, plus 2–3 pale
  leafy celery tops
2 small potatoes, chopped
1 litre/1¾ pints/4 cups stock
10–15ml/2–3 tsp ground coriander
15ml/1 tbsp chopped fresh coriander
200ml/7fl oz/⅞ cup milk
salt and freshly ground black pepper

1 Trim the carrots, peel if necessary and cut into chunks. Heat the oil and 25g/1oz of the butter in a large flame-proof casserole or heavy-based saucepan and fry the onion over a gentle heat for 3–4 minutes until slightly softened but not browned.

2 Cut the celery stalk into slices. Add the celery and potato to the onion in the pan, cook for a few minutes and then add the carrots. Fry over a gentle heat for 3–4 minutes, stirring frequently, and then cover. Reduce the heat even further and sweat for about 10 minutes. Shake the pan or stir occasionally so the vegetables do not stick to the base.

3 Add the stock, bring to the boil and then partially cover and simmer for a further 8–10 minutes until the carrots and potato are tender.

4 Remove 6–8 tiny celery leaves for garnish and finely chop the remaining celery tops (about 15ml/1 tbsp once chopped). Melt the remaining butter in a small saucepan and fry the ground coriander for about 1 minute, stirring constantly.

5 Reduce the heat and add the chopped celery tops and fresh coriander and fry for about 1 minute. Set aside.

6 Process the soup in a food processor or blender and pour into a clean saucepan. Stir in the milk, coriander mixture and seasoning. Heat gently, taste and adjust the seasoning. Serve garnished with the reserved celery leaves.

COOK'S TIP
For a more piquant flavour, add a little lemon juice just before serving.

BORSCHT

THIS CLASSIC SOUP WAS THE STAPLE DIET OF PRE-REVOLUTION RUSSIAN PEASANTS FOR HUNDREDS OF YEARS. THERE ARE MANY VARIATIONS, AND IT IS RARE TO FIND TWO RECIPES THE SAME.

SERVES SIX

INGREDIENTS

   350g/12oz whole, uncooked beetroot
   15ml/1 tbsp sunflower oil
   1 large onion, thinly sliced
   1 large carrot, cut into julienne strips
   3 celery stalks, thinly sliced
   1.5 litres/2½ pints/6¼ cups vegetable
     stock
   about 225g/8oz tomatoes, peeled,
     seeded and sliced
   about 30ml/2 tbsp lemon juice
     or wine vinegar
   30ml/2 tbsp chopped dill
   salt and freshly ground black pepper
   115g/4oz white cabbage, thinly sliced
   150ml/¼ pint/⅔ cup soured cream

1 Peel the beetroot, slice and then cut into very thin strips.

2 Heat the oil in a large, heavy-based saucepan and fry the thinly sliced onion for 2–3 minutes and then add the carrot, celery and beetroot. Cook for 4–5 minutes, stirring frequently, until the oil has been absorbed.

3 Add the stock, tomatoes, lemon juice or wine vinegar, half of the dill and seasoning. Bring to the boil and simmer for about 30–40 minutes until the vegetables are completely tender.

4 Add the cabbage and simmer for 5 minutes until tender. Adjust the seasoning and serve sprinkled with remaining dill and the soured cream.

SPICY BORSCHT

THE FLAVOUR MATURES AND IMPROVES IF THE SOUP IS MADE THE DAY BEFORE IT IS NEEDED.

SERVES SIX

INGREDIENTS
   1 onion, chopped
   450g/1lb raw beetroot, peeled and
      chopped
   1 large cooking apple, chopped
   2 celery sticks, chopped
   ½ red pepper, chopped
   115g/4oz mushrooms, chopped
   25g/1oz/2 tbsp butter
   30ml/2 tbsp sunflower oil
   2 litres/3½ pints/8 cups stock or water
   5ml/1 tsp cumin seeds
   pinch dried thyme
   1 large bay leaf
   fresh lemon juice
   salt and ground black pepper
   150ml/¼ pint/⅔ cup soured cream
   few sprigs fresh dill, to garnish

1 Place all the chopped vegetables into a large saucepan with the butter, oil and 45ml/3 tbsp of the stock or water. Cover and cook gently for about 15 minutes, shaking the pan occasionally.

2 Stir in the cumin seeds and cook for a minute, then add the remaining stock or water, dried thyme, bay leaf, lemon juice and seasoning.

3 Bring to the boil, then cover and turn down to a gentle simmer. Cook for about 30 minutes.

4 Strain the vegetables and reserve the liquid. Pass the vegetables through a food processor or blender until they are smooth and creamy.

5 Return the vegetables to the pan, stir in the reserved stock and reheat. Check the seasoning.

6 Serve the borscht with swirls of soured cream and topped with a few sprigs of fresh dill.

VARIATION
This soup can be served fairly thick, as long as the vegetables are finely chopped first.

   Beetroot is something of an under-valued vegetable, although popular in many European countries. For example, it is delicious served as a hot vegetable accompaniment with a creamy bechamel sauce and topped with crisp breadcrumbs. Alternatively, try it raw and coarsely grated, then tossed in dressing for a side salad.

ONION SQUASH RISOTTO

ONION SQUASHES ARE NOT AS SWEET AS PUMPKINS, BUT THEY HAVE A DELICATE FRUITY TASTE, WHICH IS IDEAL IN A RISOTTO. SERVE THIS AS A TASTY STARTER OR LIGHT LUNCH.

SERVES FOUR

INGREDIENTS
1 onion squash or pumpkin, about
    900g–1kg/2–2¼lb
30ml/2 tbsp olive oil
1 onion, chopped
1–2 garlic cloves, crushed
115g/4oz arborio rice
600–750ml/1–1¼ pints/2½–3 cups
    vegetable stock
salt and freshly ground black pepper
40g/1½oz Parmesan cheese, grated
15ml/1 tbsp chopped fresh parsley

1   Halve or quarter the onion squash or pumpkin, remove the seeds and skin, and then cut into chunks about 1–2cm/½–¾in in size.

2   Heat the oil in a flameproof casserole and fry the onion and garlic for about 3–4 minutes, stirring frequently. Continue frying until both the onion and garlic are lightly golden.

3   Add the squash or pumpkin, stir-fry for a few minutes. Add the rice and cook for about 2 minutes, stirring all the time.

4   Pour in about half of the stock and season. Stir well and then half cover and simmer gently for about 20 minutes, stirring occasionally. As the liquid is absorbed, add more stock and stir to prevent the mixture sticking to the base.

5   When the squash and rice are nearly tender, add a little more stock. Cook, uncovered for 5–10 minutes. Stir in the Parmesan cheese and parsley and serve.

PUMPKIN SOUP

THE SWEET FLAVOUR OF PUMPKIN IS GOOD IN SOUPS, TEAMING WELL WITH OTHER MORE SAVOURY INGREDIENTS SUCH AS ONIONS AND POTATOES TO MAKE A WARM AND COMFORTING DISH.

SERVES FOUR TO SIX

INGREDIENTS
15ml/1 tbsp sunflower oil
25g/1oz butter
1 large onion, sliced
675g/1½lb pumpkin, cut into
    large chunks
450g/1lb potatoes, sliced
600ml/1 pint/2½ cups vegetable stock
good pinch of nutmeg
5ml/1 tsp chopped fresh tarragon
salt and freshly ground black pepper
600ml/1 pint/2½ cups milk
about 5–10ml/1–2 tsp lemon juice

1   Heat the oil and butter in a heavy-based saucepan and fry the onion for 4–5 minutes over a gentle heat until soft but not browned, stirring frequently.

2   Add the pumpkin and potato, stir well and then cover and sweat over a low heat for about 10 minutes until the vegetables are almost tender, stirring occasionally to prevent them from sticking to the pan.

3   Stir in the stock, nutmeg, tarragon and seasoning. Bring to the boil and then simmer for about 10 minutes until the vegetables are completely tender.

4   Allow to cool slightly, then pour into a food processor or blender and process until smooth. Pour back into a clean saucepan and add the milk. Heat gently and then taste, adding the lemon juice and extra seasoning if necessary. Serve piping hot with crusty brown bread.

CREAM OF MUSHROOM SOUP

A GOOD MUSHROOM SOUP MAKES THE MOST OF THE SUBTLE AND SOMETIMES RATHER ELUSIVE FLAVOUR OF MUSHROOMS. BUTTON MUSHROOMS ARE USED HERE FOR THEIR PALE COLOUR; CHESTNUT OR, BETTER STILL, FIELD MUSHROOMS GIVE A FULLER FLAVOUR BUT TURN THE SOUP BROWN.

SERVES FOUR

INGREDIENTS

275g/10oz button mushrooms
15ml/1 tbsp sunflower oil
40g/1½oz butter
1 small onion, finely chopped
15ml/1 tbsp plain flour
450ml/¾ pint/1¾ cups vegetable stock
450ml/¾ pint/1¾ cups milk
pinch of dried basil
salt and freshly ground black pepper
30–45ml/2–3 tbsp single cream
  (optional)
fresh basil leaves, to garnish

1 Separate the mushroom caps from the stalks. Finely slice the caps and finely chop the stalks.

2 Heat the oil and half the butter in a heavy-based saucepan and add the onion, mushroom stalks and ½–¾ of the sliced mushroom caps. Fry for about 1–2 minutes, stirring frequently, and then cover and sweat over a gentle heat for 6–7 minutes, stirring occasionally.

3 Stir in the flour and cook for about 1 minute. Gradually add the stock and milk, to make a smooth thin sauce. Add the basil, and season with salt and pepper. Bring to the boil and then simmer, partly covered, for 15 minutes.

4 Cool slightly and then pour the soup into a food processor or blender and process until smooth. Melt the rest of the butter in a frying pan and fry the remaining mushrooms gently for 3–4 minutes until they are just tender.

5 Pour the soup into a clean saucepan and stir in the sliced mushrooms. Heat until very hot and adjust the seasoning. Add a little cream, if using. Serve sprinkled with fresh basil leaves.

SOUFFLÉ OMELETTE WITH MUSHROOM SAUCE

A SOUFFLÉ OMELETTE INVOLVES A LITTLE MORE PREPARATION THAN AN ORDINARY OMELETTE BUT THE RESULT IS LIGHT AND SPRINGY TO TOUCH. THIS DISH MAKES A DELICIOUS LIGHT LUNCH.

SERVES ONE

INGREDIENTS

2 eggs, separated
15g/½oz butter
sprig of parsley or coriander
For the mushroom sauce
15g/½oz butter
75g/3oz button mushrooms,
  thinly sliced
15ml/1 tbsp plain flour
85–120ml/3–4fl oz/½ cup milk
5ml/1 tsp chopped fresh parsley
  (optional)
salt and freshly ground black pepper

1 To make the mushroom sauce, melt the butter in a saucepan or frying pan and fry the sliced mushrooms for 4–5 minutes until tender.

2 Stir in the flour and then gradually add the milk, stirring all the time, to make a smooth sauce. Add the parsley, if using, and season with salt and pepper. Keep warm to one side.

3 Beat the egg yolks with 15ml/1 tbsp water and season with a little salt and pepper. Whisk the egg whites until stiff and then fold into the egg yolks using a metal spoon. Preheat the grill.

4 Melt the butter in a large frying pan and pour the egg mixture into the pan. Cook over a gentle heat for 2–4 minutes. Place the frying pan under the grill and cook for a further 3–4 minutes until the top is golden brown.

5 Slide the omelette on to a warmed serving plate, pour over the mushroom sauce and fold the omelette in half. Serve garnished with a sprig of parsley or fresh coriander leaves.

GAZPACHO

Gazpacho is a classic Spanish soup. It is popular all over Spain but nowhere more so than in Andalucia, where there are hundreds of variations. It is a cold soup of tomatoes, tomato juice, green pepper and garlic, which is served with a selection of garnishes.

SERVES FOUR

INGREDIENTS
  1.5kg/3–3½lb ripe tomatoes
  1 green pepper, seeded and roughly
    chopped
  2 garlic cloves, crushed
  2 slices white bread, crusts removed
  60ml/4 tbsp olive oil
  60ml/4 tbsp tarragon wine vinegar
  150ml/¼ pint/⅔ cup tomato juice
  good pinch of sugar
  salt and freshly ground black pepper
  ice cubes, to serve
For the garnishes
  30ml/2 tbsp sunflower oil
  2–3 slices white bread, diced
  1 small cucumber, peeled and
    finely diced
  1 small onion, finely chopped
  1 red pepper, seeded and finely diced
  1 green pepper, seeded and finely
    diced
  2 hard-boiled eggs, chopped

1 Skin the tomatoes, then quarter them and remove the cores.

2 Place the pepper in a food processor and process for a few seconds. Add the tomatoes, garlic, bread, olive oil and vinegar and process again. Add the tomato juice, sugar, seasoning and a little extra tomato juice or cold water and process. The consistency should be thick but not too stodgy.

3 Pour into a bowl and chill for at least 2 hours but no more than 12 hours, otherwise the textures deteriorate.

4 To prepare the bread cubes to use as a garnish, heat the oil in a frying pan and fry them over a moderate heat for 4-5 minutes until golden brown. Drain well on kitchen paper.

5 Place each garnish in a separate small dish, or alternatively arrange them in rows on a large plate.

6 Just before serving, stir a few ice cubes into the soup and then spoon into serving bowls. Serve with the garnishes.

PEAR AND WATERCRESS SOUP WITH STILTON CROÛTONS

PEARS AND STILTON TASTE VERY GOOD WHEN EATEN TOGETHER AFTER THE MAIN COURSE. HERE, FOR A CHANGE, THEY ARE COMBINED IN A STARTER.

SERVES SIX

INGREDIENTS
　1 bunch watercress
　4 medium pears, sliced
　900ml/1½ pints/3¾ cups vegetable
　　stock
　salt and pepper
　120ml/4fl oz/½ cup double cream
　juice of 1 lime
For the croûtons
　25g/1oz butter
　15ml/1 tbsp olive oil
　200g/7oz/3 cups stale bread, cubed
　140g/5oz/1 cup Stilton cheese,
　　chopped

1 Keep back about one-third of the watercress leaves. Place all the rest of the leaves and the stalks in a pan with the pears, stock and a little seasoning. Simmer for about 15–20 minutes. Reserving a few watercress leaves for garnish, add the rest and immediately blend in a food processor until smooth.

2 Put the mixture into a bowl and stir in the cream and lime juice to mix the flavours thoroughly. Season again to taste. Pour all the soup back into a pan and reheat, stirring until warmed through.

3 To make the croûtons, melt the butter and oil in a pan and fry the bread cubes until golden brown. Drain on kitchen paper. Put the cheese on top, then heat under a hot grill until bubbling.

4 Pour the reheated soup into bowls. Use the croûtons and remaining watercress leaves to garnish the soup before serving.

GARLIC MUSHROOMS

GARLIC AND MUSHROOMS MAKE A WONDERFUL COMBINATION. THEY MUST BE SERVED PIPING HOT, SO IF POSSIBLE USE A BALTI PAN OR CAST IRON FRYING PAN AND DON'T STAND ON CEREMONY – SERVE STRAIGHT FROM THE PAN.

SERVES FOUR (as a starter)

INGREDIENTS
    30ml/2 tbsp sunflower oil
    25g/1oz butter
    5 spring onions, thinly sliced
    3 garlic cloves, crushed
    450g/1lb button mushrooms
    40g/1½oz fresh white breadcrumbs
    15ml/1 tbsp chopped fresh parsley
    30ml/2 tbsp lemon juice
    salt and freshly ground black pepper

1 Heat the oil and butter in a balti pan, wok or cast iron frying pan. Add the spring onions and garlic and stir-fry over a medium heat for 1–2 minutes.

2 Add the whole button mushrooms and fry over a high heat for 4–5 minutes, stirring and tossing with a large wide spatula or wooden spoon all the time.

3 Stir in the breadcrumbs, parsley, lemon juice and seasoning. Stir-fry for a few minutes until the lemon juice has virtually evaporated and then serve.

ROAST GARLIC WITH CROÛTONS

YOUR GUESTS WILL BE ASTONISHED TO BE SERVED A WHOLE ROAST GARLIC FOR A STARTER. ROAST GARLIC HAS A HEAVENLY FLAVOUR AND IS SO IRRESISTIBLE THAT THEY WILL EVEN FORGIVE YOU THE NEXT DAY!

SERVES FOUR

INGREDIENTS
    2 garlic bulbs
    45ml/3 tbsp olive oil
    45ml/3 tbsp water
    sprig of rosemary
    sprig of thyme
    1 bay leaf
    sea salt and freshly ground
        black pepper
To serve
    slices of French bread
    olive or sunflower oil, for frying
    175g/6oz young goat's cheese or soft
        cream cheese
    10ml/2 tsp chopped fresh herbs, e.g.
        marjoram, parsley and chives

1 Preheat the oven to 190°C/375°F/ Gas 5. Place the garlic bulbs in a small ovenproof dish and pour over the oil and water. Add the rosemary, thyme and bay leaf and sprinkle with sea salt and pepper. Cover with foil and bake in the oven for 30 minutes.

2 Remove the foil, baste the garlic heads with the juices from the dish and bake for a further 15–20 minutes until they feel soft when pressed.

3 Heat a little oil in a frying pan and fry the French bread on both sides until golden. Blend the cheese with the mixed herbs and place in a serving dish.

4 Cut each garlic bulb in half and open out slightly. Serve the garlic on small plates with the croûtons and soft cheese. Each garlic clove should be squeezed out of its papery shell, spread over a croûton and eaten with the cheese.

MUSHROOMS ON TOAST

SERVE THESE ON TOAST FOR A QUICK, TASTY STARTER OR POP THEM INTO SMALL RAMEKINS AND SERVE WITH SLICES OF WARM CRUSTY BREAD. USE SOME SHITAKE MUSHROOMS, IF YOU CAN FIND THEM, FOR A RICHER FLAVOUR.

SERVES FOUR

INGREDIENTS
  450g/1lb button mushrooms, sliced if
    large
  45ml/3 tbsp olive oil
  45ml/3 tbsp stock or water
  30ml/2 tbsp dry sherry (optional)
  3 garlic cloves, crushed
  115g/4oz low fat soft cheese
  30ml/2 tbsp fresh parsley, chopped
  15ml/1 tbsp fresh chives, chopped
  salt and ground black pepper

1 Put the mushrooms into a large saucepan with the olive oil, stock or water and sherry, if using. Heat until bubbling then cover and simmer for 5 minutes.

2 Add the garlic and stir well. Cook for a further 2 minutes. Remove the mushrooms with a slotted spoon and set them aside. Cook the liquor until it reduces down to 30ml/2 tbsp. Remove from the heat and stir in the cheese and herbs.

3 Stir the mixture well until the cheese melts, then return the mushrooms to the pan so that they become coated with the cheesy mixture. Season to taste.

4 Pile the mushrooms onto thick slabs of hot toast. Alternatively, spoon them into four ramekins and serve accompanied by slices of crusty bread.

RICOTTA AND BORLOTTI BEAN PÂTÉ

FOR AN ATTRACTIVE PRESENTATION, SPOON THE PÂTÉ INTO SMALL, OILED RING MOULDS, TURN OUT AND FILL WITH WHOLE BORLOTTI BEANS, DRESSED WITH LEMON JUICE, OLIVE OIL AND FRESH HERBS.

SERVES FOUR

INGREDIENTS
  1 × 400g/14oz can borlotti beans,
    drained
  1 garlic clove, crushed
  175g/6oz ricotta cheese (or other
    cream cheese)
  50g/2oz/4 tbsp butter, melted
  juice of ½ lemon
  salt and ground black pepper
  30ml/2 tbsp fresh parsley, chopped
  15ml/1 tbsp fresh thyme or dill,
    chopped
To serve
  extra canned beans (optional)
  fresh lemon juice, olive oil and
    chopped herbs (optional)
  salad leaves, radish slices and a few
    sprigs fresh dill, to garnish

1 Blend the beans, garlic, cheese, butter, lemon juice and seasoning in a food processor until smooth.

2 Add the chopped herbs and continue to blend. Spoon into one serving dish or four lightly oiled ramekins, the bases lined with discs of non-stick baking paper. Chill the pâté so that it sets firm.

3 If serving with extra beans, dress them with lemon juice, olive oil and herbs, season well and spoon on top. Garnish with salad leaves and serve with warm crusty bread or toast.

4 If serving individually, turn each pâté out of its ramekin onto a small plate and remove the disc of paper. Top with radish slices and sprigs of dill.

VARIATION
You could try other canned pulses for this recipe, although the softer lentils would not be suitable. Butter (lima) beans are surprisingly good. For an attractive presentation, fill the centre with dark red kidney beans and chopped fresh green beans.

MEDITERRANEAN VEGETABLES WITH TAHINI

*WONDERFULLY COLOURFUL, THIS
STARTER IS EASILY PREPARED IN
ADVANCE. TAHINI IS A PASTE
MADE FROM SESAME SEEDS.*

SERVES FOUR

INGREDIENTS
    2 peppers, seeded and quartered
    2 courgettes, halved lengthways
    2 small aubergines, degorged and
       halved lengthways
    1 fennel bulb, quartered
    olive oil
    salt and ground black pepper
    115g/4oz Greek Halloumi cheese,
       sliced
For the tahini cream
    225g/8oz/1 cup tahini paste
    1 garlic cloves, crushed
    30ml/2 tbsp olive oil
    30ml/2 tbsp fresh lemon juice
    120ml/4 floz/½ cup cold water

1 Preheat the grill or barbecue until hot. Brush the vegetables with the oil and grill until just browned, turning once. (If the peppers blacken, don't worry. The skins can be peeled off.) Cook the vegetables until just softened.

2 Place the vegetables in a shallow dish and season. Allow to cool. Meanwhile, brush the cheese slices with oil and grill these on both sides until just charred. Remove them with a palette knife.

3 To make the tahini cream, place all the ingredients, except the water, in a food processor or blender. Whizz for a few seconds to mix, then, with the motor still running, pour in the water and blend until smooth.

4 Serve the vegetables and cheese on a platter and trickle over the cream. Delicious served with warm pitta or naan bread.

COOK'S TIP
To degorge aubergines, sprinkle cut slices with salt and allow the juices that form to drain away in a colander. After 30 minutes or so, rinse well and pat dry. Degorged aubergines are less bitter and easier to cook.

IMAM BAYILDI

LEGEND HAS IT THAT A MUSLIM HOLY MAN – THE IMAM – WAS SO OVERWHELMED BY THIS DISH THAT HE FAINTED IN SHEER DELIGHT! TRANSLATED, IMAM BAYILDI MEANS "THE IMAM FAINTED"

SERVES FOUR

INGREDIENTS

2 medium aubergines, degorged and
  halved lengthways
salt
60ml/4 tbsp olive oil
2 large onions, sliced thinly
2 garlic cloves, crushed
1 green pepper, seeded and sliced
1 × 400g/14oz can chopped tomatoes
30g/1½oz sugar
5ml/1 tsp ground coriander
ground black pepper
30ml/2 tbsp fresh coriander or parsley,
  chopped

1 Gently fry the aubergines, cut side down, in the oil for 5 minutes, then drain and place in a shallow ovenproof dish.

2 In the same pan gently fry the onions, garlic and green pepper, adding extra oil if necessary. Cook for about 10 minutes.

3 Add the tomatoes, sugar, ground coriander and seasoning and cook for about 5 minutes until the mixture is reduced. Stir in the coriander or parsley. Spoon this mixture on top of the aubergines.

4 Preheat the oven to 190°C/375°F/ Gas 5, cover and bake for about 30–35 minutes. When cooked, cool, then chill. Serve cold with crusty bread.

COOK'S TIP
Before cooking, slash the aubergine flesh a few times, sprinkle with salt and place in a colander for about half an hour. Rinse and pat dry.

TWICE-BAKED GOATS' CHEESE SOUFFLÉS

*A GOOD CHEF'S TRICK IS TO
REHEAT SMALL BAKED SOUFFLÉS
OUT OF THEIR RAMEKINS TO
SERVE WITH A SALAD. THEY PUFF
UP AGAIN AND THE OUTSIDES
BECOME NICE AND CRISPY.*

SERVES SIX

INGREDIENTS
    25g/1oz/2 tbsp butter
    25g/1oz/3 tbsp plain flour
    300ml/½ pint/1¼ cups hot milk
    pinch cayenne pepper
    squeeze of lemon juice
    salt and ground black pepper
    100g/3½oz semi-hard goats' cheese
    2 eggs, separated
    melted butter, for brushing
    25g/1oz/3 tbsp dried breadcrumbs
    25g/1oz/3 tbsp ground hazelnuts or
      walnuts
    2 egg whites
    salad garnish (optional)

1 Melt the butter and stir in the flour.
Cook to a roux for a minute, then
gradually whisk in the hot milk to make a
thick white sauce.

VARIATION
There is another good chef's trick –
making soufflés in advance and chilling
them unbaked. It helps to add an extra
egg white or two when whisking,
depending on the mixture. It is also
possible to freeze unbaked soufflés in
small ramekins and then to bake them
from frozen, allowing an extra 5 or 10
minutes' baking time.

2 Simmer for a minute, then season with
cayenne, lemon juice, salt and pepper.
Remove the pan from the heat and stir in
the crumbled cheese until it melts. Cool
slightly then beat in the egg yolks.

3 Brush the insides of six ramekins with
the melted butter and coat them with the
breadcrumbs and nuts mixed together.
Shake out any excess.

4 Preheat the oven to 190°C/375°F/Gas 5
and prepare a bain marie – a roasting pan
with boiling water.

5 Whisk the four egg whites to the soft
peak stage and carefully fold them into
the main mixture using a figure-of-eight
motion. Spoon into the ramekins.

6 Place the soufflés in the bain marie
and bake for about 12–15 minutes until
risen and golden brown. You can, of
course, serve them at this stage;
otherwise allow to cool then chill.

7 To serve twice-baked, reheat the oven
to the same temperature. Run a knife
round the inside of each ramekin and turn
out each soufflé onto a baking tray.

8 Bake the soufflés for about 12 minutes.
Serve on prepared plates with a dressed
salad garnish.

TRICOLOUR SALAD

*THIS CAN BE A SIMPLE STARTER
OR PART OF A LIGHT BUFFET.
LIGHTLY SALTED TOMATOES MAKE
A TASTY DRESSING WITH THEIR
OWN NATURAL JUICES.*

SERVES FOUR TO SIX

INGREDIENTS
    1 small red onion, sliced thinly
    6 large full-flavoured tomatoes
    extra virgin olive oil, to sprinkle
    50g/2oz/small bunch rocket or
        watercress, roughly chopped
    salt and ground black pepper
    175g/6oz Mozzarella cheese, thinly
        sliced or grated
    30ml/2 tbsp pine nuts (optional)

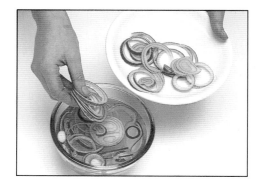

1 Soak the onion slices in a bowl of cold water for 30 minutes, then drain and pat dry. Skin the tomatoes by slashing and dipping briefly in boiling water. Remove the core and slice the flesh.

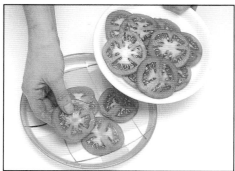

2 Slice the tomatoes and arrange half on a large platter, or divide them between small plates.

3 Sprinkle liberally with olive oil then layer with the chopped rocket or cress, and soaked onion slices, seasoning well. Add in the cheese, sprinkling over more oil and seasoning as you go.

4 Repeat with the remaining tomato slices, salad leaves, cheese and oil.

5 Season well to finish and complete with some oil and a good scattering of pine nuts. Cover the salad and chill for at least 2 hours before serving.

GUACAMOLE SALSA IN RED LEAVES

SERVE THIS LOVELY DIP WITH CHUNKS OF WARM GARLIC BREAD.

SERVES FOUR

INGREDIENTS

2 tomatoes, skinned and chopped
15ml/1 tbsp grated onion
1 garlic clove, crushed
1 green chilli, halved, seeded and chopped
2 ripe avocados
30ml/2 tbsp olive oil
2.5ml/½ tsp ground cumin
30ml/2 tbsp fresh coriander or parsley, chopped
juice of 1 lime
salt and ground black pepper
leaves from radicchio lettuce

1 Using a sharp knife, slash a small cross on the top of the tomatoes, then dip them briefly in a bowl of boiling water. The skins will slip off easily. Remove the core and chop the flesh.

2 Put the tomato flesh into a bowl together with the onion, garlic and chopped chilli. Halve the avocados, remove the stones then scoop the flesh into the bowl, mashing it with a fork.

3 Add the remaining ingredients, except for the radicchio leaves, and mix well together, seasoning to taste.

4 Lay the radicchio leaves on a platter and spoon in the salsa. Serve immediately as avocados go black when exposed to the air.

COOK'S TIP
Take care when cutting chillies. The juice can sting, so be careful not to rub your eyes until you have washed your hands.

EXTRA HOT GUACAMOLE

THIS IS QUITE A FIERY VERSION OF A POPULAR MEXICAN DISH, ALTHOUGH PROBABLY NOWHERE NEAR AS HOT AS YOU WOULD BE SERVED IN MEXICO, WHERE IT SEEMS HEAT KNOWS NO BOUNDS!

SERVES FOUR

INGREDIENTS

2 ripe avocados, peeled and stoned
2 tomatoes, peeled, seeded and finely
    chopped
6 spring onions, finely chopped
1–2 chillies, seeded and finely
    chopped
30ml/2 tbsp fresh lime or lemon juice
15ml/1 tbsp chopped fresh coriander
salt and freshly ground black pepper
coriander sprig, to garnish

1 Put the avocado halves into a large bowl and mash them roughly with a large fork.

2 Add the remaining ingredients. Mix well and season according to taste. Serve garnished with fresh coriander.

PLANTAIN APPETIZER

PLANTAINS ARE A TYPE OF COOKING BANANA WITH A LOWER SUGAR CONTENT THAN DESSERT BANANAS. THEY ARE UNSUITABLE FOR EATING RAW BUT CAN BE USED IN A WIDE RANGE OF DISHES. THIS DELICIOUS ASSORTMENT OF SWEET AND SAVOURY PLANTAINS IS A POPULAR DISH IN AFRICA.

SERVES FOUR

INGREDIENTS
2 green plantains
45ml/3 tbsp vegetable oil
1 small onion, very thinly sliced
1 yellow plantain
½ garlic clove, crushed
salt and cayenne pepper
vegetable oil, for frying

1 Peel one of the green plantains and cut into wafer-thin rounds, preferably using a swivel-headed potato peeler.

2 Heat about 15ml/1 tbsp of the oil in a large frying pan and fry the plantain slices for 2–3 minutes until golden, turning occasionally. Transfer to a plate lined with kitchen paper and keep warm.

3 Coarsely grate the other green plantain and mix with the onion.

4 Heat 15ml/1 tbsp of the remaining oil in the pan and fry the plantain and onion mixture for 2–3 minutes until golden, turning occasionally. Transfer to the plate with the plantain slices.

5 Peel the yellow plantain, cut into small chunks. Sprinkle with cayenne pepper. Heat the remaining oil and fry the yellow plantain and garlic for 4–5 minutes until brown. Drain and sprinkle with salt.

ROAST ASPARAGUS CRÊPES

ROAST ASPARAGUS IS DELICIOUS AND GOOD ENOUGH TO EAT JUST AS IT COMES. HOWEVER, FOR A REALLY SPLENDID STARTER, TRY THIS SIMPLE RECIPE. EITHER MAKE SIX LARGE OR TWICE AS MANY COCKTAIL-SIZE PANCAKES TO USE WITH SMALLER STEMS OF ASPARAGUS.

SERVES SIX (as a starter)

INGREDIENTS
  450g/1lb fresh asparagus
  90–120ml/6–8 tbsp olive oil
  175g/6oz mascarpone cheese
  60ml/4 tbsp single cream
  25g/1oz Parmesan cheese, grated
  sea salt
For the pancakes
  175g/6oz plain flour
  pinch of salt
  2 eggs
  350ml/12fl oz/1½ cups milk
  vegetable oil, for frying

1 To make the pancake batter, mix the flour with the salt in a large bowl, food processor or blender, then add the eggs and milk and beat or process to make a smooth, fairly thin, batter.

2 Heat a little oil in a large frying pan and add a small amount of batter, swirling the pan to coat the base evenly. Cook over a moderate heat for about 1 minute, then flip over and cook the other side until golden. Set aside and cook the rest of the pancakes in the same way; the mixture makes about six large or 12 smaller pancakes.

3 Preheat the oven to 180°C/350°F/ Gas 4 and lightly grease a large shallow ovenproof dish or roasting tin with some of the olive oil.

4 Trim the asparagus by placing on a board and cutting off the bases. Using a small sharp knife, peel away the woody ends, if necessary.

5 Arrange the asparagus in a single layer in the dish, trickle over the remaining olive oil, rolling the asparagus to coat each one thoroughly. Sprinkle with a little salt and then roast in the oven for about 8–12 minutes until tender (the cooking time depends on the stem thickness).

6 Blend the mascarpone cheese with the cream and Parmesan cheese and spread a generous tablespoonful over each of the pancakes, leaving a little extra for the topping. Preheat the grill.

7 Divide the asparagus spears among the pancakes, roll up and arrange in a single layer in an ovenproof dish. Spoon over the remaining cheese mixture and then place under a moderate grill for 4–5 minutes, until heated through and golden brown. Serve at once.

ARTICHOKES WITH GARLIC AND HERB BUTTER

IT IS FUN EATING ARTICHOKES AND EVEN MORE FUN TO SHARE ONE BETWEEN TWO PEOPLE. YOU CAN ALWAYS HAVE A SECOND ONE TO FOLLOW SO THAT YOU GET YOUR FAIR SHARE!

SERVES FOUR

INGREDIENTS
  2 large or 4 medium globe artichokes
  salt
For the garlic and herb butter
  75g/3oz butter
  1 garlic clove, crushed
  15ml/1 tbsp mixed chopped fresh
    tarragon, marjoram and parsley

1 Wash the artichokes well in cold water. Using a sharp knife, cut off the stalks level with the bases. Cut off the top 1cm/½in of leaves. Snip off the pointed ends of the remaining leaves with scissors.

2 Put the prepared artichokes in a large saucepan of lightly salted water. Bring to the boil, cover and cook for about 40–45 minutes or until a lower leaf comes away easily when gently pulled.

3 Drain upside down for a couple of minutes while making the sauce. Melt the butter over a low heat, add the garlic and cook for 30 seconds. Remove from the heat, stir in the herbs and then pour into one or two small serving bowls.

4 Place the artichokes on serving plates and serve with the garlic and herb butter.

COOK'S TIP
To eat an artichoke, pull off each leaf and dip into the garlic and herb butter. Scrape off the soft fleshy base with your teeth. When the centre is reached, pull out the hairy choke and discard it, as it is inedible. The base can be cut up and eaten with the remaining garlic butter.

STUFFED ARTICHOKES

*GLOBE ARTICHOKES ARE FIDDLY TO
PREPARE BUT THEIR DELICIOUS
TASTE MAKES IT WORTHWHILE. THIS
DISH CAN BE MADE IN ADVANCE
AND REHEATED BEFORE SERVING.*

SERVES FOUR

INGREDIENTS
    4 medium globe artichokes
    salt
    lemon slices
For the stuffing
    1 medium onion, chopped
    1 garlic clove, crushed
    45ml/3 tbsp olive oil
    115g/4oz mushrooms, chopped
    1 medium carrot, grated
    40g/1½oz sun-dried tomatoes in oil,
      drained and sliced
    leaves from a sprig of thyme
    about 45–60ml/3–4 tbsp water
    ground black pepper
    115g/4oz/2 cups fresh breadcrumbs
    extra olive oil, to cook
    fresh parsley, chopped, to garnish

1 Boil the artichokes in plenty of salted
water with a few slices of lemon for about
30 minutes, or until a leaf pulls easily
from the base. Strain through a colander
and cool the artichokes, laying them
upside down.

2 To make the stuffing for the artichokes,
gently fry the onion and garlic in the oil
for 5 minutes, then add the mushrooms,
carrot, sun-dried tomatoes and thyme.

3 Stir in the water, season well and cook
for a further 5 minutes, then mix in the
breadcrumbs.

VARIATION
For a lighter meal, rather than stuff the
artichokes, you could simply fill the
centres with home-made mayonnaise or
serve with a dish of vinaigrette or melted
butter for dipping the leaves.

4 Pull the artichoke leaves apart and pull
out the purple-tipped central leaves.
Using a small teaspoon, scrape out the
hairy choke, making sure that you remove
it all. Discard it.

5 Spoon the stuffing into the centre of
the artichoke, and push the leaves back
into shape. Put the artichokes into an
ovenproof dish and pour a little oil into
the centre of each one.

6 Half an hour before serving, heat the
oven to 190°C/375°F/Gas 5 and bake the
artichokes for about 20–25 minutes until
heated through. Serve garnished with a
little chopped fresh parsley on top.

BRUSCHETTA WITH GOATS' CHEESE AND TAPENADE

SIMPLE TO PREPARE IN ADVANCE,
THIS APPETIZING DISH CAN BE
SERVED AS A STARTER OR AT
FINGER BUFFETS.

SERVES FOUR TO SIX

INGREDIENTS
For the tapenade
    1 × 400g/14oz can black olives, stoned
      and finely chopped
    50g/2oz sun-dried tomatoes in oil,
      chopped
    30ml/2 tbsp capers, chopped
    15ml/1 tbsp green peppercorns, in
      brine, crushed
    45–60ml/3–4 tbsp olive oil
    2 garlic cloves, crushed
    45ml/3 tbsp fresh basil, chopped, or
    5ml/1 tsp dried basil
    salt and ground black pepper
For the bases
    12 slices ciabatta or other crusty bread
    olive oil, for brushing
    2 garlic cloves, halved
    115g/4oz soft goats' cheese
    fresh herb sprigs, to garnish

1 Mix the tapenade ingredients all together and check the seasoning. It should not need too much. Allow to marinate overnight, if possible.

2 To make the bruschetta, grill both sides of the bread lightly until golden. Brush one side with oil and then rub with a cut clove of garlic. Set aside until ready to serve.

3 Spread the bruschetta with the cheese, roughing it up with a fork, and spoon the tapenade on top. Garnish with sprigs of herbs.

COOK'S TIP
The bruschetta is nicest grilled over a open barbecue flame, if possible. Failing that, a grill will do, but avoid using a toaster – it gives too even a colour and the bruschetta is supposed to look quite rustic.

WARM AVOCADOS WITH TANGY TOPPING

LIGHTLY GRILLED WITH A TASTY
TOPPING OF RED ONIONS AND
CHEESE, THIS MAKES A DELIGHTFUL
ALTERNATIVE TO THE RATHER
HUMDRUM AVOCADO VINAIGRETTE.

SERVES FOUR

INGREDIENTS
    1 small red onion, sliced
    1 garlic clove, crushed
    15ml/1 tbsp sunflower oil
    Worcestershire sauce
    2 ripe avocados, halved and stoned
    2 small tomatoes, sliced
    15ml/1 tbsp fresh chopped basil,
      marjoram or parsley
    50g/2oz Lancashire or Mozzarella
      cheese, sliced
    salt and ground black pepper

1 Gently fry the onion and garlic in the oil for about 5 minutes until just softened. Shake in a little Worcestershire sauce.

2 Preheat a grill. Place the avocado halves on the grill pan and spoon the onions into the centre.

3 Divide the tomato slices and fresh herbs between the four halves and top each one with the cheese.

4 Season well and grill until the cheese melts and starts to brown.

VARIATION
Avocados are wonderful served in other hot dishes too. Try them chopped and tossed into hot pasta or sliced and layered in a lasagne.

TEMPURA VEGETABLES WITH DIPPING SAUCE

A Japanese favourite, thinly sliced, fresh vegetables are fried in a light crispy batter and served with flavoured soy sauce.

SERVES FOUR TO SIX

INGREDIENTS

1 medium courgette, in thin sticks
1 pepper, seeded and cut in wedges
3 large mushrooms, quartered
1 fennel bulb, cut in wedges
½ medium aubergine, thinly sliced
oil, for deep frying
For the sauce
45ml/3 tbsp soy sauce
15ml/1 tbsp medium dry sherry
5ml/1 tsp sesame seed oil
few shreds fresh ginger or spring onion
For the batter
1 egg
115g/4oz/1 cup plain flour
175ml/6fl oz/¾ cup cold water
salt and ground black pepper

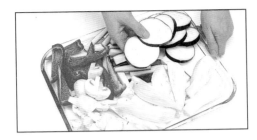

1 Prepare all the vegetables and lay them out on a tray, together with sheets of kitchen paper towel for draining the vegetables after cooking.

2 Mix the sauce ingredients together by whisking them in a jug or shaking them together in a sealed jar. Pour into a bowl.

3 Half fill a deep frying pan with oil and preheat to a temperature of about 190°C/375°F. Quickly whisk the batter ingredients together, but don't overbeat them. It doesn't matter if the batter is a little lumpy.

4 Fry the vegetables in stages by dipping a few quickly into the batter and lowering into the hot oil in a wire basket. Fry for just a minute until golden brown and crisp. Drain on the paper towel.

5 Repeat until all the vegetables are fried. Keep those you have cooked, uncovered, in a warm oven while you fry the rest. Serve the vegetables on a large platter alongside the dipping sauce.

COOK'S TIP
Successful deep frying can be quite tricky and a bit hazardous. Never leave the pan of oil unattended while the heat is turned on. If you have to leave the stove, turn off the oil. The oil will drop in temperature during cooking, so keep re-heating it between batches.

SWEETCORN BLINIS WITH DILL CREAM

A MOUTH-WATERING STARTER, THESE BLINIS ARE ALSO SUITABLE FOR COCKTAIL BUFFETS. IDEALLY, MAKE THEM AN HOUR OR TWO BEFORE SERVING, ALTHOUGH THE BATTER WILL STAND FOR LONGER.

SERVES SIX TO EIGHT

INGREDIENTS
    75g/3oz/¾ cup plain flour
    75g/3oz/⅔ cup wholemeal flour
    250ml/8fl oz/1 cup buttermilk
    4 small eggs, beaten
    2.5ml/½ tsp salt
    2.5ml/½ tsp baking powder
    25g/1oz/2 tbsp butter, melted
    good pinch bicarbonate of soda
    15ml/1 tbsp hot water
    1 × 200g/7oz can sweetcorn kernels,
      drained
    oil, for the griddle
For the dill cream
    200g/7oz crème fraîche
    30ml/2 tbsp fresh dill, chopped
    30ml/2 tbsp fresh chives, chopped
    salt and ground black pepper

1 Mix the two flours and buttermilk together until completely smooth. Cover and leave to chill for about 8 hours in the refrigerator.

2 Beat in the eggs, salt, baking powder and butter. Mix the bicarbonate with the hot water and add this too, along with the sweetcorn kernels.

3 Heat a griddle or heavy based frying pan until quite hot. Brush with a little oil and drop spoonfuls of the blinis mixture on to it. The mixture should start to sizzle immediately.

4 Cook until holes appear on the top and the mixture looks almost set. Using a palette knife flip the blinis over and cook briefly. Stack the blinis under a clean tea towel while you make the rest.

5 To make the cream, simply blend the crème fraîche with the herbs and seasoning. Serve the blinis with a few spoonfuls of cream and garnished with sliced radishes and herbs.

SWEETCORN AND CHEESE PASTIES

THESE TASTY PASTIES ARE REALLY SIMPLE TO MAKE AND EXTREMELY MORE-ISH — WHY NOT MAKE DOUBLE THE AMOUNT, AS THEY'LL GO LIKE HOT CAKES.

MAKES 18–20

INGREDIENTS
    250g/9oz sweetcorn
    115g/4oz feta cheese
    1 egg, beaten
    30ml/2 tbsp whipping cream
    15g/½oz Parmesan cheese, grated
    3 spring onions, chopped
    freshly ground black pepper
    8–10 small sheets filo pastry
    115g/4oz butter, melted

1 Preheat the oven to 190°C/375°F/ Gas 5 and butter two patty tins.

2 If using fresh sweetcorn, strip the kernels from the cob using a sharp knife and simmer in a little salted water for 3–5 minutes until tender. For canned sweetcorn, drain and rinse well under cold running water.

3 Crumble the feta cheese into a bowl and stir in the sweetcorn. Add the egg, cream, Parmesan cheese, spring onions and ground black pepper, and stir well.

4 Take one sheet of pastry and cut it in half to make a square. (Keep the remaining pastry covered with a damp cloth to prevent it drying out.) Brush with melted butter and then fold into four, to make a smaller square (about 7.5cm/3in).

5 Place a heaped teaspoon of mixture in the centre of each pastry square and then squeeze the pastry around the filling to make a "money bag" casing.

6 Continue making pasties until all the mixture is used up. Brush the outside of each "bag" with any remaining butter and then bake in the oven for about 15 minutes until golden. Serve hot.

HOT BROCCOLI TARTLETS

IN FRANCE, HOME OF THE CLASSIC QUICHE LORRAINE, YOU CAN ALSO FIND A WHOLE VARIETY OF SAVOURY TARTLETS, FILLED WITH ONIONS, LEEKS, MUSHROOMS AND BROCCOLI. THIS VERSION IS SIMPLE TO PREPARE AND WOULD MAKE AN ELEGANT START TO A MEAL.

MAKES EIGHT TO TEN

INGREDIENTS
15ml/1 tbsp oil
1 leek, finely sliced
175g/6oz broccoli, broken into florets
15g/½oz butter
15g/½oz plain flour
150ml/¼ pint/⅔ cup milk
50g/2oz goat's Cheddar or farmhouse
  Cheddar, grated
fresh chervil, to garnish
For the pastry
175g/6oz plain flour
75g/3oz butter
1 egg
pinch of salt

1 To make the pastry, place the flour and salt in a large bowl and rub in the butter and egg to make a dough. Add a little cold water if necessary, knead lightly, then wrap in clear film and leave to rest in the fridge for 1 hour.

2 Preheat the oven to 190°C/375°F/ Gas 5. Let the dough return to room temperature for 10 minutes and then roll out on a lightly floured surface and line 8–10 deep patty tins. Prick the bases with a fork and bake in the oven for about 10–15 minutes until the pastry is firm and lightly golden. Increase the oven temperature to 200°C/400°F/Gas 6.

3 Heat the oil in a small saucepan and sauté the leek for 4–5 minutes until soft. Add the broccoli, stir-fry for about 1 minute and then add a little water. Cover and steam for 3–4 minutes until the broccoli is just tender.

4 Melt the butter in a separate saucepan, stir in the flour and cook for a minute, stirring all the time. Slowly add the milk and stir to make a smooth sauce. Add half of the cheese and season with salt and pepper.

5 Spoon a little broccoli and leek into each tartlet case and then spoon over the sauce. Sprinkle each tartlet with the remaining cheese and then bake in the oven for about 10 minutes until golden.

6 Serve the tartlets as part of a buffet or as a starter, garnished with chervil.

LIGHT LUNCHES
AND
SUPPERS

*Experiment with these delicious recipes for
mouthwatering lunches that are not too filling as
well as for a sparkling selection of tasty snacks and
suppers. Ingredients range from protein-rich tofu
to irresistible cheeses and healthy pasta.*

MEXICAN BRUNCH EGGS

INSTEAD OF EGGS ON TOAST, WHY NOT TRY THEM ON FRIED CORN TORTILLAS WITH CHILLIES AND CREAMY AVOCADO? CANNED TORTILLAS ARE AVAILABLE FROM LARGER SUPERMARKETS OR DELICATESSENS.

SERVES FOUR

INGREDIENTS
oil, for frying
8 tortilla corn pancakes from a 300g/
11oz can
1 avocado
1 large tomato
50g/2oz/4 tbsp butter
8 eggs
4 jalepeno chillies, sliced
salt and ground black pepper
15ml/1 tbsp fresh coriander, chopped,
to garnish

1 Heat the oil and fry the tortillas for a few seconds each side. Remove and drain. Keep the tortillas warm.

2 Halve, stone and peel the avocado, then cut into slices. Dip the tomato into boiling water, then skin and chop roughly.

3 Melt the butter in a frying pan and fry the eggs, in batches, sunny side up.

4 Place two tortillas on four plates, slip an egg on each and top with sliced chillies, avocado and tomato. Season and serve garnished with fresh coriander.

FRIED TOMATOES WITH POLENTA CRUST

IF YOU SAW THE FILM "FRIED GREEN TOMATOES" YOU SHOULD ENJOY THIS DISH! NO NEED TO SEARCH FOR HOME-GROWN GREEN TOMATOES — ANY SLIGHTLY UNDER-RIPE ONES WILL DO.

SERVES FOUR

INGREDIENTS
4 large firm under-ripe tomatoes
115g/4oz/1 cup polenta or coarse
cornmeal
5ml//1 tsp dried oregano
2.5ml/½ tsp garlic powder
plain flour, for dredging
1 egg, beaten with seasoning
oil, for deep fat frying

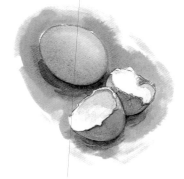

1 Cut the tomatoes into thick slices. Mix the polenta or cornmeal with the oregano and garlic powder.

2 Put the flour, egg and polenta into different bowls. Dip the tomato slices into the flour, then into the egg and finally into the polenta.

3 Fill a shallow frying pan about one-third full of oil, and heat steadily until it is quite hot.

4 Slip the tomato slices into the oil carefully, a few at a time, and fry on each side until crisp. Remove and drain. Repeat with the remaining tomatoes, reheating the oil in between. Serve with salad.

PISSALADIÈRE

A French Mediterranean classic, this is a delicious and colourful tart full of punchy flavour.

SERVES SIX

INGREDIENTS
For the pastry
    225g/8oz/2 cups plain flour
    115g/4oz/8 tbsp butter or sunflower
      margarine
    5ml/1 tsp dried mixed herbs
    pinch salt
For the filling
    2 large onions, thinly sliced
    2 garlic cloves, crushed
    45ml/3 tbsp olive oil
    fresh nutmeg, grated, to taste
    1 × 400g/14oz can chopped tomatoes
    5ml/1 tsp sugar
    leaves from small sprig of thyme
    salt and ground black pepper
    75g/3oz/⅔ cup stoned black olives,
      sliced
    30ml/2 tbsp capers
    fresh parsley, chopped, to garnish

1 Rub the flour and butter or margarine together until it forms fine crumbs, then mix in the herbs and salt. Mix to a firm dough with cold water. Preheat the oven to 190°C/375°F/Gas 5.

2 Roll out and line a 21cm/9in flan dish and bake blind, removing the baking beans and paper for the last 5 minutes for the pastry to crisp up. Cool.

3 Gently fry the onions and garlic in the oil for about 10 minutes, until quite soft, and mix in the nutmeg.

4 Stir in the tomatoes, sugar, thyme and seasoning and simmer gently for about 10 minutes until the mixture is reduced and slightly syrupy.

5 Remove from the heat and allow to cool. Mix in the olives and capers.

6 When ready to serve, spoon into the flan case, sprinkle with some fresh chopped parsley and serve at room temperature. Ideally, put the base and fillings together just before serving so that the base remains crisp.

VARIATION
To serve Pissaladière hot, top with grated cheese and grill until the cheese is golden and bubbling. The crisp-baked pastry case can be used as a base for a number of other vegetable mixtures. Try filling it with a Russian Salad – chopped, cooked root vegetables, including potato and carrot, mixed with peas, beans and onions, blended with mayonnaise and soured cream. Top with slices of hard-boiled egg and garnish with chopped fresh herbs.

TOMATO AND BASIL TART

*IN FRANCE, PÂTISSERIES DISPLAY MOUTH-WATERING SAVOURY TARTS IN THEIR WINDOWS. THIS IS A VERY
SIMPLE YET EXTREMELY TASTY TART MADE WITH RICH SHORTCRUST PASTRY, TOPPED WITH SLICES OF
MOZZARELLA CHEESE AND TOMATOES AND ENRICHED WITH OLIVE OIL AND BASIL LEAVES.*

SERVES FOUR

INGREDIENTS
    150g/5oz mozzarella,
        thinly sliced
    4 large tomatoes, thickly sliced
    about 10 basil leaves
    30ml/2 tbsp olive oil
    2 garlic cloves, thinly sliced
    sea salt and freshly ground
        black pepper
For the pastry
    115g/4oz plain flour
    50g/2oz butter or margarine
    1 egg yolk
    pinch of salt

1 To prepare the pastry, mix together
the flour and salt, then rub in the butter
and egg yolk. Add enough cold water to
make a smooth dough and knead lightly
on a floured surface. Place in a plastic
bag and chill for about 1 hour.

2 Preheat the oven to 190°C/375°F/
Gas 5. Remove the pastry from the fridge
and allow about 10 minutes for it to
return to room temperature and then roll
out into a 20cm/8in round. Press into the
base of a 20cm/8in flan dish or tin. Prick
all over with a fork and then bake in the
oven for about 10 minutes until firm but
not brown. Allow to cool slightly. Reduce
the oven temperature to 180°C/350°F/
Gas 4.

3 Arrange the mozzarella slices over the
pastry base. On top, arrange a single
layer of the sliced tomatoes, overlapping
them slightly. Dip the basil leaves in olive
oil and arrange them on the tomatoes.

4 Scatter the garlic on top, drizzle with
the remaining olive oil and season with a
little salt and a good sprinkling of black
pepper. Bake for 40–45 minutes, until
the tomatoes are well cooked. Serve hot.

HOT SOUR CHICK-PEAS

THIS DISH, KHATTE CHOLE, IS EATEN AS A SNACK ALL OVER INDIA, SOLD BY ITINERANT STREET VENDORS. THE HEAT OF THE CHILLIES IS TEMPERED PARTLY BY THE CORIANDER AND CUMIN, WHILE THE LEMON JUICE ADDS A WONDERFUL SOURNESS.

SERVES FOUR

INGREDIENTS

350g/12oz chick-peas, soaked overnight
60ml/4 tbsp vegetable oil
2 medium onions, very finely chopped
225g/8oz tomatoes, peeled and finely chopped
15ml/1 tbsp ground coriander
15ml/1 tbsp ground cumin
5ml/1 tsp ground fenugreek
5ml/1 tsp ground cinnamon
1–2 hot green chillies, seeded and finely sliced
about 2.5cm/1in fresh root ginger, grated
60ml/4 tbsp lemon juice
15ml/1 tbsp chopped fresh coriander
salt

1 Drain the chick-peas and place them in a large saucepan, cover with water and bring to the boil. Cover and simmer for 1–1¼ hours until tender, making sure the chick-peas do not boil dry. Drain, reserving the cooking liquid.

2 Heat the oil in a large flameproof casserole. Reserve about 30ml/2 tbsp of the chopped onions and fry the remainder in the casserole over a moderate heat for 4–5 minutes, stirring frequently, until tinged with brown.

3 Add the tomatoes and continue cooking over a moderately low heat for 5–6 minutes until soft. Stir frequently, mashing the tomatoes to a pulp.

4 Stir in the coriander, cumin, fenugreek and cinnamon. Cook for 30 seconds and then add the chick-peas and 350ml/12fl oz/1½ cups of the reserved cooking liquid. Season with salt, cover and simmer very gently for about 15–20 minutes, stirring occasionally and adding more cooking liquid if the mixture becomes too dry.

5 Meanwhile, mix the reserved onion with the chilli, ginger and lemon juice.

6 Just before serving, stir the onion and chilli mixture and the coriander into the chick-peas, and adjust the seasoning.

LEEKS IN EGG AND LEMON SAUCE

THE COMBINATION OF EGGS AND LEMON IN SAUCES AND SOUPS IS COMMONLY FOUND IN RECIPES FROM GREECE, TURKEY AND THE MIDDLE EAST. THIS SAUCE HAS A DELICIOUS FRESH TASTE AND BRINGS OUT THE BEST IN THE LEEKS. BE SURE TO USE TENDER BABY LEEKS FOR THIS RECIPE.

SERVES FOUR

INGREDIENTS
    675g/1½lb baby leeks
    15ml/1 tbsp cornflour
    10ml/2 tsp sugar
    2 egg yolks
    juice of 1½ lemons
    salt

1 Trim the leeks, slit them from top to bottom and rinse very well under cold water to remove any dirt.

2 Place the leeks in a large saucepan, preferably so they lie flat on the base, cover with water and add a little salt. Bring to the boil, cover and simmer for 4–5 minutes until just tender.

3 Carefully remove the leeks using a slotted spoon, drain well and arrange in a shallow serving dish. Reserve 200ml/ 7fl oz/scant 1 cup of the cooking liquid.

4 Blend the cornflour with the cooled cooking liquid and place in a small saucepan. Bring to the boil, stirring all the time, and cook over a gentle heat until the sauce thickens slightly. Stir in the sugar and then remove the saucepan from the heat and allow to cool slightly.

5 Beat the egg yolks thoroughly with the lemon juice and stir gradually into the cooled sauce. Cook over a very low heat, stirring all the time, until the sauce is fairly thick. Be careful not to overheat the sauce or it may curdle. As soon as the sauce has thickened remove the pan from the heat and continue stirring for a minute. Taste and add salt or sugar as necessary. Cool slightly.

6 Stir the cooled sauce with a wooden spoon. Pour the sauce over the leeks and then cover and chill well for at least 2 hours before serving.

LEEK SOUFFLÉ

SOME PEOPLE THINK OF A SOUFFLÉ AS A DINNER PARTY DISH, AND A RATHER TRICKY ONE AT THAT. HOWEVER, OTHERS FREQUENTLY SERVE THEM FOR FAMILY MEALS BECAUSE THEY ARE QUICK AND EASY TO MAKE, AND PROVE TO BE VERY POPULAR AND SATISFYING.

SERVES TWO TO THREE

INGREDIENTS

15ml/1 tbsp sunflower oil
40g/1½oz butter
2 leeks, thinly sliced
about 300ml/½ pint/1¼ cups milk
25g/1oz plain flour
4 eggs, separated
75g/3oz Gruyère or Emmenthal
    cheese, grated
salt and freshly ground black pepper

1 Preheat the oven to 180°C/350°F/ Gas 4 and butter a large soufflé dish. Heat the oil and 15g/½oz of the butter in a small saucepan or flameproof casserole and fry the leeks over a gentle heat for 4–5 minutes until soft but not brown, stirring occasionally.

2 Stir in the milk and bring to the boil. Cover and simmer for 4–5 minutes until the leeks are tender. Strain the liquid through a sieve into a measuring jug.

3 Melt the remaining butter in a sauce-pan, stir in the flour and cook for 1 minute. Remove pan from the heat. Make up the reserved liquid with milk to 300ml/ ½ pint/1¼ cups. Gradually stir the milk into the pan to make a smooth sauce. Return to the heat and bring to the boil, stirring. When thickened, remove from the heat. Cool slightly and then beat in the egg yolks, cheese and the leeks.

4 Whisk the egg whites until stiff and, using a large metal spoon, fold into the leek and egg mixture. Pour into the pre-pared soufflé dish and bake in the oven for about 30 minutes until golden and puffy. Serve immediately.

ASPARAGUS TART WITH RICOTTA

RICOTTA GIVES THE FILLING A DELIGHTFULLY SMOOTH, CREAMY TEXTURE, WHILE THE PARMESAN ADDS THE NECESSARY ZING.

SERVES FOUR

INGREDIENTS
For the pastry
    175g/6oz plain flour
    75g/3oz butter or margarine
    pinch of salt
For the filling
    225g/8oz asparagus
    2 eggs, beaten
    225g/8oz ricotta cheese
    30ml/2 tbsp Greek yogurt
    40g/1½oz Parmesan cheese, grated
    salt and freshly ground black pepper

1 Preheat the oven to 200°C/400°F/ Gas 6. Rub the butter or margarine into the flour and salt until the mixture resembles fine breadcrumbs. Stir in enough cold water to form a smooth dough and knead lightly on a floured surface.

2 Roll out the pastry and line a 23cm/9in flan ring. Press firmly into the tin and prick all over with a fork. Bake in the oven for about 10 minutes until the pastry is pale but firm. Remove from the oven and reduce the temperature to 180°C/350°F/Gas 4.

3 To make the filling, trim the asparagus and cut 5cm/2in from the top and chop the remaining stalks into 2.5cm/1in pieces. Add the stalks to a saucepan of boiling water and after 1 minute add the tops. Simmer for 4–5 minutes until almost tender, then drain and refresh under cold water.

4 Beat together the eggs, ricotta, yogurt, Parmesan cheese and seasoning. Stir in the asparagus stalks and pour the mixture into the pastry case. Arrange the asparagus tips on top, pressing them down slightly into the ricotta mixture.

5 Bake in the oven for 35–40 minutes until golden. Serve warm or cold.

ASPARAGUS WITH TARRAGON HOLLANDAISE

THIS IS THE IDEAL STARTER FOR AN EARLY SUMMER DINNER PARTY WHEN THE NEW SEASON'S ASPARAGUS IS JUST IN AND AT ITS BEST. MAKING HOLLANDAISE SAUCE IN A BLENDER OR FOOD PROCESSOR IS INCREDIBLY EASY AND VIRTUALLY FOOLPROOF!

SERVES FOUR

INGREDIENTS
    500g/1¼lb fresh asparagus
For the Hollandaise sauce
    2 eggs yolks
    15ml/1 tbsp lemon juice
    salt and freshly ground black pepper
    115g/4oz butter
    10ml/2 tsp finely chopped fresh
      tarragon

1 Prepare the asparagus, lay it in a steamer or in an asparagus kettle and place over a saucepan of rapidly boiling water. Cover and steam for 6–10 minutes until tender (the cooking time will depend on the thickness of the asparagus stems).

2 To make the Hollandaise sauce, place the egg yolks, lemon juice and seasoning in a blender or food processor and process briefly. Melt the butter in a small pan until foaming and then, with the blender running, pour it on to the egg mixture in a slow, steady stream.

3 Stir in the tarragon by hand or process it (for a sauce speckled with green or a pale green sauce, respectively).

4 Arrange the asparagus on small plates and pour over some of the Hollandaise sauce. Serve remaining sauce in a jug.

ARTICHOKE RÖSTI

SERVES FOUR TO SIX

INGREDIENTS
   450g/1lb Jerusalem artichokes
   juice of 1 lemon
   salt
   450g/1lb potatoes
   about 50g/2oz butter

1 Peel the Jerusalem artichokes and place in a saucepan of water together with the lemon juice and a pinch of salt. Bring to the boil and cook for about 5 minutes until barely tender.

2 Peel the potatoes and place in a separate pan of salted water. Bring to the boil and cook until barely tender – they will take slightly longer than the artichokes.

3 Drain and cool both the artichokes and potatoes, and then grate them into a bowl. Mix them with your fingers, without breaking them up too much.

4 Melt the butter in a large heavy-based frying pan. Add the artichoke mixture, spreading it out with the back of a spoon. Cook gently for about 10 minutes.

5 Invert the "cake" on to a plate and slide back into the pan. Cook for about 10 minutes until golden. Serve at once.

ARTICHOKE TIMBALES WITH SPINACH SAUCE

SERVES SIX

INGREDIENTS
   900g/2lb Jerusalem artichokes
   juice of 1 lemon
   25g/1oz butter
   15ml/1 tbsp oil
   1 onion, finely chopped
   1 garlic clove, crushed
   50g/2oz fresh white breadcrumbs
   1 egg
   60–75ml/4–5 tbsp vegetable stock
     or milk
   15ml/1 tbsp chopped fresh parsley
   5ml/1 tsp finely chopped sage
   salt and freshly ground black pepper
For the sauce
   225g/8oz fresh spinach, prepared
   15g/½oz butter
   2 shallots, finely chopped
   150ml/¼ pint/⅔ cup single cream
   175ml/6fl oz/¾ cup vegetable stock
   salt and freshly ground black pepper

1 Preheat the oven to 180°C/350°F/ Gas 4. Grease six 150ml/¼ pint/⅔ cup ramekin dishes, and then place a circle of non-stick baking paper in each base.

2 Peel the artichokes and put in a saucepan with the lemon juice and water to cover. Bring to the boil and simmer for about 10 minutes until tender. Drain and mash with the butter.

3 Heat the oil in a small frying pan and fry the onion and garlic until soft. Place in a food processor with the breadcrumbs, egg, stock or milk, parsley, sage and seasoning. Process to a smooth purée, add the artichokes and process again briefly. Do not over-process.

4 Put the mixture in the prepared dishes and smooth the tops. Cover with non-stick baking paper, place in a roasting tin half-filled with boiling water and bake for 35–40 minutes until firm.

5 To make the sauce, cook the spinach without water, in a large covered saucepan, for 2–3 minutes. Shake the pan occasionally. Strain and press out the excess liquid.

6 Melt the butter in a small saucepan and fry the shallots gently until slightly softened but not browned. Place in a food processor or blender and process to make a smooth purée. Pour back into the pan, add the cream and seasoning, and keep warm over a very low heat. Do not allow the mixture to boil.

7 Allow the timbales to stand for a few minutes after cooking and then turn out on to warmed serving plates. Spoon the warm sauce over them and serve.

COOK'S TIP
When puréeing the artichokes in a food processor or blender, use the pulse button and process for a very short time. The mixture will become cloying if it is over-processed.

YAM FRITTERS

YAMS HAVE A SLIGHTLY DRIER FLAVOUR THAN POTATOES AND ARE PARTICULARLY GOOD WHEN MIXED WITH SPICES AND THEN FRIED. THE FRITTERS CAN ALSO BE MOULDED INTO SMALL BALLS AND DEEP FRIED. THIS IS A FAVOURITE AFRICAN WAY OF SERVING YAMS.

MAKES ABOUT 18–20

INGREDIENTS
    675g/1½lb yams
    milk, for mashing
    2 small eggs, beaten
    45ml/3 tbsp chopped tomato flesh
    45ml/3 tbsp finely chopped spring
      onions
    1 green chilli, seeded and finely
      sliced
    salt and freshly ground black pepper
    flour, for shaping
    40g/1½oz white breadcrumbs
    vegetable oil, for shallow frying

1   Peel the yams and cut into chunks. Place in a saucepan of salted water and boil for 20–30 minutes until tender. Drain and mash with a little milk and about 45ml/3 tbsp of the beaten eggs.

2   Add the chopped tomato, spring onions, chilli and seasoning and stir well to mix thoroughly.

3   Using floured hands, shape the yam and vegetable mixture into round fritters, about 7.5cm/3in in diameter.

4   Dip each in the remaining beaten egg and then coat evenly with the breadcrumbs. Heat a little oil in a large frying pan and fry the yam fritters for about 4–5 minutes until golden brown. Turn the fritters over once during cooking. Drain well on kitchen paper and serve.

EDDO, CARROT AND PARSNIP MEDLEY

EDDO (TARO), LIKE YAMS, IS WIDELY EATEN IN AFRICA AND THE CARIBBEAN, OFTEN AS A PURÉE. HERE, IT IS ROASTED AND COMBINED WITH MORE COMMON ROOT VEGETABLES TO MAKE A COLOURFUL DISPLAY.

SERVES FOUR TO SIX

INGREDIENTS
    450g/1lb eddoes or taros
    350g/12oz parsnips
    450g/1lb carrots
    25g/1oz butter
    45ml/3 tbsp sunflower oil
For the dressing
    30ml/2 tbsp fresh orange juice
    30ml/2 tbsp demerara sugar
    10ml/2 tsp soft green peppercorns
    salt
    fresh parsley, to garnish

1   Preheat the oven to 200°C/400°F/ Gas 6. Peel the eddoes and cut into pieces about 5 x 2cm/2 x ¾in by 2cm/¾in, and place in a large bowl.

2   Peel the parsnips, halve lengthways and remove the inner core if necessary. Cut into the same size pieces as the eddo and add to the bowl. Blanch in boiling water for 2 minutes and then drain. Peel or scrub the carrots, and halve or quarter them according to their size.

3   Place the butter and sunflower oil in a roasting tin and heat in the oven for 3–4 minutes. Add the vegetables, turning them in the oil to coat evenly. Roast in the oven for 30 minutes.

4   Meanwhile, blend the orange juice, sugar and soft green peppercorns in a small bowl. Remove the roasting tin from the oven and allow to cool for a minute or so and then carefully pour the mixture over the vegetables, stirring to coat them all. (If the liquid is poured on immediately, the hot oil will spit.)

5   Return the tin to the oven and cook for a further 20 minutes until the vegetables are crisp and golden. Transfer to a warmed serving plate and sprinkle with salt. Garnish with parsley to serve.

Spinach Ravioli

Home-made ravioli is time-consuming, yet it is worth the effort as even the best shop-bought pasta never tastes quite as fresh. To complement this effort, make the filling exactly to your liking, tasting it for the right balance of spinach and cheese.

SERVES FOUR

INGREDIENTS
225g/8oz fresh spinach
40g/1½oz butter
1 small onion, finely chopped
25g/1oz Parmesan cheese, grated
40g/1½oz dolcellate cheese, crumbled
15ml/1 tbsp chopped fresh parsley
salt and freshly ground black pepper
shavings of Parmesan cheese, to serve
For the pasta dough
350g/12oz strong white flour
4ml/¾ tsp salt
2 eggs
15ml/1 tbsp olive oil

1 To make the pasta dough, mix together the flour and salt in a large bowl or food processor. Add the eggs, olive oil and about 45ml/3 tbsp of cold water or enough to make a pliable dough. If working by hand, mix the ingredients together and then knead the dough for about 15 minutes until very smooth. Or, process for about 1½ minutes in a food processor. Place the dough in a plastic bag and chill for at least 1 hour (or overnight if more convenient).

2 Cook the spinach in a large, covered saucepan for 3–4 minutes, until the leaves have wilted. Strain and press out the excess liquid. Set aside to cool a little and then chop finely.

3 Melt half the butter in a small saucepan and fry the onion over a gentle heat for about 5–6 minutes until soft. Place in a bowl with the chopped spinach, the Parmesan and dolcellate cheeses, and seasoning. Mix well.

4 Grease a ravioli tin. Roll out half or a quarter of the pasta dough to a thickness of about 3mm/⅛in. Lay the dough over the ravioli tin, pressing it well into each of the squares.

5 Spoon a little spinach mixture into each cavity, then roll out a second piece of dough and lay it on top. Press a rolling pin evenly over the top of the tin to seal the edges and then cut the ravioli into squares using a pastry cutter.

6 Place the ravioli in a large saucepan of boiling water and simmer for about 4–5 minutes until cooked through but *al dente*. Drain well and then toss with the remaining butter and the parsley.

7 Divide between four serving plates and serve scattered with shavings of Parmesan cheese.

COOK'S TIP
For a small ravioli tin of 32 holes, divide the dough into quarters. Roll the dough out until it covers the tin comfortably – it takes some time but the pasta needs to be thin otherwise the ravioli will be too stodgy. For a large ravioli tin of 64 holes, divide the dough in half.

THAI NOODLES WITH CHINESE CHIVES

THIS RECIPE REQUIRES A LITTLE TIME FOR PREPARATION, BUT THE COOKING TIME IS VERY FAST. EVERYTHING IS COOKED SPEEDILY IN A HOT WOK AND SHOULD BE EATEN AT ONCE.

SERVES FOUR

INGREDIENTS
   350g/12oz dried rice noodles
   1cm/½ in fresh root ginger, grated
   30ml/2 tbsp light soy sauce
   45ml/3 tbsp vegetable oil
   225g/8oz Quorn, cut into small cubes
   2 garlic cloves, crushed
   1 large onion, cut into thin wedges
   115g/4oz fried bean curd, thinly sliced
   1 green chilli, seeded
      and finely sliced
   175g/6oz beansprouts
   115g/4oz Chinese chives, cut into
      5cm/2in lengths
   50g/2oz roasted peanuts, ground
   30ml/ 2tbsp dark soy sauce
   30ml/2 tbsp chopped fresh coriander
   1 lemon, cut into wedges

1 Place the noodles in a large bowl, cover with warm water and soak for 20–30 minutes, then drain. Blend together the ginger, light soy sauce and 15ml/1 tbsp of the oil in a bowl. Stir in the Quorn and set aside for 10 minutes. Drain, reserving the marinade.

2 Heat 15ml/1 tbsp of the oil in a frying pan and fry the garlic for a few seconds. Add the Quorn and stir-fry for 3–4 minutes. Then transfer to a plate and set aside.

5 When hot, spoon on to serving plates and garnish with the remaining ground peanuts, coriander and lemon wedges.

COOK'S TIP
Quorn makes this a vegetarian meal, however thinly sliced pork or chicken could be used instead. Stir-fry it initially for 4–5 minutes.

3 Heat the remaining oil in the wok or frying pan and stir-fry the onion for 3–4 minutes until softened and tinged with brown. Add the bean curd and chilli, stir-fry briefly and then add the noodles. Stir-fry for 4–5 minutes.

4 Stir in the beansprouts, Chinese chives and most of the ground peanuts, reserving a little for the garnish. Stir well, then add the Quorn, the dark soy sauce and the reserved marinade.

CAULIFLOWER AND MUSHROOM GOUGÈRE

THIS IS AN ALL-ROUND FAVOURITE VEGETARIAN DISH.

SERVES FOUR TO SIX

INGREDIENTS
  300ml/½ pint/1¼ cups water
  115g/4oz butter or margarine
  150g/5oz plain flour
  4 eggs
  115g/4oz Gruyère or Cheddar cheese,
    finely diced
  5ml/1 tsp Dijon mustard
  salt and freshly ground black pepper
For the filling
  ½ x 400g/14oz can tomatoes
  15ml/1 tbsp sunflower oil
  15g/½oz butter or margarine
  1 onion, chopped
  115g/4oz button mushrooms, halved
    if large
  1 small cauliflower, broken into
    small florets
  sprig of thyme
  salt and freshly ground black pepper

1 Preheat the oven to 200°C/400°F/
Gas 6 and butter a large ovenproof dish.
Place the water and butter together in a
large saucepan and heat until the butter
has melted. Remove from the heat and
add all the flour at once. Beat well with a
wooden spoon for about 30 seconds until
smooth. Allow to cool slightly.

2 Beat in the eggs, one at a time, and
continue beating until the mixture is
thick and glossy. Stir in the cheese and
mustard and season with salt and
pepper. Spread the mixture around the
sides of the ovenproof dish, leaving a
hollow in the centre for the filling.

3 To make the filling, purée the
tomatoes in a blender or food processor
and then pour into a measuring jug. Add
enough water to make up to 300ml/
½ pint/1¼ cups of liquid.

4 Heat the oil and butter in a flameproof
casserole and fry the onion for about
3–4 minutes until softened but not
browned. Add the mushrooms and cook
for 2–3 minutes until they begin to be
flecked with brown. Add the cauliflower
florets and stir-fry for 1 minute.

5 Add the tomato liquid, thyme and
seasoning. Cook, uncovered, over a
gentle heat for about 5 minutes until the
cauliflower is only just tender.

6 Spoon the mixture into the hollow in
the ovenproof dish, adding all the liquid.
Bake in the oven for about 35–40
minutes, until the outer pastry is well
risen and golden brown.

DOLMADES

DOLMADES ARE STUFFED VINE LEAVES, A TRADITIONAL GREEK DISH. IF YOU CAN'T OBTAIN FRESH VINE LEAVES, USE A PACKET OF BRINED VINE LEAVES. SOAK THE LEAVES IN HOT WATER FOR 20 MINUTES THEN RINSE AND DRY WELL ON KITCHEN PAPER BEFORE USE.

MAKES 20–24

INGREDIENTS
20–30 fresh young vine leaves
30ml/2 tbsp olive oil
1 large onion, finely chopped
1 garlic clove, crushed
225g/8oz cooked long grain rice,
  or mixed white and wild rice
about 45ml/3 tbsp pine nuts
15ml/1 tbsp flaked almonds
40g/1½oz sultanas
15ml/ 1 tbsp snipped chives
15ml/ 1 tbsp finely chopped
  fresh mint
juice of ½ lemon
150ml/¼ pint/⅔ cup white wine
hot vegetable stock
salt and freshly ground black pepper
sprig of mint, to garnish
Greek yogurt, to serve

1 Bring a large pan of water to the boil and cook the vine leaves for about 2–3 minutes. They will darken and go limp after about 1 minute and simmering for a further minute or so ensures they are pliable. If using leaves from a packet, place them in a large bowl, cover with boiling water and leave for a few minutes until the leaves can be easily separated. Rinse them under cold water and drain on kitchen paper.

2 Heat the oil in a small frying pan and fry the onion and garlic for 3–4 minutes over a gentle heat until soft.

3 Spoon the onion and garlic mixture into a bowl and add the cooked rice.

4 Stir in 30ml/2 tbsp of the pine nuts, the almonds, sultanas, chives, mint, lemon juice and seasoning and mix well.

5 Lay a vine leaf on a clean work surface, veined side uppermost. Place a spoonful of filling near the stem, fold the lower part of the leaf over it and roll up, folding in the sides as you go. Continue stuffing the vine leaves in the same way.

6 Line the base of a deep frying pan with four large vine leaves. Place the stuffed vine leaves close together in the pan, seam side down, in a single layer.

7 Add the wine and enough stock to just cover the vine leaves. Place a plate directly over the leaves, then cover and simmer gently for 30 minutes, checking to make sure the pan does not boil dry.

8 Chill the vine leaves, then serve garnished with the remaining pine nuts, a sprig of mint and a little yogurt.

MUSHROOM AND CHILLI CARBONARA

DRIED PORCINI MUSHROOMS GIVE THIS QUICK EGGY SAUCE A RICHER MUSHROOM FLAVOUR.

SERVES FOUR

INGREDIENTS

1 × 15g/½oz pack dried porcini
   mushrooms
300ml/½ pint/1¼ cups hot water
225g/8oz spaghetti
1 garlic clove, crushed
25g/1oz/2 tbsp butter
15ml/1 tbsp olive oil
225g/8oz button or chestnut
   mushrooms, sliced
5ml/1 tsp dried chilli flakes
2 eggs
300ml/½ pint/1¼ cups single cream
salt and ground black pepper
fresh Parmesan cheese, grated, and
   parsley, chopped, to serve

1 Soak the dried mushrooms in the hot water for 15 minutes, drain and reserve the liquor.

2 Boil the spaghetti according to the instructions on the packet in salted water. Drain and rinse in cold water.

3 In a large saucepan, lightly sauté the garlic with the butter and oil for half a minute then add the mushrooms, including the soaked porcini ones, and the dried chilli flakes, and stir well. Cook for about 2 minutes, stirring a few times.

4 Pour in the reserved mushroom stock and boil to reduce slightly.

5 Beat the eggs with the cream and season well. Return the cooked spaghetti to the pan and toss in the eggs and cream. Reheat, without boiling, and serve hot sprinkled with Parmesan cheese and chopped parsley.

VARIATION
Instead of mushrooms, try using either finely sliced and sautéd leeks or perhaps coarsely shredded lettuce with peas. If chilli flakes are too hot and spicy for you, then try the delicious alternative of skinned and chopped tomatoes with torn, fresh basil leaves.

TAGLIATELLE WITH "HIT-THE-PAN" SALSA

IT IS POSSIBLE TO MAKE A HOT FILLING MEAL WITHIN JUST 15 MINUTES WITH THIS QUICK-COOK SALSA SAUCE.

SERVES TWO

INGREDIENTS
  115g/4oz tagliatelle
  45ml/3 tbsp olive oil, preferably extra
    virgin
  3 large tomatoes
  1 garlic clove, crushed
  4 spring onions, sliced
  1 green chilli, seeded and sliced
  juice of 1 orange (optional)
  30ml/2 tbsp fresh parsley, chopped
  salt and ground black pepper
  cheese, grated, to garnish (optional)

1 Boil the tagliatelle in plenty of salted water until it is *al dente*. Drain and toss in a little of the oil. Season well.

2 Skin the tomatoes by dipping them briefly in a bowl of boiling water. The skins should slip off easily. Chop the tomatoes roughly. If you haven't got time to peel the tomatoes, don't bother.

3 Heat the remaining oil until it is quite hot and stir-fry the garlic, onions and chilli for a minute. The pan should sizzle.

4 Add the tomatoes, orange juice (if using) and parsley. Season well and stir in the tagliatelle to reheat. Serve with the grated cheese (if used).

COOK'S TIP
You could use any pasta shape for this recipe. It would be particularly good with large rigatoni or linguini, or as a sauce for fresh ravioli or tortellini.

CHEESY BUBBLE AND SQUEAK

This London breakfast dish is enjoying something of a revival. Originally made on Mondays with leftover potatoes and cabbage from the Sunday lunch, it is suitable for any light meal. For breakfast, serve the bubble and squeak with eggs, grilled tomatoes and mushrooms.

SERVES FOUR

INGREDIENTS
about 450g/1lb/3 cups mashed potato
about 225g/8oz cooked cabbage or
  kale, shredded
1 egg, beaten
115g/4oz Cheddar cheese, grated
fresh nutmeg, grated
salt and ground black pepper
plain flour, for coating
oil, for frying

1 Mix the potatoes with the cabbage or kale, egg, cheese, nutmeg and seasoning. Divide and shape into eight patties.

2 Chill for an hour or so, if possible, as this enables the mixture to become firm and makes it easier to fry. Toss the patties in the flour. Heat about 1cm/½in of oil in a frying pan until it is quite hot.

3 Carefully slide the patties into the oil and fry on each side for about 3 minutes until golden and crisp. Drain on kitchen paper towel and serve hot and crisp.

CHEESE AND CHUTNEY TOASTIES

Quick cheese on toast can be made quite memorable with a few tasty additions. Serve these scrumptious toasties with a simple salad.

SERVES FOUR

INGREDIENTS
4 slices wholemeal bread, thickly sliced
butter or low fat spread
115g/4oz Cheddar cheese, grated
5ml/1 tsp dried thyme
ground black pepper
30ml/2 tbsp chutney or relish

1 Toast the bread slices lightly on each side then spread sparingly with butter or low-fat spread.

2 Mix the cheese and thyme together and season with pepper.

3 Spread the chutney or relish on the toast and divide the cheese between the four slices.

4 Return to the grill and cook until browned and bubbling. Cut into halves, diagonally, and serve with salad.

RISOTTO PRIMAVERA

REAL RISOTTOS SHOULD BE CREAMY AND FULL OF FLAVOUR. FOR BEST RESULTS USE A QUALITY ARBORIO RICE, WHICH HAS A GOOD AL DENTE BITE.

SERVES FOUR

INGREDIENTS

1 litre/1¾ pints/4 cups hot stock,
   preferably home-made
1 red onion, chopped
2 garlic cloves, crushed
30ml/2 tbsp olive oil
25g/1oz/2 tbsp butter
225g/8oz/1¼ cups risotto rice
   (do not rinse)
45ml/3 tbsp dry white wine
115g/4oz asparagus spears or green
   beans, sliced and blanched
2 young carrots, sliced and blanched
50g/2oz baby button mushrooms
salt and ground black pepper
50g/2oz Pecorino or Parmesan cheese,
   grated

1 It is important to follow the steps for making real risotto so that you achieve the right texture. First, heat the stock in a saucepan to simmering.

2 Next to it, in a large saucepan, sauté the onion and garlic in the oil and butter for 3 minutes.

3 Stir in the rice making sure each grain is coated well in the oil, then stir in the wine. Allow to reduce down and spoon in two ladles of hot stock, stirring continuously.

4 Allow this to bubble down then add more stock and stir again. Continue like this, ladling in the stock and stirring frequently for up to 20 minutes, by which time the rice will have swelled greatly.

5 Mix in the asparagus or beans, carrots and mushrooms, seasoning well, and cook for a minute or two more. Serve immediately in bowls with a scattering of grated cheese.

VARIATION
If you have any left-over risotto, shape it into small balls and then coat in beaten egg and dried breadcrumbs. Chill for 30 minutes before deep frying in hot oil until golden and crisp.

KITCHIRI

THIS IS THE INDIAN ORIGINAL, WHICH INSPIRED THE CLASSIC BREAKFAST DISH KNOWN AS KEDGEREE. MADE WITH BASMATI RICE AND SMALL, TASTY LENTILS, THIS WILL MAKE AN AMPLE SUPPER OR BRUNCH DISH.

SERVES FOUR

INGREDIENTS

115g/4oz/1 cup Indian masoor
   dhal or Continental green lentils
1 onion, chopped
1 garlic clove, crushed
50g/2oz/4 tbsp vegetarian ghee or
   butter
30ml/2 tbsp sunflower oil
225g/8oz/1¼ cups easy-cook basmati
   rice
10ml/2 tsp ground coriander
10ml/2 tsp cumin seeds
2 cloves
3 cardamom pods
2 bay leaves
1 stick cinnamon
1 litre/1¾ pints/4 cups stock
30ml/2 tbsp tomato purée
salt and ground black pepper
45ml/3 tbsp fresh coriander or parsley,
   chopped, to garnish

1 Cover the dhal or lentils with boiling water and soak for 30 minutes. Drain and boil in fresh water for 10 minutes. Drain once more and set aside.

2 Fry the onion and garlic in the ghee or butter and oil in a large saucepan for about 5 minutes.

3 Add the rice, stir well to coat the grains in the ghee or butter and oil, then stir in the spices. Cook gently for a minute or so.

4 Add the lentils, stock, tomato purée and seasoning. Bring to the boil, then cover and simmer for 20 minutes until the stock is absorbed and the lentils and rice are just soft. Stir in the coriander or parsley and check the seasoning. Remove the cinnamon stick and bay leaf.

MULTI-MUSHROOM STROGANOFF

A PAN FRY OF SLICED MUSHROOMS SWIRLED WITH SOURED CREAM IS MADE ESPECIALLY INTERESTING IF TWO OR THREE VARIETIES OF MUSHROOM ARE USED.

SERVES THREE TO FOUR

INGREDIENTS
45ml/3 tbsp olive oil
450g/1lb edible mushrooms (such as ceps, shiitakes or oysters), sliced
3 spring onions, sliced
2 garlic cloves, crushed
30ml/2 tbsp dry sherry or vermouth
salt and ground black pepper
300ml/½ pint/1¼ cups soured cream or crème fraîche
15ml/1 tbsp fresh marjoram or thyme leaves, chopped
fresh parsley, chopped

1 Heat the oil in a large frying pan and fry the mushrooms gently, stirring them occasionally until they are softened and just cooked.

2 Add the spring onions, garlic and sherry or vermouth and cook for a minute more. Season well.

3 Stir in the soured cream or crème fraîche and heat to just below boiling. Stir in the marjoram or thyme then scatter over the parsley. Serve with rice, pasta or boiled new potatoes.

BUTTER BEAN AND PESTO PASTA

BUY GOOD QUALITY, READY-MADE PESTO, RATHER THAN MAKING YOUR OWN. PESTO FORMS THE BASIS OF SEVERAL VERY TASTY SAUCES, AND IT IS ESPECIALLY GOOD WITH BUTTER BEANS.

SERVES FOUR

INGREDIENTS
225g/8oz pasta shapes
salt and ground black pepper
fresh nutmeg, grated
30ml/2 tbsp extra virgin olive oil
1 × 400g/14oz can butter beans, drained
45ml/3 tbsp pesto sauce
150ml/¼ pint/⅔ cup single cream
To serve
45ml/3 tbsp pine nuts
cheese, grated (optional)
sprigs of fresh basil, to garnish (optional)

1 Boil the pasta until *al dente*, then drain, leaving it a little wet. Return the pasta to the pan, season, and stir in the nutmeg and oil.

2 Heat the beans in a saucepan with the pesto and cream, stirring it until it begins to simmer. Toss the beans and pesto into the pasta and mix well.

BAKED POTATOES AND THREE FILLINGS

*POTATOES BAKED IN THEIR
JACKETS AND PACKED WITH A
VARIETY OF FILLINGS MAKE A
TASTY AND NOURISHING MEAL.*

INGREDIENTS
    4 medium size baking potatoes
    olive oil, for greasing
    sea salt, to serve

1 Preheat the oven to 200°C/400°F/
Gas 6. Score the potatoes with a cross
and rub all over with the olive oil.

2 Place on a baking sheet and cook for
45-60 minutes until a knife inserted into
the centres indicates they are cooked.

3 Cut the potatoes open along the score
lines and push up the flesh from the base
with your fingers. Season with salt and fill
with your chosen filling.

EACH FILLING IS FOR FOUR POTATOES

Red bean filling
    1 × 425g/15oz can red kidney beans
    200g/7oz low-fat cheese or cream
      cheese
    30ml/2 tbsp mild chilli sauce
    5ml/1 tsp ground cumin
Drain the beans, heat in a pan or
microwave and stir in the cream cheese,
chilli sauce and cumin.

Soy vegetables filling
    2 leeks, thinly sliced
    2 carrots, cut in sticks
    1 courgette, thinly sliced
    115g/4oz baby sweetcorn, halved
    45ml/3 tbsp groundnut or sunflower oil
    115g/4oz button mushrooms, sliced
    45ml/3 tbsp soy sauce
    30ml/2 tbsp dry sherry or vermouth
    15ml/1 tbsp sesame oil
    sesame seeds, to sprinkle
Stir-fry the leeks, carrots, courgettes and
baby corn in the oil for about 2 minutes
then add the mushrooms and cook for a
further minute. Mix together the soy
sauce, sherry and sesame oil and pour
over the vegetables. Heat until bubbling
then scatter over the sesame seeds.

Cheesy creamed corn filling
    1 × 425g/15oz can creamed corn
    115g/4oz cheese, grated
    5ml/1 tsp dried mixed herbs
Heat the corn, add the cheese and mixed
herbs.

PEANUT BUTTER FINGERS

CHILDREN LOVE THESE CRISPY CROQUETTES. FREEZE SOME READY TO FILL YOUNG TUMMIES!

MAKES 12

INGREDIENTS

1kg/2lb potatoes
1 large onion, chopped
2 large peppers, red or green, chopped
3 carrots, coarsely grated
45ml/3 tbsp sunflower oil
2 courgettes, coarsely grated
125g/4oz mushrooms, chopped
15ml/1 tbsp dried mixed herbs
115g/4oz Cheddar cheese, grated
75g/3oz/½ cup crunchy peanut butter
salt and ground black pepper
2 eggs, beaten
50g/2oz/½ cup dried breadcrumbs
45ml/3 tbsp dried Parmesan cheese
oil, for deep fat frying

1 Boil the potatoes until tender, then drain well and mash. Set aside.

2 Fry the onion, pepper and carrot in the oil for about 5 minutes, add the courgettes and mushrooms and cook for 5 minutes.

3 Mix the potato with the dried mixed herbs, grated cheese and peanut butter. Season, allow to cool for 30 minutes then stir in one of the eggs.

4 Spread out on a large plate, cool and chill, then divide into 12 portions and shape. Dip your hands in cold water if the mixture sticks.

5 Put the second egg in a bowl and dip the potato fingers into the egg first, then into the crumbs and Parmesan cheese until coated evenly. Put in fridge to set.

6 Heat oil in a deep fat frier to 190°C/375°F then fry the fingers in batches for about 3 minutes until golden. Drain well on kitchen paper towel. Serve hot.

COOK'S TIP
To reheat, thaw for about 1 hour, then grill or oven bake at 190°C/375°F/Gas 5 for 15 minutes.

AUBERGINE AND COURGETTE BAKE

AUBERGINES AND COURGETTES ALWAYS GO WELL TOGETHER, AND
THEY FORM THE BASIS OF MANY DELICIOUS RECIPES.

SERVES FOUR TO SIX

INGREDIENTS
  1 large aubergine
  30ml/2 tbsp olive oil
  1 large onion, chopped
  1–2 garlic cloves, crushed
  900g/2lb tomatoes, peeled
    and chopped
  a handful of basil leaves, shredded
    or 5ml/1tsp dried basil
  15ml/1 tbsp chopped fresh parsley
  2 courgettes, sliced lengthways
  plain flour, for coating
  75–90ml/5–6 tbsp sunflower oil
  350g/12oz mozzarella, sliced
  25g/1oz Parmesan cheese, grated

1 Slice the aubergine, sprinkle with salt
and set aside for 45–60 minutes.

2 Heat the olive oil in a large frying pan.
Fry the onion and garlic for 3–4 minutes
until softened. Stir in the tomatoes, half
the basil, the parsley and seasoning.
Bring to the boil. Reduce the heat and
cook, stirring, for 25–35 minutes until
thickened. Mash the tomatoes to a pulp.

3 Rinse and dry the aubergine. Dust the
aubergine and courgettes with flour.

4 Heat the sunflower oil in another
frying pan and fry the aubergine and
courgettes until golden brown. Set aside.

5 Preheat the oven to 180°C/350°F/
Gas 4. Butter an ovenproof dish. Put a
layer of aubergine and then courgettes in
the dish, pour over half the sauce and
scatter with half the mozzarella. Sprinkle
over most of the remaining basil and a
little parsley. Repeat the layers, ending
with mozzarella. Sprinkle the Parmesan
cheese and remaining herbs on top and
bake for 30–35 minutes. Serve at once.

AUBERGINES WITH TZATZIKI

SERVES FOUR

INGREDIENTS
  2 medium-sized aubergines
  oil, for deep frying
  salt
For the batter
  75g/3oz plain flour
  1 egg
  120–150ml/4–5fl oz/½–⅔ cup milk,
    or half milk, half water
  pinch of salt
For the tzatziki
  ½ cucumber, peeled and diced
  150ml/¼ pint/⅔ cup natural yogurt
  1 garlic clove, crushed
  15ml/1 tbsp chopped fresh mint

1 To make the tzatziki, place the
cucumber in a colander, sprinkle with
salt and leave for 30 minutes. Rinse,
drain well and pat dry on kitchen paper.
Mix the yogurt, garlic, mint and
cucumber in a bowl. Cover and chill.
Slice the aubergine lengthways. Sprinkle
with salt. Leave for 1 hour.

2 To make the batter, sift the flour and
salt into a large bowl, add the egg and
milk and beat until smooth.

3 Rinse the aubergine slices and pat
dry. Heat 1cm/½in of oil in a large frying
pan. Dip the aubergine slices in the
batter and fry them for 3–4 minutes until
golden, turning once. Drain on kitchen
paper and serve with the tzatziki.

EGGS FLAMENCO

A VARIATION OF THE POPULAR BASQUE DISH PIPERADE, THE EGGS ARE COOKED WHOLE INSTEAD OF BEATING THEM BEFORE ADDING TO THE PEPPER MIXTURE. THE RECIPE IS KNOWN AS CHAKCHOUKA IN NORTH AFRICA AND MAKES A GOOD LUNCH OR SUPPER DISH.

SERVES FOUR

INGREDIENTS
2 red peppers
1 green pepper
30ml/2 tbsp olive oil
1 large onion, finely sliced
2 garlic cloves, crushed
5–6 tomatoes, peeled and chopped
120ml/4fl oz/½ cup puréed canned
  tomatoes or tomato juice
good pinch of dried basil
4 eggs
40ml/8 tsp single cream
pinch of cayenne pepper (optional)
salt and freshly ground black pepper

1 Preheat the oven to 180°C/350°F/ Gas 4. Seed and thinly slice the peppers. Heat the olive oil in a large frying pan. Fry the onion and garlic gently for about 5 minutes, stirring, until softened.

2 Add the peppers to the onion and fry for 10 minutes. Stir in the tomatoes and tomato purée or juice, the basil and seasoning. Cook gently for a further 10 minutes until the peppers are soft.

3 Spoon the mixture into four ovenproof dishes, preferably earthenware. Make a hole in the centre and break an egg into each. Spoon 10ml/2 tsp cream over the yolk of each egg and sprinkle with a little black pepper or cayenne, as preferred.

4 Bake in the oven for 12–15 minutes until the white of the egg is lightly set. Serve at once with chunks of crusty warm French or Spanish bread.

SPINACH AND PEPPER PIZZA

MAKES TWO 30cm/12in PIZZAS

INGREDIENTS
   450g/1lb fresh spinach
   60ml/4 tbsp single cream
   25g/1oz Parmesan cheese, grated
   15ml/1 tbsp olive oil
   1 large onion, chopped
   1 garlic clove, crushed
   ½ green pepper, seeded and thinly
      sliced
   ½ red pepper, seeded and thinly sliced
   175–250ml/6–8fl oz/¾-1 cup passata
      sauce or puréed tomatoes
   50g/2oz black olives, pitted and
      chopped
   15ml/1 tbsp chopped fresh basil
   175g/6oz mozzarella, grated
   175g/6oz Cheddar cheese, grated
   salt
For the dough
   25g/1oz fresh yeast or 15ml/1 tbsp
      dried yeast and 5ml/1 tsp sugar
   about 225ml/7fl oz/⅞ cup warm water
   350g/12oz strong white flour
   30ml/2 tbsp olive oil
   5ml/1 tsp salt

1 To make the dough, cream together the fresh yeast and 150ml/¼ pint/⅔ cup of the water and set aside until frothy. If using dried yeast, stir the sugar into 150ml/¼ pint/⅔ cup water, sprinkle over the yeast and leave until frothy.

2 Place the flour and salt in a large bowl, make a well in the centre and pour in the olive oil and yeast mixture. Add the remaining water, mix to make a stiff but pliable dough. Knead on a lightly floured surface for about 10 minutes until smooth and elastic.

3 Shape the dough into a ball and place in a lightly oiled bowl, cover with clear film and leave in a warm place for about 1 hour until it has doubled in size.

4 To prepare the topping, cook the spinach over a moderate heat for 4–5 minutes until the leaves have wilted. Strain and press out the excess liquid. Place in a bowl and mix with the cream, Parmesan cheese and salt to taste.

5 Heat the oil in a frying pan and fry the onion and garlic over a moderate heat for 3–4 minutes until the onion has slightly softened. Add the peppers and continue cooking until the onion is lightly golden, stirring regularly.

6 Preheat the oven to 220°C/425°F/ Gas 7. Knead the dough briefly on a lightly floured surface. Divide the dough and roll out into two 30cm/12in rounds.

7 Spread each base with the passata sauce or puréed tomatoes. Add the onions and peppers and then spread over the spinach mixture. Scatter the olives and basil leaves and sprinkle with the mozzarella and Cheddar cheese.

8 Bake in the oven for 15–20 minutes, or until the crust is lightly browned and the top is beginning to turn golden. Allow to cool slightly before serving.

RADICCHIO PIZZA

THIS UNUSUAL PIZZA TOPPING CONSISTS OF CHOPPED RADICCHIO WITH LEEKS, TOMATOES AND PARMESAN AND MOZZARELLA CHEESES. THE BASE IS A SCONE DOUGH, MAKING THIS A QUICK AND EASY SUPPER DISH TO PREPARE. SERVE WITH A CRISP GREEN SALAD.

SERVES TWO

INGREDIENTS
    ½ x 400g/14oz can chopped tomatoes
    2 garlic cloves, crushed
    pinch of dried basil
    25ml/1½ tbsp olive oil, plus extra
       for dipping
    2 leeks, sliced
    100g/3½oz radicchio, roughly
       chopped
    20g/¾oz Parmesan, grated
    115g/4oz mozzarella cheese, sliced
    10–12 black olives, pitted
    basil leaves, to garnish
    salt and freshly ground black pepper
For the dough
    225g/8oz self-raising flour
    2.5ml/½ tsp salt
    50g/2oz butter or margarine
    about 120ml/4fl oz/½ cup milk

1 Preheat the oven to 220°C/425°F/
Gas 7 and grease a baking sheet. Mix the
flour and salt in a bowl, rub in the butter
or margarine and gradually stir in the
milk and mix to a soft dough.

2 Roll the dough out on a lightly floured
surface to make a 25–28cm/10–11in
round. Place on the baking sheet.

3 Purée the tomatoes and then pour
into a small saucepan. Stir in one of the
crushed garlic cloves, together with the
dried basil and seasoning, and simmer
over a moderate heat until the mixture is
thick and reduced by about half.

4 Heat the olive oil in a large frying pan
and fry the leeks and remaining garlic for
4–5 minutes until slightly softened. Add
the radicchio and cook, stirring
continuously for a few minutes, and then
cover and simmer gently for about
5–10 minutes. Stir in the Parmesan
cheese and season with salt and pepper.

5 Cover the dough base with the tomato
mixture and then spoon the leek and
radicchio mixture on top. Arrange the
mozzarella slices on top and scatter over
the black olives. Dip a few basil leaves in
olive oil, arrange on top and then bake
the pizza for 15–20 minutes until the
scone base and top are golden brown.

ITALIAN ROAST PEPPERS

SIMPLE AND EFFECTIVE, THIS DISH WILL DELIGHT ANYONE WHO LIKES PEPPERS. IT CAN BE EATEN EITHER AS A STARTER SERVED WITH FRENCH BREAD, OR AS A LIGHT LUNCH WITH COUSCOUS OR RICE.

SERVES FOUR

INGREDIENTS

4 small red peppers, halved, cored
  and seeded
30–45ml/2–3 tbsp capers, chopped
10–12 black olives, pitted
  and chopped
2 garlic cloves, finely chopped
50–75g/2–3 oz mozzarella, grated
25–40g/1–1½oz fresh white
  breadcrumbs
120ml/4fl oz/½ cup white wine
45ml/3 tbsp olive oil
5ml/1 tsp finely chopped fresh mint
5ml/1 tsp chopped fresh parsley
freshly ground black pepper

1 Preheat the oven to 180°C/350°F/
Gas 4 and butter a shallow ovenproof
dish. Place the peppers tightly together
in the dish and sprinkle over the
chopped capers, black olives, garlic,
mozzarella and breadcrumbs.

2 Pour over the wine and olive oil and
then sprinkle with the mint, parsley and
freshly ground black pepper.

3 Bake for 30–40 minutes until the
topping is crisp and golden brown.

STUFFED MUSHROOMS

THIS IS A CLASSIC MUSHROOM DISH, STRONGLY FLAVOURED WITH GARLIC. USE FLAT MUSHROOMS OR FIELD MUSHROOMS THAT ARE SOMETIMES AVAILABLE FROM FARM SHOPS.

SERVES FOUR

INGREDIENTS
    450g/1lb large flat mushrooms
    butter, for greasing
    about 75ml/5 tbsp olive oil
    2 garlic cloves, minced or very
      finely chopped
    45ml/3 tbsp finely chopped fresh
      parsley
    40–50g/1½–2oz fresh white
      breadcrumbs
    salt and freshly ground black pepper
    sprig of flat leaf parsley, to garnish

1 Preheat the oven to 180°C/350°F/ Gas 4. Cut off the mushroom stalks and reserve on one side.

2 Arrange the mushroom caps in a buttered shallow dish, gill side upwards.

COOK'S TIP
The cooking time for the mushrooms depends on their size and thickness. If they are fairly thin, cook for slightly less time. They should be tender but not too soft when cooked. If a stronger garlic flavour is preferred, do not cook the garlic before adding it to the breadcrumb mixture.

3 Heat 15ml/1 tbsp of oil in a frying pan and fry the garlic briefly. Finely chop the mushroom stalks and mix with the parsley and breadcrumbs. Add the garlic, seasoning and 15ml/1tbsp of the oil. Pile a little of the mixture on each mushroom.

4 Add the remaining oil to the dish and cover the mushrooms with buttered greaseproof paper. Bake for about 15–20 minutes, removing the paper for the last 5 minutes to brown the tops. Garnish with a sprig of flat leaf parsley.

TAGLIATELLE FUNGI

THE MUSHROOM SAUCE IS QUICK TO MAKE AND THE PASTA COOKS VERY QUICKLY; BOTH NEED TO BE COOKED AS NEAR TO SERVING AS POSSIBLE SO CAREFUL COORDINATION IS REQUIRED. PUT THE PASTA IN TO COOK WHEN THE MASCARPONE CHEESE IS ADDED TO THE SAUCE.

SERVES FOUR (as a snack or starter)

INGREDIENTS

    about 50g/2oz butter
    225–350g/8–12oz chanterelles
        or other wild mushrooms
    15ml/1 tbsp plain flour
    150ml/¼ pint/⅔ cup milk
    90ml/6 tbsp crème fraîche
    15ml/1 tbsp chopped fresh parsley
    275g/10oz fresh tagliatelle,
        preferably multi-coloured
    olive oil
    salt and freshly ground black pepper

1 Melt 40g/1½oz of the butter in a frying pan and fry the mushrooms for about 2–3 minutes over a gentle heat until the juices begin to run, then increase the heat and cook until the liquid has almost evaporated. Transfer the mushrooms to a bowl using a slotted spoon.

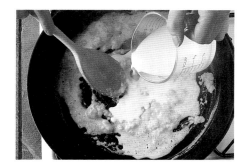

2 Stir in the flour, adding a little more butter if necessary, and cook for about 1 minute, and then gradually stir in the milk to make a smooth sauce.

3 Add the crème fraîche, parsley, mushrooms and seasoning and stir well. Cook very gently to heat through and then keep warm while cooking the pasta.

4 Cook the pasta in a large saucepan of boiling water for 4–5 minutes (or according to the instructions on the packet). Drain well, toss in a little olive oil and then turn on to a warmed serving plate. Pour the mushroom sauce over and serve immediately.

COOK'S TIP

Chanterelles are a little tricky to wash, as they are so delicate. However, since these are woodland mushrooms, it's important to clean them thoroughly. Hold each one by the stalk and let cold water run under the gills to dislodge hidden dirt. Shake gently to dry.

WILD MUSHROOMS <u>IN</u> BRIOCHE

SERVES FOUR

INGREDIENTS
    4 small brioches
    olive oil, for glazing
    20ml/4 tsp lemon juice
    sprigs of parsley, to garnish
For the mushroom filling
    25g/1oz butter
    2 shallots
    1 garlic clove, crushed
    175–225g/6–8oz assorted wild
      mushrooms, halved if large
    45ml/3 tbsp white wine
    45ml/3 tbsp double cream
    5ml/1 tsp chopped fresh basil
    5ml/1 tsp chopped fresh parsley
    salt and freshly ground black pepper

1 Preheat the oven to 180°C/350°F/ Gas 4. Using a serrated or grapefruit knife, cut a circle out of the top of each brioche and reserve. Scoop out the bread inside to make a small cavity.

2 Place the brioches and the tops on a baking sheet and brush inside and out with olive oil. Bake for 7–10 minutes until golden and crisp. Squeeze 5ml/ 1 tsp of lemon juice inside each brioche.

3 To make the filling, melt the butter in a frying pan and fry the shallots and garlic for 2–3 minutes until softened.

4 Add the mushrooms and cook gently for about 4–5 minutes, stirring.

5 When the juices begin to run, reduce the heat and continue cooking for about 3–4 minutes, stirring occasionally, until the pan is fairly dry.

6 Stir in the wine. Cook for a few more minutes and then stir in the cream, basil, parsley and seasoning to taste.

7 Pile the mushroom mixture into the brioche shells and return to the oven and reheat for about 5–6 minutes. Serve as a starter, garnished with a sprig of parsley.

WILD MUSHROOMS <u>WITH</u> PANCAKES

SERVES SIX

INGREDIENTS
    225–275g/8–10oz assorted wild
      mushrooms
    50g/2oz butter
    1–2 garlic cloves
    splash of brandy (optional)
    freshly ground black pepper
    soured cream, to serve
For the pancakes
    115g/4oz self-raising flour
    20g/¾oz buckwheat flour
    2.5ml/½ tsp baking powder
    2 eggs
    about 250ml/8fl oz/1 cup milk
    pinch of salt
    oil, for frying

1 To make the pancakes, mix together the flours, baking powder and salt in a large bowl or food processor. Add the eggs and milk and beat or process to make a smooth batter, about the consistency of single cream.

2 Grease a large griddle or frying pan with a little oil and when hot, pour small amounts of batter (about 15–30ml/ 1–2 tbsp per pancake) on to the griddle, well spaced apart.

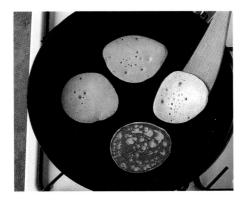

3 Fry for a few minutes until bubbles begin to appear on the surface and the underside is golden, and then flip over. Cook for about 1 minute until golden. Keep warm, wrapped in a clean dish towel. (Makes about 18–20 pancakes.)

4 If the mushrooms are large, cut them in half. Melt the butter in a frying pan and add the garlic and mushrooms. Fry over a moderate heat for a few minutes until the juices begin to run and then increase the heat and cook, stirring frequently, until nearly all the juices have evaporated. Stir in the brandy, if using, and season with a little black pepper.

5 Arrange the warm pancakes on a serving plate and spoon over a little soured cream. Top with the hot mushrooms and serve immediately.

COOK'S TIP
This makes a delicious and elegant starter for a dinner party. Alternatively, make cocktail-size pancakes and serve as part of a buffet supper.

PASTA SALADE TIÈDE

BOIL A PAN OF PASTA SHAPES
AND TOSS WITH VINAIGRETTE
DRESSING AND SOME FRESHLY
PREPARED SALAD VEGETABLES AND
YOU HAVE THE BASIS FOR A
DELICIOUS WARM SALAD.

SERVES TWO

INGREDIENTS
   115g/4oz pasta shapes (e.g. shells)
   45ml/3 tbsp vinaigrette dressing
   3 sun-dried tomatoes in oil, snipped
   2 spring onions, sliced
   25g/1oz/2 or 3 sprigs watercress or
      rocket, chopped
   ¼ cucumber, halved, seeded and sliced
   salt and ground black pepper
   about 40g/1½oz Pecorino cheese,
      coarsely grated

1 Boil the pasta according to the
instructions on the packet. Drain and toss
in the dressing.

2 Mix in the tomatoes, onions, cress or
rocket and cucumber. Season to taste.

3 Divide between two plates and sprinkle
over the cheese. Eat at room temperature,
if possible.

PENNE WITH "CAN CAN" SAUCE

THE QUALITY OF CANNED PULSES
AND TOMATOES IS SO GOOD THAT
IT IS POSSIBLE TO TRANSFORM
THEM INTO A VERY FRESH
TASTING PASTA SAUCE IN
MINUTES. AGAIN, CHOOSE
WHATEVER PASTA YOU LIKE.

SERVES THREE TO FOUR

INGREDIENTS
   225g/8oz penne pasta
   1 onion, sliced
   1 red pepper, seeded and sliced
   30ml/2 tbsp olive oil
   1 × 400g/14oz can chopped tomatoes
   1 × 425g/15oz can chick peas
   30ml/2 tbsp dry vermouth (optional)
   5ml/1 tsp dried oregano
   1 large bay leaf
   30ml/2 tbsp capers
   salt and ground black pepper

1 Boil the pasta as instructed on the
package, then drain. In a saucepan,
gently fry the onion and pepper in the oil
for about 5 minutes, stirring occasionally,
until softened.

2 Add the tomatoes, chick peas with their
liquor, vermouth (if liked), herbs and
capers.

3 Season and bring to a boil then simmer
for about 10 minutes. Remove the bay
leaf and mix in the pasta, reheat and
serve hot.

TOFU AND CRUNCHY VEGETABLES

TOFU IS NICEST IF MARINATED LIGHTLY BEFORE COOKING. SMOKED TOFU IS EVEN TASTIER.

SERVES FOUR

INGREDIENTS
- 2 × 225g/8oz cartons smoked tofu
- 45ml/3 tbsp soy sauce
- 30ml/2 tbsp dry sherry or vermouth
- 15ml/1 tbsp sesame oil
- 45ml/3 tbsp groundnut or sunflower oil
- 2 leeks, thinly sliced
- 2 carrots, cut in sticks
- 1 large courgette, thinly sliced
- 115g/4oz baby sweetcorn, halved
- 115g/4oz button or shiitake
  mushrooms, sliced
- 15ml/1 tbsp sesame seeds
- 1 packet of egg noodles, cooked

1 Cut the tofu into cubes, and marinate it in the soy sauce, sherry or vermouth and sesame oil for at least half an hour. Drain and reserve the marinade.

2 Heat the groundnut or sunflower oil in a wok and stir-fry the tofu cubes until browned all over. Remove and reserve.

3 Stir-fry the leeks, carrots, courgette and baby corn, stirring and tossing for about 2 minutes. Add the mushrooms and cook for a further minute.

4 Return the tofu to the wok and pour in the marinade. Heat until bubbling, then scatter over the sesame seeds.

5 Serve as soon as possible with the hot cooked noodles, dressed in a little sesame oil, if liked.

VARIATION
Tofu is also excellent marinated and skewered, then lightly grilled. Push the tofu off the skewers into pockets of pitta bread. Fill with lemon-dressed salad and serve with a final trickle of tahini cream.

EGG FOO YUNG

A GREAT WAY OF TURNING A BOWL OF LEFTOVER COOKED RICE INTO A MEAL FOR FOUR, THIS ORIENTAL DISH IS TASTY AND FULL OF TEXTURE. USE BOUGHT BEANSPROUTS OR GROW YOUR OWN — IT'S EASY AND FUN.

SERVES FOUR

INGREDIENTS
  salt and ground black pepper
  3 eggs, beaten
  good pinch five spice powder (optional)
  45ml/3 tbsp groundnut or sunflower oil
  4 spring onions, sliced
  1 garlic clove, crushed
  1 small green pepper, seeded and chopped
  115g/4oz fresh bean sprouts
  225g/8oz/3 cups cooked white rice
  45ml/3 tbsp light soy sauce
  15ml/1 tbsp sesame oil

1 Season the eggs and beat in the five spice powder, if using.

2 In a wok or large frying pan, heat one tablespoon of the oil and when quite hot, add the egg. Cook like an omelette, pulling the mixture away from the sides and allowing the rest to slip underneath.

3 Cook the egg until firm then tip out. Chop the "omelette" into small strips.

4 Heat the remaining oil and stir-fry the onion, garlic, pepper and bean sprouts for about 2 minutes, stirring and tossing continuously.

5 Mix in the rice and heat thoroughly, stirring well. Add the soy sauce and sesame oil then return the egg and mix in well. Serve immediately, piping hot.

EGGS BENEDICT WITH QUICK HOLLANDAISE

To make traditional Hollandaise quickly, yet still achieve a thick, creamy sauce, use a blender.

SERVES FOUR

INGREDIENTS

2 egg yolks
5ml/1 tsp dry mustard
good pinch each salt and ground black pepper
15ml/1 tbsp wine vinegar or lemon juice
175g/6oz/¾ cup butter
4 muffins, split
butter or low fat spread
4 large eggs
30ml/2 tbsp capers
a little fresh parsley, chopped, to garnish

2 Heat the butter until it is on the point of bubbling then, with the machine still running, slowly pour the butter onto the egg yolks.

3 The mixture should emulsify instantly and become thick and creamy. Switch off the blender and set the sauce aside.

5 Poach the eggs either in gently simmering water or in an egg poacher. Drain well and slip carefully onto the uncut muffin halves.

6 Spoon the sauce over the muffins and then sprinkle with capers and parsley. Serve immediately with the buttered muffin quarters.

1 Blend the egg yolks with the mustard and seasoning in a blender or food processor for a few seconds until well mixed. Mix in the vinegar or lemon juice.

4 Toast the split muffins. Cut four of the halves in two and lightly butter. Place the four uncut halves on warmed plates and leave unbuttered.

VARIATION
This classic American brunch dish is said to have originated in New York, and is ideal to serve on a special occasion such as a birthday treat or New Year's day.

Instead of the toasted muffin, you could make more of a main meal by serving the dish on a bed of lightly steamed or blanched spinach mixed with quick fried sliced mushrooms and onions. The quick Hollandaise sauce is, of course, ideal as an all-purpose serving sauce for vegetables, baked potatoes, cauliflower and broccoli.

LIGHT RATATOUILLE

*THIS LIGHTLY COOKED MEDLEY IS
COOKED WITH SIMPLE POACHED
EGGS AND TOPPED WITH CRISP
BREADCRUMBS.*

SERVES FOUR

INGREDIENTS
   45ml/3 tbsp olive oil
   50g/2oz/1 cup fresh breadcrumbs
   1 pepper, seeded and thinly sliced
   2 garlic cloves, crushed
   2 leeks, thinly sliced
   2 courgettes, thinly sliced
   2 tomatoes, skinned and sliced
   5ml/1 tsp dried rosemary, crushed
   4 eggs
   salt and ground black pepper

1 Heat half the oil in a shallow fireproof dish (or frying pan with a lid) and fry the breadcrumbs until they are golden and crisp. Drain on kitchen paper towel.

2 Add the remaining oil and fry the pepper, garlic and leeks in the same pan for about 10 minutes until softened.

3 Add the courgettes, tomatoes and rosemary and cook for a further 5 minutes. Season well.

4 Using the back of a spoon, make four wells in the vegetable mixture and break an egg into each one. Lightly season the eggs then cover and cook on a gentle heat for about 3 minutes until they are just set.

5 Sprinkle over the crisp breadcrumbs and serve immediately, piping hot.

MACARONI SOUFFLÉ

THIS IS GENERALLY A GREAT
FAVOURITE WITH CHILDREN, AND
IS RATHER LIKE A LIGHT AND
FLUFFY MACARONI CHEESE.

SERVES THREE TO FOUR

INGREDIENTS
    75g/3oz short cut macaroni
    melted butter, to coat
    25g/1oz/3 tbsp dried breadcrumbs
    50g/2oz/4 tbsp butter
    5ml/1 tsp ground paprika
    40g/1½oz/⅓ cup plain flour
    300ml/½ pint/1¼ cups milk
    75g/3oz Cheddar or Gruyère cheese,
      grated
    50g/2oz Parmesan cheese, grated
    salt and ground black pepper
    3 eggs, separated

1 Boil the macaroni according to the instructions on the pack. Drain well and then set aside. Preheat the oven to 150°C/300°F/Gas 2.

2 Brush the insides of a 1.2 litre/2 pint soufflé dish with melted butter, then coat evenly with the breadcrumbs, shaking out any excess.

3 Put the butter, paprika, flour and milk into a saucepan and bring to the boil slowly, whisking it constantly until it is smooth and thick.

4 Simmer the sauce for a minute, then take off the heat and stir in the cheeses until they melt. Season well and mix with the macaroni.

5 Beat in the egg yolks. Whisk the egg whites until they form soft peaks and spoon a quarter into the sauce mixture, beating it gently to loosen it up.

6 Using a large metal spoon, carefully fold in the rest of the egg whites and transfer to the prepared soufflé dish.

7 Bake in the centre of the oven for about 40–45 minutes until the soufflé has risen and is golden brown. The middle should wobble very slightly and the soufflé should be lightly creamy inside.

COWBOY HOT POT

A GREAT DISH TO SERVE AS A CHILDREN'S MAIN MEAL. YOU CAN USE ANY VEGETABLE MIXTURE YOU LIKE — ALTHOUGH BEANS ARE A MUST FOR EVERY SELF-RESPECTING COWBOY!

SERVES FOUR TO SIX

INGREDIENTS
    1 onion, sliced
    1 red pepper, sliced
    1 sweet potato or 2 carrots, chopped
    45ml/3 tbsp sunflower oil
    115g/4oz green beans, chopped
    1 × 400g/14oz can baked beans
    1 × 200g/7oz can sweet corn
    15ml/1 tbsp tomato purée
    5ml/1 tsp barbecue spice seasoning
    115g/4oz cheese (preferably smoked)
    450g/1lb potatoes, thinly sliced
    25g/1oz/2 tbsp butter, melted
    salt and ground black pepper

1 Fry the onion, pepper and sweet potato or carrots gently in the oil until softened but not browned.

2 Add the green beans, baked beans, sweetcorn (and liquor), tomato purée and barbecue spice seasoning. Bring to a boil then simmer for 5 minutes.

3 Transfer the vegetables to a shallow ovenproof dish and then scatter with the cubed cheese.

4 Cover the vegetable and cheese mixture with the sliced potato, brush with butter, season and bake at 190°C/375°F/Gas 5 for 30–40 minutes until golden brown on top and the potato is cooked.

STIR-FRY RICE AND VEGETABLES

LEFT-OVER COOKED RICE AND A FEW VEGETABLES FROM THE BOTTOM OF THE REFRIGERATOR ARE THE BASIS FOR THIS QUICK AND TASTY MEAL.

SERVES FOUR

INGREDIENTS
    ½ cucumber
    2 spring onions, sliced
    1 garlic clove, crushed
    2 carrots, thinly sliced
    1 small red or yellow pepper, seeded and sliced
    45ml/3 tbsps sunflower or groundnut oil
    ¼ small green cabbage, shredded
    225g/8oz/4 cups cooked long grain rice
    30ml/2 tbsp light soy sauce
    15ml/1 tbsp sesame oil
    salt and ground black pepper
    fresh parsley or coriander, chopped (optional)
    115g/4oz unsalted cashew nuts, almonds or peanuts

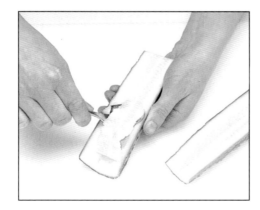

1 Halve the cucumber lengthways and scoop out the seeds with a teaspoon. Slice the flesh diagonally. Set aside.

2 In a wok or large frying pan, stir-fry the onions, garlic, carrots and pepper in the oil for about 3 minutes until they are just soft.

3 Add the cabbage and cucumber and fry for another minute or two until the leaves begin to just wilt. Mix in the rice, soy sauce, sesame oil and seasoning. Reheat the mixture thoroughly, stirring and tossing all the time.

4 Add the herbs, if using, and nuts. Check the seasoning and serve piping hot.

PEPPER AND POTATO TORTILLA

TRADITIONALLY A SPANISH DISH, TORTILLA IS BEST EATEN COLD IN CHUNKY WEDGES. USE A HARD SPANISH CHEESE, SUCH AS MAHON, OR A GOATS' CHEESE, IF YOU CAN.

SERVES FOUR

INGREDIENTS
   2 medium size potatoes
   45ml/3 tbsp olive oil
   1 large onion, thinly sliced
   2 garlic cloves, crushed
   2 peppers, one green and one red,
     thinly sliced
   6 eggs, beaten
   115g/4oz mature cheese, grated
   salt and ground black pepper

1 Do not peel the potatoes, but wash them thoroughly. Par-boil them as they are for about 10 minutes, then drain and slice them thickly. Switch on the grill so that it has time to warm up while you prepare the tortilla.

2 In a large non-stick or well seasoned frying pan, heat the oil and fry the onion, garlic and peppers on a moderate heat for 5 minutes until softened.

3 Add the potatoes and continue frying, stirring occasionally until the potatoes are completely cooked and the vegetables are soft. Add a little extra oil if the pan seems rather dry.

4 Pour in half the eggs, then sprinkle over half the cheese then the rest of the egg, seasoning as you go. Finish with a layer of cheese.

5 Continue to cook on a low heat, without stirring, half covering the pan with a lid to help set the eggs.

6 When the mixture is firm, flash the pan under the hot grill to just lightly seal the top. Leave the tortilla in the pan to cool. This helps it firm up further and makes it easier to turn out.

VARIATION
You can add any sliced and lightly cooked vegetables, such as mushrooms, courgettes or broccoli, to this tortilla dish instead of peppers. Cooked pasta or brown rice are excellent alternatives, too.

CAULIFLOWER AND EGG CHEESE

A QUICK ALL-IN-ONE SAUCE CAN BE MADE IN MINUTES, WHILE A SMALL PACK OF SOUP CROÛTONS GIVES THE DISH A DELICIOUS CRUNCHY TOPPING.

SERVES FOUR

INGREDIENTS

1 medium size cauliflower, in florets
1 medium onion, sliced
2 eggs, hard boiled, peeled and chopped
40g/1½oz/3 tbsp wholemeal flour
5ml/1 tsp mild curry powder
25g/1oz/2 tbsp sunflower margarine or low-fat spread
450ml/¾ pint/2 cups milk
2.5ml/½ tsp dried thyme
salt and ground black pepper
115g/4oz mature cheese, grated
small packet of soup croûtons

1 Boil the cauliflower and onion in enough salted water to cover until they are just tender. Be careful not to overcook them. Drain well.

2 Arrange the cauliflower and onion in a shallow ovenproof dish and scatter over the chopped egg.

3 Put the flour, curry powder, fat and milk in a saucepan all together. Bring slowly to the boil, stirring well until thickened and smooth. Stir in the thyme and seasoning and allow the sauce to simmer for a minute or two. Remove the pan from the heat and stir in about three quarters of the cheese.

4 Pour the sauce over the cauliflower, scatter over the croûtons and sprinkle with the remaining cheese. Brown under a hot grill until golden and serve. Delicious with thick crusty bread.

QUICK BASMATI AND NUT PILAFF

LIGHT AND FRAGRANT BASMATI RICE FROM THE FOOTHILLS OF THE HIMALAYAS COOKS PERFECTLY USING THIS SIMPLE PILAFF METHOD. USE WHATEVER NUTS ARE YOUR FAVOURITE — EVEN UNSALTED PEANUTS ARE GOOD, ALTHOUGH ALMONDS, CASHEWS OR PISTACHIOS ARE MORE EXOTIC.

SERVES FOUR TO SIX

INGREDIENTS
    225g/8oz/1¼ cups basmati rice
    1 onion, chopped
    1 garlic clove, crushed
    1 large carrot, coarsely grated
    15–30ml/1–2 tbsp sunflower oil
    5ml/1 tsp cumin seeds
    10ml/2 tsp ground coriander
    10ml/2 tsp black mustard seeds
      (optional)
    4 cardamom pods
    450ml/¾ pint/2 cups vegetable stock or
      water
    1 bay leaf
    salt and ground black pepper
    75g/3oz/½ cup unsalted nuts
    fresh chopped parsley or coriander,
      to garnish

RINSING BASMATI
For light, fluffy grains basmati rice is best rinsed before cooking to remove any surface starch. The traditional method is to put the rice into a large bowl of cold water. Swill the grains around with your hands then tip out the cloudy water. The rice will quickly sink to the bottom. Ideally, leave the rice to soak for 30 minutes in the last rinsing water. This ensures a lighter, fluffier grain.

1 Wash the rice either by the traditional Indian method (see note) or in a sieve under a running tap. If there is time, soak the rice for 30 minutes, then drain well in a sieve.

2 In a large shallow pan, gently fry the onion, garlic and carrot in the oil for a few minutes.

3 Stir in the rice and spices and cook for a further 1–2 minutes so that that the grains are coated in oil.

4 Pour in the stock or water, add the bay leaf and season well. Bring to a boil, cover and simmer very gently for about 10 minutes.

5 Remove from the heat without lifting the lid – this helps the rice to firm up and cook further. Leave for about 5 minutes.

6 If the rice is cooked, there will be small steam holes in the centre. Discard the bay leaf and cardamom pods.

7 Stir in the nuts and check the seasoning. Scatter the mixture with the chopped parsley or coriander. This whole dish can be made ahead and reheated.

QUORN WITH GINGER, CHILLI AND LEEKS

QUORN EASILY ABSORBS DIFFERENT FLAVOURS AND RETAINS A FIRM TEXTURE, WHICH IS IDEAL FOR STIR-FRYING.

SERVES FOUR

INGREDIENTS
    225g/8oz Quorn cubes
    45ml/3 tbsp soy sauce
    30ml/2 tbsp dry sherry or vermouth
    10ml/2 tsp clear honey
    150ml/¼ pint/⅔ cup stock
    10ml/2 tsp cornflour
    45ml/3 tbsp sunflower or groundnut oil
    3 leeks, thinly sliced
    1 red chilli, seeded and sliced
    2.5cm/1in piece fresh root ginger,
      peeled and shredded
    salt and ground black pepper

1 Toss the Quorn in the soy sauce and sherry or vermouth until well coated and leave to marinate for about 30 minutes.

2 Strain the Quorn from the marinade and reserve the juices in a jug. Mix the marinade with the honey, stock and cornflour to make a paste.

3 Heat the oil in a wok or large frying pan and when hot, stir-fry the Quorn until it is crisp on the outside. Remove the Quorn and set aside.

4 Reheat the oil and stir-fry the leeks, chilli and ginger for about 2 minutes until they are just soft. Season lightly.

5 Return the Quorn to the pan, together with the marinade, and stir well until the liquid is thick and glossy. Serve hot with rice or egg noodles.

CHINESE POTATOES <u>WITH</u> CHILLI BEANS

AN AMERICAN-STYLE DISH WITH A CHINESE FLAVOUR. TRY IT AS A QUICK SUPPER DISH WHEN YOU FANCY A MEAL WITH ZING!

SERVES FOUR

INGREDIENTS

    4 medium potatoes, cut in thick chunks
    3 spring onions, sliced
    1 large fresh chilli, seeded and sliced
    30ml/2 tbsp sunflower or groundnut oil
    2 garlic cloves, crushed
    1 × 400g/14oz can red kidney beans,
      drained
    30ml/2 tbsp soy sauce
    15ml/1 tbsp sesame oil
To serve
    salt and ground black pepper
    15ml/1 tbsp sesame seeds
    fresh coriander or parsley, chopped,
      to garnish

1 Boil the potatoes until they are just tender. Take care not to overcook them. Drain and reserve.

2 In a large frying pan or wok, stir-fry the onions and chilli in the oil for about 1 minute then add the garlic and fry for a few seconds longer.

3 Add the potatoes, stirring well, then the beans and finally the soy sauce and sesame oil.

4 Season to taste and cook the vegetables until they are well heated through. Sprinkle with the sesame seeds and the coriander or parsley.

TABBOULEH

ALMOST THE ULTIMATE QUICK GRAIN SALAD — IT SIMPLY NEEDS SOAKING, DRAINING AND MIXING. BULGUR IS PAR-BOILED WHEAT. MAKE THIS A DAY AHEAD, IF POSSIBLE, SO THAT THE FLAVOURS HAVE TIME TO DEVELOP.

SERVES FOUR

INGREDIENTS
115g/4oz/¾ cup bulgur wheat
90ml/6 tbsp fresh lemon juice
75ml/5 tbsp extra virgin olive oil
90ml/6 tbsp fresh parsley, chopped
60ml/4 tbsp fresh mint, chopped
3 spring onions, finely chopped
4 firm tomatoes, skinned and chopped
salt and ground black pepper

1 Cover the bulgur with cold water and soak for 20 minutes, then drain well and squeeze out even more water from it with your hands.

2 Put the bulgur into another bowl and add all the other ingredients, stirring and seasoning well.

3 Cover and chill for a few hours, overnight, if possible.

CRUDITÉS WITH HUMMUS

ALWAYS A GREAT FAMILY FAVOURITE, HOME-MADE HUMMUS IS SPEEDILY MADE WITH THE HELP OF A BLENDER. THE TAHINI PASTE IS THE SECRET OF HUMMUS, AND IT IS READILY AVAILABLE IN DELICATESSENS OR LARGER SUPERMARKETS.

SERVES TWO TO THREE

INGREDIENTS
1 × 425g/15oz can chick-peas, drained
30ml/2 tbsp tahini paste
30ml/2 tbsp fresh lemon juice
1 garlic clove, crushed
salt and ground black pepper
olive oil and paprika pepper, to garnish
To serve
selection of salad vegetables (e.g. cucumber, chicory, baby carrots, pepper strips, radishes)
bite size chunks of bread (e.g. pitta, walnut, naan, bruscheta or grissini sticks)

1 Put the chick-peas, tahini paste, lemon juice, garlic and plenty of seasoning into a food processor or blender and mix to a smooth paste.

2 Spoon the hummus into a bowl and swirl the top with the back of a spoon. Trickle over a little olive oil and sprinkle with paprika.

3 Prepare a selection of fresh salad vegetables and chunks of your favourite fresh bread or grissini sticks into finger size pieces.

4 Set out in a colourful jumble on a large plate with the bowl of hummus in the centre. Then dip and eat!

PITTA PIZZAS

*PITTA BREADS MAKE VERY GOOD
BASES FOR QUICK, THIN AND
CRISPY PIZZAS.*

SERVES FOUR

INGREDIENTS
Basic pizzas
    4 pitta breads, ideally wholemeal
    small jar of pasta sauce
    225g/8oz Mozzarella cheese, sliced or
      grated
    dried oregano or thyme, to sprinkle
    salt and ground black pepper
Extra toppings – choose from
    1 small red onion, thinly sliced and
      lightly fried
    mushrooms, sliced and fried
    1 × 200g/7oz can sweetcorn, drained
    jalapeno chillis, sliced
    black or green olives, stoned and sliced
    capers, drained

1 Prepare two or three toppings of your choice for the pizzas.

2 Preheat the grill and lightly toast the pitta breads on each side.

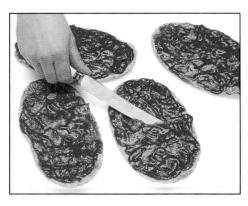

3 Spread pasta sauce on each pitta, right to the edge. This prevents the edges of the pitta from burning.

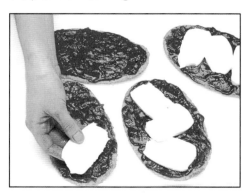

4 Arrange cheese slices or grated cheese on top of each pitta and sprinkle with herbs and seasoning.

5 Add the toppings of your choice and then grill the pizzas for about 5–8 minutes until they are golden brown and bubbling. Serve immediately.

TAGLIATELLE <u>WITH</u> SPINACH <u>AND</u> SOY GARLIC CHEESE

ITALIAN PASTA AND SPINACH COMBINE WITH CHINESE SOY AND FRENCH GARLIC CREAM CHEESE TO CREATE THIS WONDERFULLY MOUTH-WATERING DISH.

SERVES FOUR

INGREDIENTS
225g/8oz tagliatelle
225g/8oz fresh leaf spinach
30ml/2 tbsp light soy sauce
75g/3oz garlic and herb cheese
45ml/3 tbsp milk
salt and ground black pepper

1 Boil the tagliatelle according to the instructions on the pack and drain. Return the pasta to the pan.

2 Meanwhile, blanch the spinach in a tiny amount of water until just wilted, then drain very well, squeezing dry with the back of a wooden spoon. Chop roughly with kitchen scissors.

3 Return the spinach to its pan and stir in the soy sauce, garlic and herb cheese and milk. Bring slowly to a boil, stirring until smooth. Season to taste.

4 When the sauce is ready, pour it over the pasta. Toss the pasta and sauce together well and serve hot.

SPAGHETTI WITH FETA

WE THINK OF PASTA AS BEING ESSENTIALLY ITALIAN, BUT THE GREEKS HAVE A GREAT APPETITE FOR IT TOO, AND IT COMPLEMENTS BEAUTIFULLY THE TANGY, FULL-FLAVOURED FETA CHEESE.

SERVES TWO TO THREE

INGREDIENTS
    115g/4oz spaghetti
    1 garlic clove
    30ml/2 tbsp extra virgin olive oil
    8 cherry tomatoes, halved
    a little freshly grated nutmeg
    salt and ground black pepper
    75g/3oz feta cheese, crumbled
    15ml/1 tbsp chopped fresh basil
    a few black olives (optional), to serve

1 Boil the spaghetti in plenty of lightly salted water according to the instructions on the pack, then drain.

2 In the same pan gently heat the garlic clove in the oil for a minute or two then add the cherry tomatoes.

3 Increase the heat to fry the tomatoes lightly for a minute, then remove the garlic and discard.

4 Toss in the spaghetti, season with the nutmeg and seasoning to taste then stir in the crumbled feta and basil.

5 Check the seasoning, remembering that feta can be quite salty, and serve hot topped with olives if liked.

POTATOES WITH BLUE CHEESE AND WALNUTS

WE ARE SO USED TO EATING POTATOES AS A SIDE DISH, WE FORGET THEY CAN BE A GOOD MAIN MEAL TOO. THIS DISH IS SO VERSATILE IT CAN BE SERVED AS EITHER. USE STILTON, DANISH BLUE, BLUE BRIE OR ANY OTHER BLUE CHEESE.

SERVES FOUR

INGREDIENTS
    450g/1lb small new potatoes
    small head of celery, sliced
    small red onion, sliced
    115g/4oz blue cheese, mashed
    150ml/¼ pint/⅔ cup single cream
    salt and ground black pepper
    100g/3½oz/½ cup walnut pieces
    30ml/2 tbsp fresh parsley, chopped

1 Cover the potatoes with water and boil for about 15 minutes, adding the sliced celery and onion to the pan for the last 5 minutes or so.

2 Drain the vegetables and put them into a shallow serving dish.

3 In a small saucepan melt the cheese in the cream, slowly, stirring occasionally. Do not allow the mixture to boil but heat it until it scalds.

4 Season the sauce to taste. Pour it over the vegetables and scatter over the walnuts and parsley. Serve hot.

ADUKI BEAN BURGERS

ALTHOUGH NOT QUICK TO MAKE, THESE ARE A DELICIOUS ALTERNATIVE TO SHOP-BOUGHT BURGERS.

MAKES 12

INGREDIENTS
    200g/7oz/1 cup brown rice
    1 onion, chopped
    2 garlic cloves, crushed
    30ml/2 tbsp sunflower oil
    50g/2oz/4 tbsp butter
    1 small green pepper, seeded and
        chopped
    1 carrot, coarsely grated
    1 × 400g/14oz can aduki beans,
        drained (or 125g/4oz dried weight,
        soaked and cooked)
    1 egg, beaten
    125g/4oz mature cheese, grated
    5ml/1 tsp dried thyme
    50g/2oz/½ cup roasted hazelnuts or
        toasted flaked almonds
    salt and ground black pepper
    wholemeal flour or cornmeal, for
        coating
    oil, for deep frying

1 Cook the rice according to the instructions on the pack, allowing it to slightly overcook so that it is softer. Strain the rice and transfer it to a large bowl.

2 Fry the onion and garlic in the oil and butter together with the green pepper and carrot for about 10 minutes until the vegetables are softened.

3 Mix this vegetable mixture into the rice, together with the aduki beans, egg, cheese, thyme, nuts or almonds and plenty of seasoning. Chill until quite firm.

4 Shape into 12 patties, using wet hands if the mixture sticks. Coat the patties in flour or cornmeal and set aside.

5 Heat 1cm/½in oil in a large, shallow frying pan and fry the burgers in batches until browned on each side, about five minutes in total. Remove and drain on kitchen paper. Eat some burgers freshly cooked, and freeze the rest for later. Serve in buns with salad and relish.

COOK'S TIP
To freeze the burgers, cool them after cooking, then open freeze them before wrapping and bagging. Use within six weeks. Cook from frozen by baking in a pre-heated moderately hot oven for 20–25 minutes.

COURGETTE QUICHE

If possible, use a hard goats' cheese for this flan as its flavour complements the courgettes nicely. Bake the pastry base blind for a crisp crust.

SERVES SIX

INGREDIENTS
For the pastry
   115g/4oz/scant 1 cup wholemeal flour
   115g/4oz/1 cup plain flour
   115g/4oz/1/2 cup sunflower margarine
For the filling
   1 red onion, thinly sliced
   30ml/2 tbsp olive oil
   2 large courgettes, sliced
   175g/6oz cheese, grated
   30ml/2 tbsp fresh basil, chopped
   3 eggs, beaten
   300ml/½ pint/1¼ cups milk
   salt and ground black pepper

1 Preheat the oven to 200°C/400°F/Gas 6. Mix the flours together and rub in the margarine until it resembles crumbs then mix to a firm dough with cold water.

2 Roll out the pastry and use it to line a 23–25cm/9–10in flan tin, ideally at least 2.5cm/1in deep. Prick the base, chill for 30 minutes then fill with paper or foil and baking beans.

3 Bake the pastry case blind on a baking sheet for 20 minutes, uncovering it for the last 5 minutes so that it can crisp up.

4 Meanwhile, sweat the onion in the oil for 5 minutes, until it is soft. Add the courgettes and fry for another 5 minutes.

5 Spoon the onions and courgettes into the pastry case. Scatter over most of the cheese and all of the basil.

6 Beat together the eggs, milk and seasoning and pour over the filling. Top with the remaining cheese.

7 Turn the oven down to 180°C/350°F/Gas 4 and return the flan for about 40 minutes until risen and just firm to the touch in the centre. Allow to cool slightly before serving.

SPRING ROLLS

BAMBOO SHOOTS AND BEANSPROUTS ARE PERFECT COMPANIONS IN THIS POPULAR SNACK, PROVIDING THE CONTRAST IN TEXTURE THAT IS THE PRINCIPAL ELEMENT IN CHINESE COOKING. THE BAMBOO SHOOTS RETAIN THEIR CRISPNESS, WHILE THE BEANSPROUTS BECOME MORE CHEWY WHEN COOKED.

MAKES ABOUT TWENTY

INGREDIENTS
60ml/4 tbsp vegetable oil
30ml/2 tbsp dark soy sauce
30ml/2 tbsp medium dry sherry
about 1cm/½in fresh root ginger,
   finely grated
225g/8oz minced Quorn
50g/2oz rice vermicelli
4–5 shiitake mushrooms
4–5 spring onions
200g/7oz can bamboo shoots
1 garlic clove, crushed
1 carrot, grated
75g/3oz beansprouts, roughly
   chopped
15ml/1 tbsp cornflour, blended
   with 30ml/2 tbsp water
about 20 15cm/6in spring roll
   wrappers
vegetable oil, for deep frying

1 Blend together 30ml/2 tbsp of the oil, the soy sauce, sherry and ginger in a medium-size bowl. Add the Quorn, stir well and set aside for 10–15 minutes.

2 Place the rice vermicelli in a large bowl, cover with boiling water and leave to stand for 15–20 minutes. Drain well and then chop roughly.

3 Wipe the mushrooms, remove the stalks and slice the caps thinly, halving them if the mushrooms are large.

4 Using a sharp knife, cut the spring onions into diagonal slices, including all but the tips of the green parts.

5 Drain the bamboo shoots and rinse them very well under cold running water. Cut in half if they are large.

6 Heat 15ml/1 tbsp of the remaining oil in a wok or large frying pan and cook the garlic for a few seconds. Add the spring onions and stir-fry for 2–3 minutes. Add the mushrooms and stir-fry for a further 3–4 minutes. Transfer the vegetables to a plate, using a slotted spoon.

7 Heat the remaining oil in the wok. Drain the Quorn, reserving the marinade, and then stir-fry for 4–5 minutes.

8 Add the bamboo shoots to the Quorn together with the carrot, vermicelli, the mushroom and onion mixture and the reserved marinade, and stir well. Add the beansprouts, stir well and then remove from the heat and cool.

9 Place a level tablespoon of mixture at one corner of a spring roll sheet. Brush the edges of the pastry with the cornflour mixture and roll up, folding the left and right corners inwards as you roll. Continue making spring rolls in this way, until all the mixture is used up.

10 Heat some oil in a large wok or deep-fryer and fry two or three rolls at a time for 3–4 minutes until golden, turning them so that they cook evenly. Drain on kitchen paper and keep warm. Serve with extra soy sauce.

LOOFAH AND AUBERGINE RATATOUILLE

LOOFAHS HAVE A SIMILAR FLAVOUR TO COURGETTES AND CONSEQUENTLY TASTE EXCELLENT WITH AUBERGINES AND TOMATOES. BE SURE TO USE VERY YOUNG LOOFAHS AND ALSO ENSURE THAT YOU PEEL AWAY THE ROUGH SKIN, AS IT CAN BE SHARP.

SERVES FOUR

INGREDIENTS

   1 large or 2 medium aubergines
   450g/1lb young loofahs or
     sponge gourds
   1 large red pepper, cut into
     large chunks
   225g/8oz cherry tomatoes
   225g/8oz shallots, peeled
   10ml/2 tsp ground coriander
   60ml/4 tbsp olive oil
   2 garlic cloves, finely chopped
   a few coriander leaves
   salt and freshly ground black pepper

1 Cut the aubergine into thick chunks and sprinkle the pieces with salt. Set aside in a colander for about 45 minutes and then rinse well under cold running water and pat dry.

2 Preheat the oven to 220°C/425°F/ Gas 7. Peel and slice the loofahs into 2cm/¾in pieces. Place the aubergine, loofah and pepper pieces, together with the tomatoes and shallots in a roasting pan which is large enough to take all the vegetables in a single layer.

3 Sprinkle with the ground coriander and olive oil and then scatter the chopped garlic and coriander leaves on top. Season to taste.

4 Roast for about 25 minutes, stirring the vegetables occasionally, until the loofah is golden brown and the peppers are beginning to char at the edges.

MAIN COURSES

Create healthy and satisfying meals that make
the most of a wonderful variety of herbs and
spices as well as nutritious ingredients such as
lentils, rice and every possible type of vegetable.

ARABIAN SPINACH

Stir-fry spinach with onions and spices; then mix in a can of chick-peas and you have a delicious family main course meal in next to no time.

SERVES FOUR

INGREDIENTS

1 onion, sliced
30ml/2 tbsp olive or sunflower oil
2 garlic cloves, crushed
400g/14oz spinach, washed and shredded
5ml/1 tsp cumin seeds
1 × 425g/15oz can chick-peas, drained
knob of butter
salt and ground black pepper

1 In a large frying pan or wok, fry the onion in the oil for about 5 minutes until softened. Add the garlic and cumin seeds, then fry for another minute.

2 Add the spinach, in stages, stirring it until the leaves begin to wilt. Fresh spinach condenses down dramatically on cooking and it will all fit into the pan.

3 Stir in the chick-peas, butter and seasoning. Reheat until just bubbling, then serve hot. Drain off any pan juices, if you like, but this dish is rather nice served slightly saucy.

VEGETABLE MEDLEY WITH LENTIL BOLOGNESE

Instead of a white or cheese sauce, it makes a nice change to top a selection of lightly steamed vegetables with a healthy and delicious lentil sauce.

SERVES 6

INGREDIENTS

1 small cauliflower, in florets
225g/8oz broccoli florets
2 leeks, thickly sliced
225g/8oz Brussels sprouts, halved if large
Lentil Bolognese Sauce

1 Make up the sauce and keep warm.

2 Place all the vegetables in a steamer over a pan of boiling water and cook for 8–10 minutes until just tender.

3 Drain and place in a shallow serving dish. Spoon the sauce on top, stirring slightly to mix. Serve hot.

FALAFELS

Made with ground chick-peas, herbs and spices, falafels are a Middle Eastern street food, normally served tucked into warm pitta breads with scoopfuls of salad. They are delicious served with tahini cream or dollops of natural yogurt.

MAKES EIGHT

INGREDIENTS
    1 × 425g/15oz can chick-peas, drained
    1 garlic clove, crushed
    30ml/2 tbsp fresh parsley, chopped
    30ml/2 tbsp fresh coriander, chopped
    15ml/1 tbsp fresh mint, chopped
    5ml/1 tsp cumin seeds
    30ml/2 tbsp fresh breadcrumbs
    5ml/1 tsp salt
    ground black pepper
    oil, for deep frying

1 Grind the chick-peas in a food processor until they are just smooth, then mix them with all the other ingredients until you have a thick, creamy paste. Add pepper to taste.

2 Using wet hands, shape the chick-pea mixture into 8 rounds and chill for 30 minutes so that they become firm.

3 Meanwhile, heat about 5mm/¼in of oil in a shallow frying pan and fry the falafels a few at a time. Cook each one for about 8 minutes, turning the falafels carefully just once.

4 Drain the falafels on kitchen paper towel and fry the rest in batches, reheating the oil in between. Serve tucked inside warm pitta breads with sliced salad, tomatoes and tahini cream or yogurt.

COOK'S TIP
Falafels can be made in batches and frozen. Allow them to cool, spread out on wire cooling racks and open freeze until solid. Tip into a freezer-proof plastic container. To reheat, bake in a moderate oven for 10–15 minutes.

CARIBBEAN RICE AND PEAS

A great family favourite in West Indian culture, this dish is not only tasty, but nutritionally well balanced.

SERVES FOUR

INGREDIENTS
    225g/8oz/1¼ cups easy-cook long grain rice
    115g/4oz/¾ cup dried gunga peas or red kidney beans, soaked and cooked
    750ml/1¼ pints/3⅔ cups water
    50g/2oz creamed coconut, chopped
    5ml/1 tsp dried thyme or 15ml/1 tbsp fresh thyme leaves
    1 small onion stuck with 6 whole cloves
    salt and ground black pepper

1 Put the rice and peas or kidney beans into a large saucepan with the water, coconut, thyme, onion and seasoning.

2 Bring to the boil, stirring until the coconut melts, then cover and simmer gently for 20 minutes.

3 Remove the lid and allow to cook uncovered for 5 minutes to reduce down any excess liquid. Remove from the heat and stir occasionally to separate the grains. The rice should be quite dry.

GREEK STUFFED VEGETABLES

VEGETABLES SUCH AS PEPPERS MAKE WONDERFUL CONTAINERS FOR SAVOURY FILLINGS. THICK, CREAMY GREEK YOGURT IS THE IDEAL ACCOMPANIMENT.

SERVES THREE TO SIX

INGREDIENTS

    1 medium aubergine
    1 large green pepper
    2 large tomatoes
    1 large onion, chopped
    2 garlic cloves, crushed
    45ml/3 tbsp olive oil
    200g/7oz/1 cup brown rice
    600ml/1 pint/2½ cups stock
    75g/3oz/¾ cup pine nuts
    50g/2oz/⅓ cup currants
    salt and ground black pepper
    45ml/3 tbsp fresh dill, chopped
    45ml/3 tbsp fresh parsley, chopped
    15ml/1 tbsp fresh mint, chopped
    extra olive oil, to sprinkle
    natural Greek yogurt, to serve
    fresh sprigs of dill

1 Halve the aubergine, scoop out the flesh with a sharp knife and chop finely. Salt the insides and leave to drain upside down for 20 minutes while you prepare the other ingredients.

2 Halve the pepper, seed and core. Cut the tops from the tomatoes, scoop out the insides and chop roughly along with the tomato tops.

3 Fry the onion, garlic and chopped aubergine in the oil for 10 minutes, then stir in the rice and cook for 2 minutes.

4 Add the tomato flesh, stock, pine nuts, currants and seasoning. Bring to the boil, cover and simmer for 15 minutes then stir in the fresh herbs.

5 Blanch the aubergines and green pepper halves in boiling water for about 3 minutes, then drain them upside down.

6 Spoon the rice filling into all six vegetable 'containers' and place on a lightly greased ovenproof shallow dish.

7 Heat the oven to 190°C/375°F/Gas 5, drizzle over some olive oil and bake the vegetables for 25–30 minutes. Serve hot, topped with spoonfuls of natural yogurt and dill sprigs.

RED ONION AND COURGETTE PIZZA

IT'S EASY TO MAKE A HOME-MADE PIZZA USING ONE OF THE NEW FAST-ACTION YEASTS. YOU CAN EITHER ADD THE TRADITIONAL CHEESE AND TOMATO TOPPING OR TRY SOMETHING DIFFERENT, SUCH AS THE ONE DESCRIBED HERE.

SERVES FOUR

INGREDIENTS
  350g/12oz/3 cups plain flour
  1 sachet fast action/easy blend yeast
  10ml/2 tsp salt
  lukewarm water to mix
For the topping
  2 red onions, thinly sliced
  60ml/4 tbsp olive oil
  2 courgettes, thinly sliced
  salt and ground black pepper
  fresh nutmeg, grated
  about 115g/4oz semi-soft goats' cheese
  6 sun-dried tomatoes in oil, snipped
  dried oregano
  extra olive oil, to sprinkle

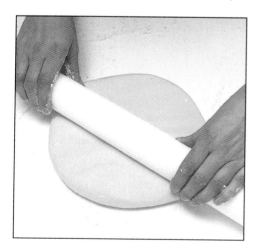

1 Preheat the oven to 200°C/400°F/ Gas 6. Mix the flour, yeast and salt together, then mix to a firm dough with warm water.

2 Knead the dough for about 5 minutes until it is smooth and elastic then roll it out to a large circle and place on a lightly greased baking sheet. Set the base aside somewhere warm to rise slightly while you make the topping.

3 Gently fry the onions in half the oil for 5 minutes then add the courgettes and fry for a further 2 minutes. Season and add nutmeg to taste.

4 Spread the pizza base with fried vegetable mixture and dot with the cheese, tomatoes and oregano. Sprinkle over the rest of the olive oil and bake for 12–15 minutes until golden and crisp.

SPROUTING BEANS AND PAK-CHOI

SUPERMARKETS ARE FAST BECOMING COSMOPOLITAN AND MANY STOCK ETHNIC VEGETABLES.

SERVES FOUR

INGREDIENTS
  45ml/3 tbsp groundnut oil
  3 spring onions, sliced
  2 garlic cloves, cut in slivers
  2.5cm/1in cube fresh root
    ginger, cut in slivers
  1 carrot, cut in thin sticks
  150g/5oz sprouting beans
  1 × 200g/7oz pak-choi, shredded
  50g/2oz/½ cup unsalted cashew nuts or
    halved almonds
For the sauce
  45ml/3 tbsp light soy sauce
  30ml/2 tbsp dry sherry
  15ml/1 tbsp sesame oil
  150ml/¼ pint/⅔ cup cold water
  5ml/1 tsp cornflour
  5ml/1 tsp clear honey
  ground black pepper

1 Heat the oil in a large wok and stir-fry the onions, garlic, ginger and carrot for 2 minutes. Add the sprouting beans and fry for another 2 minutes, stirring and tossing them together.

2 Add the pak-choi and nuts or almonds and stir-fry until the cabbage leaves are just wilting. Quickly mix all the sauce ingredients together in a jug and pour them into the wok, stirring immediately.

3 The vegetables will be coated in a thin, glossy sauce. Season and serve as soon as possible.

THAI TOFU CURRY

THAI FOOD IS A MARVELLOUS MIXTURE OF CHINESE AND INDIAN STYLES.

SERVES FOUR

INGREDIENTS
  2 × 200g/7oz packs tofu curd, cubed
  30ml/2 tbsp light soy sauce
  30ml/2 tbsp groundnut oil
For the paste
  1 small onion, chopped
  2 green chillies, seeded and chopped
  2 garlic cloves, chopped
  5ml/1 tsp grated fresh ginger
  5ml/1 tsp lime rind, grated
  10ml/2 tsp coriander berries, crushed
  10ml/2 tsp cumin seeds, crushed
  45ml/3 tbsp fresh coriander, chopped
  15ml/1 tbsp Thai fish sauce
  juice of 1 lime or small lemon
  5ml/1 tsp sugar
  25g/1oz creamed coconut dissolved in
    150ml/¼ pint/⅔ cup boiling water
For the garnish
  thin slices fresh red chilli or red pepper
  fresh coriander leaves

1 Toss the tofu cubes in soy sauce and leave to marinate for 15 minutes or so while you prepare the paste.

2 Put all the paste ingredients into a food processor and grind until smooth.

3 To cook, heat the oil in a wok until quite hot. Drain the tofu cubes and stir fry at a high temperature until well browned on all sides and just firm. Drain on kitchen paper towel.

4 Wipe out the wok. Pour in the paste and stir well. Return the tofu to the wok and mix it into the paste, reheating the ingredients as you stir.

5 Serve this dish on a flat platter garnished with red chilli or pepper and chopped coriander. Bowls of Thai fragrant or jasmine rice are the perfect accompaniment to the curry.

FESTIVE JALOUSIE

An excellent pie to serve during the holiday period. Use Chinese dried chestnuts, soaked and cooked, instead of fresh ones.

SERVES SIX

INGREDIENTS
    450g/1lb puff pastry, thawed if frozen
    450g/1lb Brussels sprouts, trimmed
    16 whole chestnuts, peeled if fresh
    1 large red pepper, sliced
    1 large onion, sliced
    45ml/3 tbsp sunflower oil
    1 egg yolk, beaten with 15ml/1 tbsp
      water
For the sauce
    40g/1½oz/scant ½ cup plain flour
    40g/1½oz/3 tbsp butter
    300ml/½ pint milk
    75g/3oz Cheddar cheese, grated
    30ml/2 tbsp dry sherry
    good pinch dried sage
    salt and ground black pepper
    45ml/3 tbsp fresh parsley, chopped

2 Blanch the Brussels sprouts for 4 minutes in 300ml/½ pint/1¼ cups boiling water then drain, reserving the water. Refresh the sprouts under cold running water.

3 Cut each chestnut in half. Lightly fry the red pepper and onion in the oil for 5 minutes. Set aside till later.

6 Fit the larger piece of pastry into your pie dish and layer the sprouts, chestnuts, peppers and onions on top. Trickle over the sauce, making sure it seeps through to wet the vegetables.

7 Brush the pastry edges with beaten egg yolk and fit the second pastry sheet on top, pressing the edges well to seal them.

8 Crimp, knock up the edges then mark the centre. Glaze well all over with egg yolk. Set aside to rest somewhere cool while you preheat the oven to 200°C/400°F/Gas 6. Bake for 30–40 minutes until golden brown and crisp.

1 Roll out the pastry to make two large rectangles, roughly the size of your dish. The pastry should be about 6mm/¼in thick and one rectangle should be slightly larger than the other. Set the pastry aside in the refrigerator.

4 Make up the sauce by beating the flour, butter and milk together over a medium heat. Beat the sauce continuously, bringing it to the boil, stirring until it is thickened and smooth.

5 Stir in the reserved sprout water, and the cheese, sherry, sage and seasoning. Simmer for 3 minutes to reduce and mix in the parsley.

SHEPHERDESS PIE

*A NO-MEAT VERSION OF THE
TIMELESS CLASSIC, THIS DISH
CONTAINS NO DAIRY PRODUCTS,
SO IS SUITABLE FOR VEGANS.*

SERVES SIX TO EIGHT

INGREDIENTS
    1kg/2lb potatoes
    45ml/3 tbsp extra virgin olive oil
    salt and ground black pepper
    1 large onion, chopped
    1 green pepper, chopped
    2 carrots, coarsely grated
    2 garlic cloves
    45ml/3 tbsp sunflower oil or margarine
    115g/4oz mushrooms, chopped
    2 × 400g/14oz cans aduki beans
    600ml/1 pint/2½ cups stock
    5ml/1 tsp vegetable yeast extract
    2 bay leaves
    5ml/1 tsp dried mixed herbs
    dried breadcrumbs or chopped nuts,
        to sprinkle

1 Boil the potatoes in the skins until tender, then drain, reserving a little of the water to moisten them.

2 Mash well, mixing in the olive oil and seasoning until you have a smooth purée. (Potatoes are easier to peel when boiled in their skins. This also preserves vitamins.)

3 Gently fry the onion, pepper, carrots and garlic in the sunflower oil or margarine for about 5 minutes until they are soft. Preheat the grill.

4 Stir in the mushrooms and drained beans and cook for a further 2 minutes, then add the stock, yeast extract, bay leaves and mixed herbs. Simmer for 15 minutes.

5 Remove the bay leaves and empty the vegetables into a shallow ovenproof dish. Spoon on the potatoes in dollops and sprinkle over the crumbs or nuts. Grill until golden brown.

MAGNIFICENT MARROW

At autumn time, marrows — with their wonderful green and cream stripes — look so attractive and tempting. They make delicious and inexpensive main courses, just right for a satisfying family Sunday lunch.

SERVES FOUR TO SIX

INGREDIENTS

250g/8oz/3 cups pasta shells
1.5–1.75kg/3–4lb marrow
1 onion, chopped
1 pepper, seeded and chopped
15ml/1 tbsp fresh root ginger, grated
2 garlic cloves, crushed
45ml/3 tbsp sunflower oil
4 large tomatoes, skinned and chopped
salt and ground black pepper
50g/2oz/½ cup pine nuts
15ml/1 tbsp fresh basil, chopped
cheese, grated, to serve (optional)

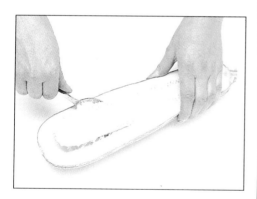

1 Preheat the oven to 190°C/375°F/ Gas 5. Boil the pasta according to the instructions on the pack, slightly overcooking it so it is just a little soft. Drain and reserve.

2 Cut the marrow in half lengthways and scoop out the seeds. These can be discarded. Use a small sharp knife and tablespoon to scoop out the marrow flesh. Chop the flesh roughly.

3 Gently fry the onion, pepper, ginger and garlic in the oil for 5 minutes.

4 Add the marrow flesh, tomatoes and seasoning. Cover and cook for 10–12 minutes until the vegetables are soft. Add to the pan the pasta, pine nuts and basil, stir well and set aside.

5 Meanwhile, place the marrow halves in a roasting pan, season lightly and pour a little water around the marrow, taking care it does not spill inside. Cover with foil and bake for 15 minutes.

6 Remove the foil, discard the water and fill the shells with the vegetable mixture. Recover with foil and return to the oven for a further 20–25 minutes.

7 If you wish, serve this dish topped with grated cheese. The marrow can either be served cut into sections or scooped out of the "shell".

GREEK SPINACH PIES

THESE LITTLE HORNS OF FILO PASTRY ARE STUFFED WITH A SIMPLE SPINACH AND FETA CHEESE FILLING TO MAKE A QUICK AND EASY MAIN COURSE.

SERVES EIGHT

INGREDIENTS
225g/8oz fresh leaf spinach
2 spring onions, chopped
175g/6oz feta cheese, crumbled
1 egg, beaten
15ml/1 tbsp fresh dill, chopped
ground black pepper
4 large sheets or 8 small sheets of filo
    pastry
olive oil, for brushing

1 Preheat the oven to 190°C/375°F/ Gas 5. Blanch the spinach in the tiniest amount of water until just wilted, then drain very well, pressing it through a sieve with the back of a wooden spoon.

2 Chop the spinach finely and mix with the onions, feta, egg, dill and ground black pepper.

3 Lay out a sheet of filo pastry and brush with olive oil. If large, cut the pieces in two and sandwich them together. If small, fit another sheet on top and brush with olive oil.

4 Spread a quarter of the filling on one edge of the filo at the bottom, then roll it up firmly, but not too tightly. Shape into a crescent and place on a baking sheet.

5 Brush the pastry well with oil and bake for about 20–25 minutes in the preheated oven until golden and crisp. Cool slightly then remove to a wire rack to cool further.

CHUNKY VEGETABLE PAELLA

THE SPANISH RICE DISH HAS BECOME A FIRM FAMILY FAVOURITE THE WORLD OVER.

SERVES SIX

INGREDIENTS
good pinch saffron strands
1 aubergine, cut in thick chunks
salt
90ml/6 tbsp olive oil
1 large onion, sliced
3 garlic cloves, crushed
1 yellow pepper, sliced
1 red pepper, sliced
10ml/2 tsp paprika
225g/8oz/1¼ cups risotto rice
600ml/1 pint/2½ cups stock
450g/1lb fresh tomatoes
ground black pepper
115g/4oz sliced mushrooms
115g/4oz cut green beans
1 × 400g/14oz can chick peas

1 Steep the saffron in 45ml/3 tbsp hot water. Sprinkle the aubergine with salt, leave to drain in a colander for 30 minutes, then rinse and dry.

2 In a large paella or frying pan, heat the oil and fry the onion, garlic, peppers and aubergine for about 5 minutes, stirring occasionally. Sprinkle in the paprika and stir again.

3 Mix in the rice, then pour in the stock, tomatoes, skinned and chopped, saffron and seasoning. Bring to a boil then simmer for 15 minutes, uncovered, shaking the pan frequently and stirring occasionally.

4 Stir in the mushrooms, green beans and chick peas (with the liquor). Continue cooking for a further 10 minutes, then serve hot from the pan.

GREEN LENTIL KULBYAKA

This traditional Russian fish dish can be adapted to make a light, crisp vegetarian centrepiece.

<u>SERVES SIX</u>

INGREDIENTS
175g/6oz/1 cup green lentils
2 bay leaves
2 onions, sliced
1.2 litres/2 pints/5 cups stock
175g/6oz/¾ cup butter, melted
225g/8oz/1¼ cups long grain rice
salt and ground black pepper
60ml/4 tbsp fresh parsley, chopped
30ml/2 tbsp fresh dill, chopped
1 egg, beaten
225g/8oz mushrooms, sliced
about 8 sheets filo pastry
3 eggs, hard boiled and sliced

1 Soak the lentils for 30 minutes, drain them, then simmer with the bay leaves, one onion and half the stock for 25 minutes until cooked and thick. Season well, cool and set aside.

2 Gently fry the remaining onion in another saucepan with 30ml/2 tbsp of the butter for 5 minutes. Stir in the rice then the rest of the stock.

3 Season, bring to the boil, then cover and cook gently for 12 minutes (for basmati) or 15 minutes (for long grain). Leave to stand, uncovered for 5 minutes, then stir in the fresh herbs. Cool, then beat in the raw egg.

4 Fry the mushrooms in 45ml/3 tbsp of the butter for 5 minutes until they are just soft. Cool and set aside.

5 Brush the inside of a large, shallow ovenproof dish with more butter. Lay the sheets of filo in it, covering the base and making sure most of the pastry overhangs the sides. Brush well with butter in between and overlapping the pastry as required. Make sure there is a lot of pastry to fold over the mounded filling.

6 Into the pastry lining, layer rice, lentils and mushrooms, repeating the layers at least once and tucking the sliced egg in between. Season as you layer and form an even mound of filling.

7 Bring up the sheets of pastry over the filling, scrunching the top into attractive folds. Brush all over with the rest of the butter and set aside to chill and firm up.

8 Preheat the oven to 190°C/375°F/Gas 5. When ready, bake the Kulbyaka for about 45 minutes until golden and crisp. Allow to stand for 10 minutes before you cut it and serve.

TURNIP AND CHICK-PEA COBBLER

A GOOD MID-WEEK MEAL.

SERVES FOUR TO SIX

INGREDIENTS
  1 onion, sliced
  2 carrots, chopped
  3 medium size turnips, chopped
  1 small sweet potato or swede, chopped
  2 celery sticks, sliced thinly
  45ml/3 tbsp sunflower oil
  2.5ml/½ tsp ground coriander
  2.5ml/½ tsp dried mixed herbs
  1 × 425g/15oz can chopped tomatoes
  1 × 400g/14oz can chick-peas
  1 vegetable stock cube
  salt and ground black pepper
For the topping
  225g/8oz/2 cups self-raising flour
  5ml/1 tsp baking powder
  50g/2oz/4 tbsp margarine
  45ml/3 tbsp sunflower seeds
  30ml/2 tbsp Parmesan cheese, grated
  150ml/¼ pint/⅔ cup milk

1 Fry all the vegetables in the oil for about 10 minutes until they are soft. Add the coriander, herbs, tomatoes, chick peas with their liquor and stock cube. Season well and simmer for 20 minutes.

2 Pour the vegetables into a shallow casserole dish while you make the topping. Preheat the oven to 190°C/375°F/Gas 5.

3 Mix together the flour and baking powder then rub in the margarine until it resembles fine crumbs. Stir in the seeds and Parmesan cheese. Add the milk and mix to a firm dough.

4 Lightly roll out the topping to a thickness of 1cm/½ in and stamp out star shapes or rounds, or simply cut it into small squares.

5 Place the shapes on top of the vegetable mixture and brush with a little extra milk. Bake for 12–15 minutes until risen and golden brown. Serve hot with green, leafy vegetables.

TANGY FRICASSEE

VEGETABLES IN A LIGHT TANGY SAUCE AND COVERED WITH A CRISPY CRUMB TOPPING MAKE A SIMPLE AND EASY MAIN COURSE TO SERVE WITH CRUSTY BREAD AND SALAD.

<u>SERVES FOUR</u>

INGREDIENTS
   4 courgettes, sliced
   115g/4oz green beans, sliced
   4 large tomatoes, skinned and sliced
   1 onion, sliced
   50g/2oz/4 tbsp butter or sunflower
      margarine
   40g/1½oz/⅓ cup plain flour
   10ml/2 tsp coarse grain mustard
   450ml/¾ pint/2 cups milk
   150ml/¼ pint/⅔ cup natural yogurt
   5ml/1 tsp dried thyme
   115g/4oz mature cheese, grated
   salt and ground black pepper
   60ml/4 tbsp fresh wholemeal
      breadcrumbs tossed with 15ml/1 tbsp
      sunflower oil

1 Blanch the courgettes and beans in a small amount of boiling water for just 5 minutes, then drain and arrange in a shallow ovenproof dish. Arrange all but three slices of tomato on top. Put the onion into a saucepan with the butter or margarine and fry gently for 5 minutes.

2 Stir in the flour and mustard, cook for a minute then add the milk gradually until the sauce has thickened. Simmer for a further 2 minutes.

3 Remove the pan from the heat, add the yogurt, thyme and cheese, stirring until melted. Season to taste. Reheat gently if you wish, but do not allow the sauce to boil or it will curdle.

4 Pour the sauce over the vegetables and scatter the breadcrumbs on top. Brown under a preheated grill until golden and crisp, taking care not to let them burn. Garnish with the reserved tomato slices if desired.

CHILLI CON QUESO

THIS CLASSIC MEXICAN DISH IS JUST AS TASTY WHEN MADE WITH ALL-RED BEANS. FOR AN EXTRA GOOD FLAVOUR, USE SMALL CUBES OF SMOKED CHEESE, AND SERVE WITH RICE.

SERVES FOUR

INGREDIENTS
  225g/8oz/2 cups red kidney beans, soaked and drained
  45ml/3 tbsp sunflower oil
  1 onion, chopped
  1 red pepper, chopped
  2 garlic cloves, crushed
  1 fresh red chilli, chopped (optional)
  15ml/1 tbsp chilli powder (mild or hot)
  5ml/1 tsp ground cumin
  1 litre/1¾ pints/4 cups stock or water
  5ml/1 tsp crushed dried epazote leaves (optional)
  ground black pepper and salt
  15g/½oz granulated sugar
  115g/4oz cheese, cubed, to serve

1 Rinse the beans. In a large saucepan heat the oil and fry the onion, pepper, garlic and fresh chilli for about 5 minutes.

2 Stir in the spices and cook for another minute, then add the beans, stock or water, epazote (if using) and a grinding of pepper. Don't add salt at this stage. Boil for 10 minutes, cover and turn down to a gentle simmer. Cook for about 50 minutes checking the water level and adding extra if necessary.

3 When the beans are tender, season them well with salt. Remove about a quarter of the mixture and mash to a pulp.

4 Return the purée to the pan and stir well. Add sugar and serve hot with the cheese sprinkled on top. Great with plain boiled long grain rice.

COOK'S TIP
Epazote is a traditional Mexican herb, found in specialist stores.

BIG BARLEY BOWL

BARLEY SEEMS TO HAVE SLIPPED FROM FASHION IN RECENT YEARS — A PITY AS IT HAS A DELICIOUS, NUTTY TEXTURE.

SERVES SIX

INGREDIENTS
  1 red onion, sliced
  ½ fennel bulb, sliced
  2 carrots, cut in sticks
  1 parsnip, sliced
  45ml/3 tbsp sunflower oil
  115g/4oz/1 cup pearl barley
  1 litre/1.2 pints/4 cups stock
  5ml/1 tsp dried thyme
  fresh parsley, chopped, to garnish
  salt and ground black pepper
  115g/4oz/⅔ cup green beans, sliced
  1 × 425g/15oz can pinto beans
For the croûtes
  1 medium sized baguette, sliced
  olive oil, for brushing
  1 garlic clove, cut in half
  60ml/4 tbsp grated Parmesan cheese

1 In a large, heatproof casserole, sauté the onion, fennel, carrots and parsnip gently in the oil for 10 minutes.

2 Stir in the barley and stock. Bring to a boil, add the herbs and seasoning, then cover and simmer gently for 40 minutes.

3 Stir in the green beans and drained pinto beans and continue cooking – covered – for a further 20 minutes.

4 Meanwhile, preheat the oven to 190°C/375°F/Gas 5. Brush the baguette slices with olive oil and place them on a baking sheet.

5 Bake for about 15 minutes until light golden and crisp. Remove from the oven and quickly rub each croûte with the garlic halves. Sprinkle over the cheese and return to the oven to melt.

6 Ladle the barley into warm bowls and serve sprinkled with parsley, accompanied by the cheesy croûtes. This dish is nicest eaten with a spoon.

VEGETABLES JULIENNE WITH A RED PEPPER COULIS

JUST THE RIGHT COURSE FOR
THOSE WATCHING THEIR WEIGHT.
CHOOSE A SELECTION OF AS
MANY VEGETABLES AS YOU FEEL
YOU CAN EAT.

SERVES TWO

INGREDIENTS
A selection of vegetables (choose from:
carrots, turnips, asparagus, parsnips,
courgettes, green beans, broccoli,
salsify, cauliflower, mangetouts)
For the red pepper coulis
1 small onion, chopped
1 garlic clove, crushed
15ml/1 tbsp sunflower oil
15ml/1 tbsp water
3 red peppers, roasted and skinned
120ml/8 tbsp fromage frais
squeeze of fresh lemon juice
salt and ground black pepper
sprigs of fresh rosemary and thyme
2 bay leaves
fresh green herbs, to garnish

2 Make the coulis. Lightly sauté the onion and garlic in the oil and water for 3 minutes then add the peppers and cook for a further 2 minutes.

3 Pass the coulis through a food processor, then work in the fromage frais, lemon juice and seasoning.

1 Prepare the vegetables by cutting them into thin fingers or small, bite size pieces.

4 Boil some salted water with the fresh rosemary, thyme and bay leaves, and fit a steamer over the top.

5 Arrange the prepared vegetables on the steamer, placing the harder root vegetables at the bottom and steaming these for about 3 minutes.

6 Add the other vegetables according to their natural tenderness and cook for a further 2–4 minutes.

7 Serve the vegetables on plates with the sauce to one side. Garnish with fresh green herbs, if you wish.

VARIATION
The red pepper coulis makes a wonderful sauce for many other dishes. Try it spooned over fresh pasta with lightly steamed or fried courgettes, or use it as a pouring sauce for savoury filled crêpes.

PISTACHIO PILAFF IN A SPINACH CROWN

SAFFRON AND GINGER ARE
DELICIOUS WHEN MIXED WITH
FRESH PISTACHIO NUTS. THIS IS
A GOOD, LIGHT MAIN COURSE.

SERVES FOUR

INGREDIENTS
3 onions
60ml/4 tbsp olive oil
2 garlic cloves, crushed
2.5cm/1in cube fresh root ginger
1 fresh green chilli, chopped
2 carrots, coarsely grated
225g/8oz/1¼ cups basmati rice
1.25ml/¼ tsp saffron strands, crushed
450ml/¾ pint/2 cups stock
1 cinnamon stick
5ml/1 tsp ground coriander
salt and ground black pepper
5g/3oz/¼ cup fresh pistachio nuts
450g/1 lb fresh leaf spinach
5ml/1 tsp garam masala powder

1 Roughly chop two of the onions. Heat half the oil in a large saucepan and fry the onion with half the garlic, the grated ginger and the chilli for 5 minutes.

2 Mix in the carrots and rinsed rice, cook for 1 more minute and then add the saffron, stock, cinnamon and coriander. Season well. Bring to a boil, then cover and simmer gently for 10 minutes, without lifting the lid.

3 Remove from the heat and leave to stand, uncovered, for 5 minutes. Add the pistachio nuts, mixing them in with a fork. Remove the cinnamon stick and keep the rice warm.

4 Thinly slice the third onion and fry in the remaining oil for about 3 minutes. Stir in the spinach. Cover and cook for another 2 minutes.

5 Add the garam masala powder. Cook until just tender then drain and roughly chop the spinach.

6 Spoon the spinach round the edge of a round serving dish and pile the pilaff in the centre. Serve hot, accompanied by a simple tomato salad.

SPINACH BREAD AND BUTTER BAKE

IDEALLY USE CIABATTA BREAD FOR THIS, OR FAILING THAT, A FRENCH-STYLE BAGUETTE.

SERVES FOUR TO SIX

INGREDIENTS

400g/14oz fresh leaf spinach
1 ciabatta loaf, thinly sliced
50g/2oz/4 tbsp softened butter, olive oil or low fat spread
1 red onion, thinly sliced
115g/4oz mushrooms, thinly sliced
30ml/2 tbsp olive oil
5ml/1 tsp cumin seeds
salt and ground black pepper
115g/4oz Gruyère cheese, grated
3 eggs
500ml/18fl oz/2¼ cups milk
fresh nutmeg, grated

1 Rinse the spinach well and blanch it in the tiniest amount of water for 2 minutes. Drain well, pressing out any excess water and chop roughly.

2 Spread the bread slices thinly with the butter or low fat spread. Grease a large shallow ovenproof dish and line the base and sides with bread.

3 Fry the onion and mushrooms lightly in the oil for 5 minutes then add the cumin seeds and spinach. Season well.

4 Layer the spinach mixture with the remaining bread and half the cheese. For the top, mix everything together and sprinkle over the remaining cheese.

5 Beat the eggs with the milk, adding seasoning and nutmeg to taste. Pour slowly over the whole dish and set aside for a good hour to allow the custard to be absorbed into the bread.

6 Preheat the oven to 190°C/375°F/Gas 5. Stand the dish in a roasting pan and pour around some boiling water for a bain marie. Bake for 40–45 minutes until risen, golden brown and crispy on top.

CURRIED PARSNIP PIE

SWEET, CREAMY PARSNIPS ARE
BEAUTIFULLY COMPLEMENTED BY
THE CURRY SPICES AND CHEESE.

SERVES FOUR

INGREDIENTS
For the pastry
    115g/4oz/8 tbsp butter or margarine
    225g/8oz/1 cup plain flour
    salt and ground black pepper
    5ml/1 tsp dried thyme or oregano
    cold water, to mix
For the filling
    8 baby onions, or shallots, peeled
    2 large parsnips, thinly sliced
    2 carrots, thinly sliced
    25g/1oz/2 tbsp butter or margarine
    30ml/2 tbsp wholemeal flour
    15ml/1 tbsp mild curry or tikka paste
    300ml/½ pint/1¼ cups milk
    115g/4oz mature cheese, grated
    salt and ground black pepper
    45ml/3 tbsp fresh coriander or parsley
    1 egg yolk, beaten with 10ml/2 tsp
      water

2 Blanch the baby onions or shallots with the parsnips and carrots in just enough water to cover, for about 5 minutes. Drain, reserving about 300ml/½ pint/1¼ cups of the liquid.

3 In a clean pan, melt the butter or margarine, and stir in the flour and spice paste to make a roux. Gradually whisk in the reserved stock and milk until smooth. Simmer for a minute or two.

1 Make the pastry by rubbing the butter or margarine into the flour until it resembles fine breadcrumbs. Season and stir in the thyme or oregano, then mix to a firm dough with cold water.

4 Take the pan off the heat, stir in the cheese and seasoning, then mix in the chopped coriander or parsley.

5 Pour into a pie dish, fix a pie funnel in the centre and allow to cool.

6 Roll out the pastry, large enough to fit the top of the pie dish. Re-roll the trimmings into long strips.

7 Brush the pastry edges with egg yolk wash and fit on the pastry strips. Brush again with egg yolk wash.

8 Using a rolling pin, lift the rolled out pastry over the pie top and fit over the funnel, pressing it down well onto the strips underneath.

9 Cut off the overhanging pastry and crimp the edges. Cut a hole for the funnel, brush all over well with the remaining egg yolk wash and make decorations with the trimmings, glazing them too.

10 Place the pie dish on a baking sheet and chill for 30 minutes while you preheat the oven to 200°C/400°F/Gas 6. Bake the pie for about 25–30 minutes until golden brown and crisp on top.

COOK'S TIP
This pie freezes well and makes a good stand-by for a mid-week meal. For best results, make up to the final stage and freeze unbaked. Open freeze until solid, then wrap well in freezer cling film, seal and label. Use within one month.
    Cauliflower, broccoli or any other favourite vegetable can be added to the filling to give a variety of flavours and textures.

VEGETABLES UNDER A LIGHT CREAMY CRUST

THE SUBTLE FLAVOURS OF LEEKS, COURGETTES AND MUSHROOMS ARE TOPPED BY FROMAGE FRAIS, CHEESE AND BREADCRUMBS.

SERVES FOUR

INGREDIENTS
    2 leeks, thinly sliced
    3 courgettes, thickly sliced
    350g/12oz sliced mushrooms
    2 garlic cloves, crushed
    30ml/2 tbsp olive oil
    25g/1oz/2 tbsp butter
    15ml/1 tbsp plain flour
    300ml/½ pint/1¼ cups stock
    5ml/1 tsp dried thyme
    salt and ground black pepper
    30ml/2 tbsp fromage frais
For the topping
    450g/1lb fromage frais
    25g/1oz/2 tbsp butter, melted
    3 eggs, beaten
    salt and ground black pepper
    fresh nutmeg, grated
    freshly grated cheese and dried
      breadcrumbs, to sprinkle

1 Preheat the oven to 190°C/375°F/ Gas 5. In a saucepan, gently fry the leeks, courgettes, mushrooms and garlic in the oil and butter for about 7 minutes, stirring occasionally, until the vegetables are just soft.

2 Stir in the flour then gradually mix in the stock. Bring to a boil, stirring until thickened. Add the thyme and seasoning. Take the pan off the heat and stir in the fromage frais. Pour the vegetable mixture into a shallow ovenproof dish.

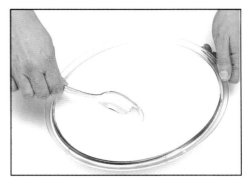

3 Beat the topping ingredients together, seasoning them well and adding the nutmeg to taste. Spoon on top of the vegetables and sprinkle with cheese and breadcrumbs.

4 Bake for about 30 minutes until a light golden, firm crust forms. Serve hot with pasta or crusty bread.

POTATO AND PARSNIP AMANDINE

SHELLS OF BAKED POTATOES, FILLED WITH A PARSNIP AND ALMOND MIX, MAKE AN UNUSUAL ALTERNATIVE TO PLAIN JACKET POTATOES.

SERVES FOUR

INGREDIENTS
    4 large baking potatoes
    olive oil, for greasing
    225g/8oz parsnips, diced
    25g/1oz/2 tbsp butter
    5ml/1 tsp cumin seeds
    5ml/1 tsp ground coriander
    30ml/2 tbsp single cream or natural
      yogurt
    salt and ground black pepper
    115g/4oz Gruyère or Cheddar cheese,
      grated
    1 egg, beaten
    50g/2oz/¼ cup flaked almonds

1 Rub the potatoes all over with oil, score in half, then bake at 200°C/400°F/Gas 6 for about 1 hour until cooked.

2 Meanwhile, boil the parsnips until tender, then drain well, mash and mix with the butter, spices and cream or natural yogurt.

3 When the potatoes are cooked, halve, scoop out and mash the flesh then mix with the parsnip, seasoning well.

4 Stir in the cheese, egg and three-quarters of the almonds. Fill the potato shells with the mixture and sprinkle over the remaining almonds.

5 Return to the oven and bake for about 15–20 minutes until golden brown and the filling has set lightly. Serve hot with a side salad.

AUTUMN GLORY

PUMPKIN AND PASTA MAKE
MARVELLOUS PARTNERS,
ESPECIALLY AS A MAIN COURSE
SERVED FROM THE BAKED SHELL.

SERVES FOUR

INGREDIENTS
    1 × 2kg/4lb pumpkin
    1 onion, sliced
    2.5cm/1in cube fresh root ginger
    45ml/3 tbsp extra virgin olive oil
    1 courgette, sliced
    115g/4oz sliced mushrooms
    1 × 400g/14oz can chopped tomatoes
    75g/3oz/1 cup pasta shells
    450ml/¾ pint/2 cups stock
    salt and ground black pepper
    60ml/4 tbsp fromage frais
    30ml/2 tbsp fresh basil, chopped

1 Preheat the oven to 180°C/350°F/Gas 4. Cut the top off the pumpkin with a large, sharp knife and scoop out and discard the seeds.

2 Using a small sharp knife and a sturdy tablespoon extract as much of the pumpkin flesh as possible, then chop it into chunks.

3 Bake the pumpkin with its lid on for 45 minutes to one hour until the inside begins to soften.

4 Meanwhile, make the filling. Gently fry the onion, ginger and pumpkin flesh in the olive oil for about 10 minutes, stirring occasionally.

5 Add the courgette and mushrooms and cook for a further 3 minutes, then stir in the tomatoes, pasta shells and stock. Season well, bring to a boil, then cover and simmer gently for 10 minutes.

6 Stir the fromage frais and basil into the pasta and spoon the mixture into the pumpkin. (It may not be possible to fit all the filling into the pumpkin shell, so serve the rest separately if this is the case.)

FESTIVE LENTIL AND NUT ROAST

SERVE WITH VEGETARIAN GRAVY, CRANBERRIES AND FRENCH PARSLEY.

<u>SERVES SIX TO EIGHT</u>

INGREDIENTS
115g/4oz/⅔ cup red lentils
115g/4oz/1 cup hazelnuts
115g/4oz/1 cup walnuts
1 large carrot
2 celery sticks
1 large onion, sliced
115g/4oz mushrooms
50g/2oz/4 tbsp butter
10ml/2 tsp mild curry powder
30ml/2 tbsp tomato ketchup
30ml/2 tbsp Worcestershire sauce
1 egg, beaten
10ml/2 tsp salt
60ml/4 tbsp fresh parsley, chopped
150ml/¼ pint/⅔ cup water

1 Soak the lentils for 1 hour in cold water then drain well. Grind the nuts in a food processor until quite fine but not too smooth. Set the nuts aside.

2 Chop the carrot, celery, onion and mushrooms into small chunks, then pass them through a food processor or blender until they are quite finely chopped.

3 Fry the vegetables gently in the butter for 5 minutes then stir in the curry powder and cook for a minute. Cool.

4 Mix the lentils with the nuts, vegetables and remaining ingredients.

5 Grease and line the base and sides of a long 1 kg/2 lb loaf tin with greaseproof paper or a sheet of foil. Press the mixture into the tin. Preheat the oven to 190°C/375°F/Gas 5.

6 Bake for about 1–1¼ hours until just firm, covering the top with a butter paper or piece of foil if it starts to burn. Allow the mixture to stand for about 15 minutes before you turn it out and peel off the paper. It will be fairly soft when cut as it is a moist loaf.

VEGETARIAN GRAVY

MAKE UP A LARGE BATCH AND FREEZE IT IN SMALL CONTAINERS READY TO REHEAT AND SERVE.

<u>MAKES ABOUT 1 LITRE/1¾ PINTS</u>

INGREDIENTS
1 large red onion, sliced
3 turnips, sliced
3 celery sticks, sliced
115g/4oz open cut mushrooms, halved
2 whole garlic cloves
90ml/6 tbsp sunflower oil
1.5 litres/2½ pints/6 cups vegetable
   stock or water
45ml/3 tbsp soy sauce
good pinch of granulated sugar
salt and ground black pepper

1 Cook the vegetables and garlic on a moderately high heat with the oil in a large saucepan, stirring occasionally until nicely browned but not singed. This should take about 15-20 minutes.

2 Add the stock or water and soy sauce, bring to the boil then cover and simmer for another 20 minutes.

3 Purée the vegetables, adding a little of the stock, and return them to the pan by rubbing the pulp through a sieve with the back of a ladle or wooden spoon.

4 Taste for seasoning and add the sugar. Freeze at least half of the gravy to use later and reheat the rest to serve with the rice and peas or lentil and nut roast.

HOME-MADE RAVIOLI

IT IS A PLEASURE TO MAKE YOUR OWN FRESH PASTA AND YOU WILL BE SURPRISED AT HOW EASY IT IS TO FILL AND SHAPE RAVIOLI. ALLOW A LITTLE MORE TIME THAN YOU WOULD FOR READY-MADE OR DRIED PASTA.

SERVES SIX

INGREDIENTS
  200g/7oz/1½ cups strong plain flour
  5ml/½ tsp salt
  15ml/1 tbsp olive oil
  2 eggs, beaten
For the filling
  1 small red onion, finely chopped
  1 small green pepper, finely chopped
  1 carrot, coarsely grated
  15ml/1 tbsp olive oil
  50g/2oz/½ cup walnuts, chopped
  115g/4oz ricotta cheese
  30ml/2 tbsp fresh Parmesan or
    Pecorino cheese, grated
  15ml/1 tbsp fresh marjoram or basil
  salt and ground black pepper
  extra oil or melted butter, to serve

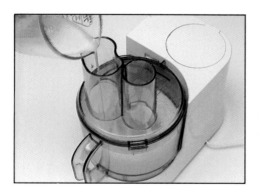

1 Sift the flour and salt into a food processor. With the machine running, trickle in the oil and eggs and blend to a stiff but smooth dough.

2 Allow the machine to run for at least a minute if possible, otherwise remove the dough and knead it by hand for 5 minutes.

3 If you are using a pasta machine, break off small balls of dough and then feed them through the rollers, several times, according to the machine manufacturer's instructions.

4 If rolling the pasta by hand, divide the dough into two and roll out on a lightly floured surface to a thickness of about 6mm/¼in.

5 Fold the pasta into three and re-roll. Repeat this up to six times until the dough is smooth and no longer sticky. Roll the pasta slightly more thinly each time.

6 Keep the rolled dough under clean, dry tea towels while you complete the rest and make the filling. You should aim to have an even number of pasta sheets, all the same size if rolling by machine.

7 Fry the onion, pepper and carrot in the oil for 5 minutes, then allow to cool. Mix with the walnuts, cheeses, chopped herbs and seasoning.

COOK'S TIP
A food processor will save you time and effort in making and kneading the dough. A pasta rolling machine helps with the rolling out, but both these jobs can be done by hand if necessary.

8 Lay out a pasta sheet and place small scoops of the filling in neat rows about 5cm/2in apart. Brush in between with a little water and then place another pasta sheet on the top.

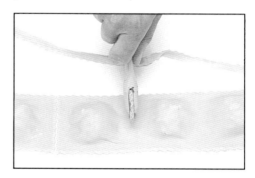

9 Press down well in between the rows then, using a ravioli or pastry cutter, cut into squares. If the edges pop open occasionally, press them back gently with your fingers.

10 Leave the ravioli to dry in the refrigerator, then boil in plenty of lightly salted water for just 5 minutes.

11 Toss the cooked ravioli in a little oil or melted butter before serving with either homemade tomato sauce or extra cheese.

MUSHROOM PUFF PASTIES

IF POSSIBLE, USE FULL-
FLAVOURED CHESTNUT, OR
BROWN, MUSHROOMS.

MAKES EIGHT

INGREDIENTS
    2 × 225g/8oz blocks frozen puff pastry
    1 egg, beaten
For the filling
    1 onion, chopped
    1 carrot, coarsely grated
    1 medium potato, coarsely grated
    45ml/3 tbsp sunflower oil
    225g/8oz sliced mushrooms
    30ml/2 tbsp soy sauce
    15ml/1 tbsp tomato ketchup
    15ml/1 tbsp dry sherry (optional)
    good pinch dried thyme
    salt and ground black pepper

1 Roll out the thawed pastry blocks until they are 6mm/¼in thick and cut each block into four 15cm/6in squares. Reserve a little pastry for decoration. Cover the rolled pastry squares and oddments and set aside in a cool place to rest.

2 Make the filling by gently frying the onion, carrot and potato in the oil for 5 minutes, then add the mushrooms, soy sauce, ketchup, sherry (if using), thyme and seasoning.

3 Cook, stirring occasionally, until the mushrooms and vegetables have softened and feel quite tender. Cool.

4 Divide the filling between the eight squares, placing it to one side across the diagonal. Brush the pastry edges with egg then fold over into triangles and press well to seal. From the pastry scraps, cut out little shapes, such as mushrooms, to decorate the pasties.

5 Crimp each pastie edge and top with the cut-out shapes. Set on two baking sheets. Preheat the oven to 200°C/400°F/ Gas 6 and in the meantime allow the pasties to rest somewhere cool.

6 Glaze the pasties with beaten egg, then bake for 15–20 minutes until golden brown and crisp.

WINTER CASSEROLE WITH HERB DUMPLINGS

When the cold weather draws in, gather together a good selection of vegetables and make this comforting casserole with some hearty old-fashioned dumplings.

SERVES SIX

INGREDIENTS

2 potatoes
2 carrots
1 small fennel bulb
1 small swede
2 leeks
2 courgettes
50g/2oz/4 tbsp butter or margarine
30ml/2 tbsp plain flour
1 × 425g/15oz can butter beans, with liquor
600ml/1 pint/2½ cups stock
30ml/2 tbsp tomato purée
1 cinnamon stick
10ml/2 tsp ground coriander
2.5ml/½ tsp ground ginger
2 bay leaves
salt and ground black pepper
For the dumplings
200g/7oz/1½ cups plain flour
115g/4oz vegetable suet, shredded, or chilled butter, grated
5ml/1 tsp dried thyme
5ml/1 tsp salt
120ml/4fl oz/½ cup milk

1 Cut all the vegetables into even, bite size chunks, then fry gently in the butter or margarine for about 10 minutes.

2 Stir in the flour then the liquor from the beans, the stock, tomato purée, spices, bay leaves and seasoning. Bring to the boil, stirring.

3 Cover and simmer for 10 minutes then add the beans and cook for a further 5 minutes.

4 Meanwhile, to make the dumplings, simply mix the flour, suet or butter, thyme and salt to a firm but moist dough with the milk and knead with your hands until it is smooth.

5 Divide the dough into 12 pieces, rolling each one into a ball with your fingers. Uncover the simmering stew and then add the dumplings, allowing space between each one for expansion.

6 Replace the lid and cook on a gentle simmer for a further 15 minutes. Do not peek – or you will let out all the steam. Neither should you cook dumplings too fast, or they will break up. Remove the cinnamon stick and bay leaves before you serve this dish, steaming hot.

LASAGNE ROLLS

PERHAPS A MORE ELEGANT PRESENTATION THAN ORDINARY LASAGNE, BUT JUST AS TASTY AND POPULAR. YOU WILL NEED TO BOIL "NO-NEED-TO-COOK" LASAGNE AS IT NEEDS TO BE SOFT ENOUGH TO ROLL!

SERVES FOUR

INGREDIENTS
    8–10 lasagne sheets
    lentil bolognese sauce (see below)
    225g/8oz fresh leaf spinach
    115g/4oz mushrooms, sliced
    115g/4oz Mozzarella cheese, thinly
      sliced
For the béchamel sauce
    40g/1½oz/scant ½ cup plain flour
    40g/1½oz/3 tbsp butter or margarine
    600ml/1 pint/2½ cups milk
    bay leaf
    salt and ground black pepper
    fresh nutmeg, grated
    freshly grated Parmesan or Pecorino
      cheese, to serve

1 Cook the lasagne sheets according to the instructions on the pack, or for about 10 minutes. Drain and allow to cool.

2 Cook the spinach in the tiniest amount of water for 2 minutes then add the sliced mushrooms and cook for a further 2 minutes. Drain very well, pressing out all the excess liquor, and chop roughly.

3 Put all the béchamel ingredients into a saucepan and bring slowly to a boil, stirring continuously until the sauce is thick and smooth. Simmer for 2 minutes with the bay leaf, then season well and stir in grated nutmeg to taste.

4 Lay out the pasta sheets and spread with the bolognese, spinach and mushrooms and mozzarella. Roll up each one and place in a large shallow casserole dish with the join face down.

5 Remove and discard the bay leaf and then pour the sauce over the pasta. Sprinkle over the cheese and place under a hot grill to brown.

LENTIL BOLOGNESE SAUCE

THIS IS A REALLY USEFUL SAUCE TO SERVE WITH PASTA, AS A PANCAKE STUFFING OR EVEN AS A PROTEIN-PACKED SAUCE FOR VEGETABLES.

SERVES SIX

INGREDIENTS
    1 onion, chopped
    2 garlic cloves, crushed
    2 carrots, coarsely grated
    2 celery sticks, chopped
    45ml/3 tbsp olive oil
    115g/4oz/⅔ cup red lentils
    1 × 400g/14oz can chopped tomatoes
    30ml/2 tbsp tomato purée
    450ml/¾ pint/2 cups stock
    15ml/1 tbsp fresh marjoram, chopped,
      or 5ml/1 tsp dried marjoram
    salt and ground black pepper

1 In a large saucepan, gently fry the onion, garlic, carrots and celery in the oil for about 5 minutes, until they are soft.

2 Add the lentils, tomatoes, tomato purée, stock, marjoram and seasoning.

3 Bring the mixture to a boil then partially cover with a lid and simmer for 20 minutes until thick and soft. Use the bolognese as required.

ARTICHOKE AND LEEK CRÊPES

THIN PANCAKES WITH A MOUTH-WATERING SOUFFLÉ MIXTURE OF ARTICHOKES AND LEEKS.

SERVES FOUR

INGREDIENTS
    115g/4oz/1 cup plain flour
    pinch of salt
    1 egg
    300ml/½ pint/1¼ cup milk
    oil, for brushing
For the soufflé filling
    450g/1lb Jerusalem artichokes
    1 large leek, sliced thinly
    50g/2oz/4 tbsp butter
    30ml/2 tbsp self-raising flour
    30ml/2 tbsp single cream
    75g/3oz mature Cheddar cheese
    30ml/2 tbsp fresh parsley, chopped
    fresh nutmeg, grated
    2 eggs, separated
    salt and ground black pepper

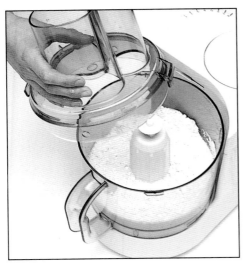

1 Make the crêpe batter by blending the flour, salt, egg and milk to a smooth batter in a food processor or blender.

2 Using a crêpe or omelette pan with a diameter of about 20cm/8in, make a batch of thin pancakes. you will need about 30ml/2 tbsp of batter for each one.

3 Stack the pancakes under a clean tea towel as you make them. Reserve eight for this dish and freeze the rest.

4 Cook the artichokes and leek with the butter in a covered saucepan on a gentle heat for about 12 minutes until very soft. Mash with the back of a wooden spoon. Season well.

5 Stir the flour into the vegetables and cook for 1 minute. Take the pan off the heat and beat in the cream, grated cheese, parsley and nutmeg to taste. Cool, then add the egg yolks.

6 Whisk the egg whites until they form soft peaks and carefully fold them into the leek/artichoke mixture.

7 Lightly grease a small ovenproof dish and preheat the oven to 190°C/375°F/ Gas 5. Fold each pancake in four, hold the top open and spoon the mixture into the centre.

8 Arrange the crêpes in the prepared dish with the filling uppermost if possible. Bake for about 15 minutes until risen and golden. Eat immediately!

COOK'S TIP
Make sure the pan is at a good steady heat and is well oiled before you pour in the batter. It should sizzle as it hits the pan. Swirl the batter round to coat the pan, and then cook quickly.

BROCCOLI RISOTTO TORTE

LIKE A SPANISH OMELETTE, THIS IS A SAVOURY CAKE SERVED IN WEDGES. IT IS GOOD COLD OR HOT, AND NEEDS ONLY A SALAD AS AN ACCOMPANIMENT.

SERVES SIX

INGREDIENTS
225g/8oz broccoli, cut into very small florets
1 onion, chopped
2 garlic cloves, crushed
1 large yellow pepper, sliced
30ml/2 tbsp olive oil
50g/2oz/4 tbsp butter
225g/8oz/1¼ cups risotto rice
120ml/4 floz/½ cup dry white wine
1 litre/1¾ pints/4½ cups stock
salt and ground black pepper
115g/4oz fresh or Parmesan cheese, coarsely grated
4 eggs, separated
oil, for greasing
sliced tomato and chopped parsley, to garnish

1 Blanch the broccoli for 3 minutes then drain and reserve.

2 In a large saucepan, gently fry the onion, garlic and pepper in the oil and butter for 5 minutes until they are soft.

3 Stir in the rice, cook for a minute then pour in the wine. Cook, stirring the mixture until the liquid is absorbed.

4 Pour in the stock, season well, bring to the boil then lower to a simmer. Cook for 20 minutes, stirring occasionally.

5 Meanwhile, grease a 25cm/10in round deep cake tin and line the base with a disc of greaseproof paper. Preheat the oven to 180°C/350°F/Gas 4.

6 Stir the cheese into the rice, allow the mixture to cool for 5 minutes, then beat in the egg yolks.

7 Whisk the egg whites until they form soft peaks and carefully fold into the rice. Turn into the prepared tin and bake for about 1 hour until risen, golden brown and slightly wobbly in the centre.

8 Allow the torte to cool in the tin, then chill if serving cold. Run a knife round the edge of the tin and shake out onto a serving plate. If liked, garnish with sliced tomato and chopped parsley.

LEEK AND CHÈVRE LASAGNE

An unusual and lighter than average lasagne using a soft French goats' cheese. The pasta sheets are not so chewy if boiled briefly first.

SERVES SIX

INGREDIENTS

6–8 lasagne pasta sheets
salt
1 large aubergine
3 leeks, thinly sliced
30ml/2 tbsp olive oil
2 red peppers, roasted and peeled
200g/7oz chèvre, broken into pieces
50g/2oz Pecorino or Parmesan cheese,
   freshly grated
For the sauce
60g/2½oz/½ cup plain flour
60g/2½oz/5 tbsp butter
900ml/1½ pints/3¾ cups milk
2.5ml/½ tsp ground bay leaves
fresh nutmeg, grated
ground black pepper

1 Blanch the pasta sheets in plenty of boiling water for just 2 minutes. Drain and place on a clean tea towel.

2 Lightly salt the aubergine and lay in a colander to drain for 30 minutes, then rinse and pat dry with kitchen paper.

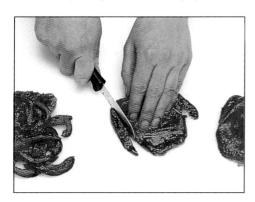

3 Lightly fry the leeks in the oil for about 5 minutes until softened. Peel the roasted peppers and cut into strips.

4 Make the sauce. Put the flour, butter and milk into a saucepan and bring to the boil, stirring constantly until it has thickened. Add the ground bay leaves, nutmeg and seasoning. Simmer for a further 2 minutes.

5 In a greased shallow casserole, layer the leeks, pasta, aubergine, chèvre and Pecorino or Parmesan. Trickle the sauce over the layers, ensuring that plenty goes in between.

6 Finish with a layer of sauce and grated cheese. Bake in the oven at 190°C/375°F/ Gas 5 for 30 minutes or until browned on top. Serve immediately.

GLAMORGAN SAUSAGES

This old Welsh recipe tastes particularly good served with creamy mashed potatoes and lightly cooked green cabbage.

SERVES FOUR

INGREDIENTS
   115g/4oz/2 cups fresh wholemeal
      breadcrumbs
   150g/6oz mature Cheddar or Caerphilly
      cheese, grated
   30ml/2 tbsp leek or spring onion, finely
      chopped
   30ml/2 tbsp fresh parsley, chopped
   15ml/1 tbsp fresh marjoram, chopped
   15ml/1 tbsp coarse grain mustard
   2 eggs, 1 separated
   ground black pepper
   50g/2oz/½ cup dried breadcrumbs
   oil, for deep fat frying

1 Mix the fresh breadcrumbs with the cheese, leek or onion, parsley, marjoram, mustard, whole egg, one egg yolk and ground black pepper to taste. The mixture may appear dry at first but knead it lightly with your fingers and it will come together. Make 8 small sausage shapes.

2 Whisk the egg white until lightly frothy and put the dried breadcrumbs into a bowl. Dip the sausages first into egg white, and then coat them evenly in breadcrumbs, shaking off any excess.

3 Heat a deep fat frying pan a third full of oil and carefully fry four sausages at a time for 2 minutes each. Drain on kitchen paper towel and reheat the oil to repeat.

4 Keep the sausages warm in the oven, uncovered. Alternatively, open freeze, bag and seal, then to reheat, thaw for 1 hour and cook in a moderately hot oven for 10–15 minutes.

BRAZILIAN STUFFED PEPPERS

Colourful and full of flavour, these stuffed peppers are easy to make in advance. They can be reheated quickly in a microwave oven and browned under a grill.

SERVES FOUR

INGREDIENTS
   4 peppers, halved and seeded
   1 aubergine, cut in chunks
   1 onion, sliced
   1 garlic clove, crushed
   30ml/2 tbsp olive oil
   1 × 400g/14oz can chopped tomatoes
   5ml/1 tsp ground coriander
   salt and ground black pepper
   15ml/1 tbsp fresh basil, chopped
   115g/4oz goats' cheese, coarsely grated
   30ml/2 tbsp dried breadcrumbs

1 Blanch the pepper halves in boiling water for 3 minutes, then drain well.

2 Sprinkle the aubergine chunks with salt, place in a colander and leave to drain for 20 minutes. Rinse and pat dry.

3 Fry the onion and garlic in the oil for 5 minutes until they are soft, then add the aubergine and cook for a further 5 minutes, stirring occasionally.

VARIATION
There are all sorts of delicious stuffings for peppers. Rice or pasta make a good base, mixed with some lightly fried onion, garlic and spices. Mixed nuts, finely chopped, can be added and a beaten egg or grated cheese helps to bind it all together. Vegans can leave out the last two ingredients.

4 Pour in the tomatoes, coriander and seasoning. Bring to the boil, then simmer for 10 minutes until the mixture is thick. Cool slightly, stir in the basil and half of the cheese.

5 Spoon into the pepper halves and place on a shallow heatproof serving dish. Sprinkle with cheese and breadcrumbs, then brown lightly under the grill. Serve with rice and salad.

MUSHROOM GOUGÈRE

A savoury choux pastry ring makes a marvellous main course dish that can be made ahead, then baked when required. Why not try it for a dinner party? It looks so very special.

SERVES FOUR

INGREDIENTS

   115g/4oz/½ cup strong plain flour
   2.5ml/½ tsp salt
   75g/3oz/6 tbsp butter
   200ml/7fl oz/¾ cup cold water
   3 eggs, beaten
   75g/3oz/¾ cup diced Gruyère or mature
     Gouda cheese
For the filling
   1 small onion, sliced
   1 carrot, coarsely grated
   225g/8oz button mushrooms, sliced
   40g/1½oz/3 tbsp butter or margarine
   5ml/1 tsp tikka or mild curry paste
   25g/1oz/2 tbsp plain flour
   300ml/½ pint/1¼ cups milk
   30ml/2 tbsp fresh parsley, chopped
   salt and ground black pepper
   30ml/2 tbsp flaked almonds

1 Preheat the oven to 200°C/400°F/ Gas 6. Grease a shallow ovenproof dish about 23cm/9in long.

2 To make the choux pastry, first sift the flour and salt onto a sheet of greaseproof paper. In a large saucepan, heat the butter and water until the butter melts. Do not let the water boil. Fold the paper and shoot the flour into the pan all at once.

3 With a wooden spoon, beat the mixture rapidly until the lumps become smooth and the mixture comes away from the sides of the pan. Cool for 10 minutes.

4 Beat the eggs gradually into the mixture until you have a soft, but still quite stiff, dropping consistency. You may not need all the egg.

5 Stir in the cheese, then spoon the mixture round the sides of the greased ovenproof dish.

6 To make the filling, sauté the onion, carrot and mushrooms in the butter or margarine for 5 minutes. Stir in the curry paste then the flour.

7 Gradually stir in the milk and heat until thickened. Mix in the parsley, season well, then pour into the centre of the choux pastry.

8 Bake for 35–40 minutes until risen and golden brown, sprinkling on the almonds for the last 5 minutes or so. Serve at once.

COOK'S TIP
Choux pastry is remarkably easy to make, as no rolling out is required. The secret of success is to let the flour and butter mixture cool before beating in the eggs, to prevent them from setting.

COUSCOUS-STUFFED CABBAGE

CUT INTO WEDGES AND SERVE ACCOMPANIED BY A FRESH TOMATO OR CHEESE SAUCE OR EVEN A VEGETARIAN GRAVY.

SERVES FOUR

INGREDIENTS
   1 medium size cabbage
   115g/4oz/1 cup couscous grains
   1 onion, chopped
   1 small red pepper, chopped
   2 garlic cloves, crushed
   30ml/2 tbsp olive oil
   5ml/1 tsp ground coriander
   2.5ml/½ tsp ground cumin
   good pinch ground cinnamon
   115g/4oz/½ cup green lentils, soaked
   600ml/1 pint/2½ cups stock
   30ml/2 tbsp tomato purée
   salt and ground black pepper
   30ml/2 tbsp fresh parsley, chopped
   30ml/2 tbsp pine nuts
   75g/3oz mature Cheddar cheese
   1 egg, beaten

1 Cut the top quarter off the cabbage and remove any loose outer leaves. Using a small sharp knife, cut out as much of the middle as you can. Reserve a few larger leaves for later.

2 Blanch the cabbage in a pan of boiling water for 5 minutes, then drain it well, upside down.

COOK'S TIP
A whole stuffed cabbage makes a wonderful main dish, especially for a Sunday lunch. It can be made ahead and steamed when required.

3 Steam the couscous according to the instructions on the pack, ensuring they are light and fluffy.

4 Lightly fry the onion, pepper and garlic in the oil for 5 minutes until soft then stir in the spices and cook for a further 2 minutes.

5 Add the lentils and pour in the stock and tomato purée. Bring to the boil, season and simmer for 25 minutes until the lentils are cooked.

6 Mix in the couscous, parsley, pine nuts, grated cheese and egg. Check the seasoning again. Open up the cabbage and spoon in the stuffing.

7 Blanch the leftover outer cabbage leaves and place these over the top of the stuffing, then wrap the whole thing in a sheet of buttered foil.

8 Place in a steamer over simmering water and cook for about 45 minutes. Remove from the foil and serve cut into wedges.

COUSCOUS AROMATIQUE

*MOROCCO AND TUNISIA HAVE
MANY WONDERFUL DISHES USING
COUSCOUS, WHICH IS STEAMED
OVER SIMMERING SPICY STEWS.*

SERVES FOUR TO SIX

INGREDIENTS

450g/1lb couscous grains
60ml/4 tbsp olive oil
1 onion, cut in chunks
2 carrots, cut in thick slices
4 baby turnips, halved
8 small new potatoes, halved
1 green pepper, cut in chunks
115g/4oz green beans, halved
1 small fennel bulb, sliced thickly
2.5cm/1in cube fresh root ginger
2 garlic cloves, crushed
5ml/1 tsp ground turmeric
15ml/1 tbsp ground coriander
5ml/1 tsp cumin seeds
5ml/1 tsp ground cinnamon
45ml/3 tbsp red lentils
1 × 400g/14oz can chopped tomatoes
1 litre/1¾ pints/4½ cups stock
60ml/4 tbsp raisins
salt and ground black pepper
rind and juice of 1 lemon
harissa paste, to serve (optional)

1 Cover the couscous with cold water
and soak for 10 minutes. Drain and
spread out on a tray for 20 minutes,
stirring it occasionally with your fingers.

2 Meanwhile, in a large saucepan, heat
the oil and fry the vegetables for about
10 minutes, stirring from time to time.

3 Add the grated ginger, garlic and
spices, stir well and cook for 2 minutes.
Pour in the lentils, tomatoes, stock and
raisins, and add seasoning.

4 Bring to a boil, then turn down to a
simmer. By this time the couscous should
be ready for steaming. Place in a steamer
and fit this on top of the stew.

5 Cover and steam gently for about 20
minutes. The grains should be swollen
and soft. Fork through and season well.
Spoon into a serving dish.

6 Add the lemon rind and juice to the
stew and check the seasoning. If liked,
add harissa paste to taste; it is quite hot
so beware! Serve the stew from a
casserole dish separately. Spoon the
couscous onto a plate and ladle the stew
on top.

CABBAGE ROULADES WITH LEMON SAUCE

CABBAGE OR CHARD LEAVES FILLED WITH RICE AND LENTILS MAKE A TASTY MAIN COURSE.

SERVES FOUR TO SIX

INGREDIENTS
    12 large cabbage or chard leaves
    salt
    30ml/2 tbsp sunflower oil
    1 onion, chopped
    1 large carrot, grated
    115g/4oz sliced mushrooms
    600ml/1 pint/2½ cups stock
    115g/4oz/½ cup long grain rice
    60ml/4 tbsp red lentils
    5ml/1 tsp dried oregano or marjoram
    ground black pepper
    100g/3½oz soft cheese with garlic
For the sauce
    30g/1oz/3 tbsp plain flour
    juice of 1 lemon
    3 eggs, beaten

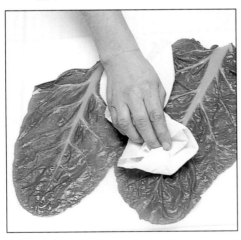

1 Remove the stalks and blanch the leaves in boiling, salted water until they begin to wilt. Drain, reserve the water and pat the leaves dry with kitchen paper.

2 Heat the oil and lightly fry the onion, carrot and mushrooms for 5 minutes, and then pour in the stock.

3 Add the rice, lentils, herbs and seasoning. Bring to a boil, cover and simmer gently for 15 minutes. Remove, then stir in the cheese. Preheat the oven to 190°C/375°F/Gas 5.

4 Lay out the chard or cabbage leaves rib side down, and spoon on the filling at the stalk end. Fold the sides in and roll up.

5 Place the join side down in a small roasting pan and pour in the reserved cabbage water. Cover with lightly greased foil and bake for 30–45 minutes until the leaves are tender.

6 Remove the cabbage rolls from the oven, drain and place on a serving dish. Strain 600ml/1 pint/2½ cups of the cooking water into a saucepan and bring to a boil.

7 Blend the flour to a runny paste with a little cold water and whisk into the boiling stock, together with the lemon juice.

8 Beat the eggs in a heatproof jug and slowly pour on the hot stock, whisking well as you go.

9 Return to the stove and on the lowest heat, stir until smooth and thick. Do not allow the sauce to boil or it will curdle. Serve the rolls with some of the sauce poured over and the rest handed separately.

IRISH COLCANNON

THIS LOVELY WARMING WINTER'S DISH IS QUITE LIKE EGGS FLORENTINE. HERE, BAKED EGGS NESTLE AMONG CREAMY POTATOES WITH CURLY KALE OR CABBAGE AND A TOPPING OF GRATED CHEESE.

SERVES FOUR

INGREDIENTS

   1kg/2lb potatoes, cut in even pieces
   225g/8oz curly kale or crisp green
     cabbage, shredded
   2 spring onions, chopped
   butter or margarine, to taste
   fresh nutmeg, grated
   salt and ground black pepper
   4 large eggs
   75g/3oz mature cheese, grated

1 Boil the potatoes until just tender, then drain and mash well.

2 Lightly cook the kale or cabbage until just tender but still crisp. Preheat the oven to 190°C/375°F/Gas 5.

3 Drain the greens and mix them into the potato with the onions, butter or margarine and nutmeg. Season to taste.

4 Spoon the mixture into a shallow ovenproof dish and make four hollows in the mixture. Crack an egg into each and season well.

5 Bake for about 12 minutes or until the eggs are just set, then serve sprinkled with the cheese.

PASTA WITH CAPONATA

THIS EXCELLENT SWEET AND SOUR VEGETABLE DISH GOES WONDERFULLY WELL WITH PASTA.

SERVES FOUR

INGREDIENTS

   1 medium aubergine, cut into sticks
   2 medium courgettes, cut into sticks
   8 baby onions, peeled or 1 large onion,
     sliced
   2 garlic cloves, crushed
   1 large red pepper, sliced
   60ml/4 tbsp olive oil
   450ml/¾ pint/scant 2 cups tomato
     juice or 1 × 500ml/17fl oz carton
     creamed tomatoes
   150ml/¼ pint/⅔ cup water
   30ml/2 tbsp balsamic vinegar
   juice of 1 lemon
   15ml/1 tbsp sugar
   30ml/2 tbsp sliced black olives
   30ml/2 tbsp capers
   salt and ground black pepper
   400g/14oz tagliatelle or pasta ribbons

1 Lightly salt the aubergine and courgettes and leave them to drain in a colander for 30 minutes. Rinse and pat dry with kitchen paper towel.

2 In a large saucepan, lightly fry the onions, garlic and pepper in the oil for 5 minutes, then stir in the aubergine and courgettes and fry for a further 5 minutes.

3 Stir in the tomato juice or creamed tomatoes plus the water. Stir well, bringing the mixture to the boil, then add all the rest of the ingredients except the pasta. Season to taste and then simmer for 10 minutes.

4 Meanwhile, boil the pasta according to the instructions on the pack, then drain. Serve the caponata with the pasta.

SPINACH GNOCCHI

THIS WHOLESOME ITALIAN DISH IS IDEAL FOR MAKING IN ADVANCE. SERVE IT WITH A FRESH TOMATO SAUCE.

SERVES FOUR TO SIX

INGREDIENTS

400g/14oz fresh leaf spinach or 150g/
  6oz frozen leaf spinach, thawed
750ml/1¼ pints/good 3 cups milk
200g/7oz/1¼ cups semolina
50g/2oz/4 tbsp butter, melted
50g/2oz Parmesan cheese, freshly
  grated, plus extra to serve
fresh nutmeg, grated
salt and ground black pepper
2 eggs, beaten

2 In a large saucepan, heat the milk and when just on the point of boiling, sprinkle in the semolina in a steady stream, stirring it briskly with a wooden spoon.

5 Stamp out shapes using a plain round cutter with a diameter of about 4cm/1½in. Reserve the trimmings.

1 Blanch the spinach in the tiniest amount of water then drain and squeeze dry through a sieve with the back of a ladle. Chop the spinach roughly.

3 Simmer the semolina for 2 minutes then remove from the heat and stir in half the butter, most of the cheese, nutmeg and seasoning to taste and the spinach. Allow to cool for 5 minutes.

6 Grease a shallow ovenproof dish. Place the trimmings on the base and arrange the gnocchi rounds on top with each one overlapping.

7 Brush the tops with the remaining butter and sprinkle over the last of the cheese.

8 Preheat the oven when ready to bake to 190°C/375°F/Gas 5 and cook for about 35 minutes until golden and crisp on top. Serve hot with fresh tomato sauce and extra cheese.

VARIATION
For a special occasion, make half plain and half spinach gnocchi and arrange in an attractive pattern to serve. Use the same recipe as above but halve the amount of spinach and add to half the mixture in a separate bowl to make two batches. Stamp out and cook the gnocchi as normal. For a more substantial, healthy meal, make a tasty vegetable base of lightly sautéed peppers, courgettes and mushrooms and place the gnocchi on top.

4 Stir in the eggs then tip the mixture out onto a shallow baking sheet, spreading it to a 1cm/½in thickness. Allow to cool completely then chill until solid.

RED RICE RISSOLES

RISOTTO RICE CHILLS TO A FIRM TEXTURE YET REMAINS LIGHT AND CREAMY WHEN REHEATED AS CRISP CRUMBED RISSOLES.

<u>SERVES ABOUT EIGHT</u>

INGREDIENTS

1 large red onion, chopped
1 red pepper, chopped
2 garlic cloves, crushed
1 red chilli, finely chopped
30ml/2 tbsp olive oil
25g/1oz/2 tbsp butter
225g/8oz/1¼ cups risotto rice
1 litre/1¾ pints/4½ cups stock
4 sun-dried tomatoes, chopped
30ml/2 tbsp tomato purée
10ml/2 tsp dried oregano
salt and ground black pepper
45ml/3 tbsp fresh parsley, chopped
150g/6oz cheese (e.g. Red Leicester)
1 egg, beaten
115g/4oz/1 cup dried breadcrumbs
oil, for deep frying

1 Fry the onion, pepper, garlic and chilli in the oil and butter for 5 minutes. Stir in the rice and fry for a further 2 minutes.

2 Pour in the stock and add the tomatoes, purée, oregano and seasoning. Bring to the boil, stirring occasionally, then cover and simmer for 20 minutes.

3 Stir in the parsley then turn into a shallow dish, cool and chill until firm. When cold, divide into 12 and shape into balls.

4 Cut the cheese into 12 pieces and press a nugget into the centre of each of the rissoles.

5 Put the beaten egg in one bowl and the breadcrumbs into another. Dip the rissoles first into the egg then into the breadcrumbs, coating each one evenly.

6 Lay the coated rissoles on a plate and chill again for 30 minutes. Fill a deep fat frying pan one-third full of oil and heat until a cube of day-old bread browns in under a minute.

7 Fry the rissoles in batches, reheating the oil in between, for about 3–4 minutes. Drain on kitchen paper and keep warm, uncovered, before serving.

BROAD BEAN AND CAULIFLOWER CURRY

A TASTY MID-WEEK CURRY TO SERVE WITH RICE, SMALL POPADUMS AND MAYBE A CUCUMBER RAITA.

SERVES FOUR

INGREDIENTS

    2 garlic cloves, chopped
    2.5cm/1in cube fresh root ginger
    1 fresh green chilli, seeded and
      chopped
    15ml/1 tbsp oil
    1 onion, sliced
    1 large potato, chopped
    30ml/2 tbsp ghee or softened butter
    15ml/1 tbsp curry powder, mild or hot
    1 medium size cauliflower, cut into
      small florets
    600ml/1 pint/2½ cups stock
    25g/1oz/2 tbsp creamed coconut
    salt and ground black pepper
    1 × 285g/10oz can broad beans
    juice of half a lemon (optional)
    fresh coriander or parsley, to serve

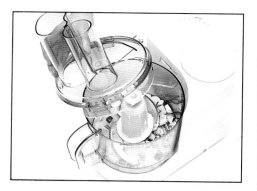

1 Blend the garlic, ginger, chilli and oil in a food processor until they form a smooth paste.

2 In a large saucepan, fry the onion and potato in the ghee or butter for 5 minutes then stir in the spice paste and curry powder. Cook for 1 minute.

3 Add the cauliflower florets and stir well into the spicy mixture, then pour in the stock. Bring to a boil and mix in the coconut, stirring until it melts.

4 Season well then cover and simmer for 10 minutes. Add the beans and their liquor and cook uncovered for a further 10 minutes.

5 Check the seasoning and add a good squeeze of lemon juice if liked. Serve hot garnished with chopped coriander or parsley.

BAKED ONIONS STUFFED WITH FETA

RED ONIONS HAVE A MILD FLAVOUR AND ATTRACTIVE APPEARANCE. THE RED FLESH LOOKS ESPECIALLY APPEALING WITH THIS FETA AND BREADCRUMB STUFFING.

SERVES FOUR

INGREDIENTS
    4 large red onions
    15ml/1 tbsp olive oil
    25g/1oz pine nuts
    115g/4oz feta cheese, crumbled
    25g/1oz fresh white breadcrumbs
    15ml/1 tbsp chopped fresh coriander
    salt and freshly ground black pepper

1 Preheat the oven to 180°C/350°F/ Gas 4 and lightly grease a shallow oven-proof dish. Peel the onions and cut a thin slice from the top and base of each. Place in a large saucepan of boiling water and cook for 10–12 minutes until just tender. Remove with a slotted spoon. Drain on kitchen paper and leave to cool slightly.

2 Using a small knife or your fingers, remove the inner sections of the onions, leaving about two or three outer rings. Finely chop the inner sections and place the shells in an ovenproof dish.

3 Heat the oil in a medium-sized frying pan and fry the chopped onions for 4–5 minutes until golden, then add the pine nuts and stir-fry for a few minutes.

4 Place the feta cheese in a small bowl and stir in the onions and pine nuts, the breadcrumbs and coriander. Season well with salt and pepper and then spoon the mixture into the onion shells. Cover loosely with foil and bake in the oven for about 30 minutes and remove the foil for the last 10 minutes.

5 Serve as a starter or as a light lunch with warm olive bread.

ONION TARTS WITH GOAT'S CHEESE

A VARIATION OF A CLASSIC FRENCH DISH, TARTE A L'OIGNON, THIS DISH USES YOUNG GOAT'S CHEESE INSTEAD OF CREAM, AS IT IS MILD AND CREAMY AND COMPLEMENTS THE FLAVOUR OF THE ONIONS. THIS RECIPE MAKES EITHER EIGHT INDIVIDUAL TARTS OR ONE LARGE 23CM/9IN TART.

SERVES EIGHT

INGREDIENTS
For the pastry
    175g/6oz plain flour
    65g/2½oz butter
    25g/1oz goat's cheddar or Cheddar
      cheese, grated
For the filling
    15–25ml/1–1½ tbsp olive or
      sunflower oil
    3 onions, finely sliced
    175g/6oz young goat's cheese
    2 eggs, beaten
    15ml/1 tbsp single cream
    50g/2oz goat's cheddar, grated
    15ml/1 tbsp chopped fresh tarragon
    salt and freshly ground black pepper

1 To make the pastry, sift the flour into a bowl and rub in the butter until the mix-ture resembles fine breadcrumbs. Stir in the grated cheese and add enough cold water to make a dough. Knead lightly, put in a polythene bag and chill. Preheat the oven to 190°C/375°F/Gas 5.

2 Roll out the dough on a lightly floured surface, and then cut into eight rounds using a 11.5cm/4½in pastry cutter and line eight 10cm/4in patty tins. Prick the bases with a fork and bake in the oven for 10–15 minutes until firm but not browned. Reduce the oven temperature to 180°C/350°F/Gas 4.

3 Heat the olive or sunflower oil in a large frying pan and fry the onions over a low heat for 20–25 minutes until they are a deep golden brown. Stir occasionally to prevent them burning.

4 Beat the goat's cheese with the eggs, cream, goat's cheddar and tarragon. Season with salt and pepper and then stir in the fried onions.

5 Pour the mixture into the part-baked pastry cases and bake in the oven for 20–25 minutes until golden. Serve warm or cold with a green salad.

PARSNIP AND CHESTNUT CROQUETTES

THE SWEET NUTTY TASTE OF CHESTNUTS BLENDS PERFECTLY WITH THE SIMILARLY SWEET BUT EARTHY FLAVOUR OF PARSNIPS. FRESH CHESTNUTS NEED TO BE PEELED BUT FROZEN CHESTNUTS ARE EASY TO USE AND ARE NEARLY AS GOOD AS FRESH FOR THIS RECIPE.

MAKES TEN TO TWELVE

INGREDIENTS

    450g/1lb parsnips, cut roughly into
      small pieces
    115g/4oz frozen chestnuts
    25g/1oz butter
    1 garlic clove, crushed
    15ml/1tbsp chopped fresh coriander
    1 egg, beaten
    40–50g/1½–2oz fresh white
      breadcrumbs
    vegetable oil, for frying
    salt and freshly ground black pepper
    sprig of coriander, to garnish

1 Place the parsnips in a saucepan with enough water to cover. Bring to the boil, cover and simmer for 15–20 minutes until completely tender.

2 Place the frozen chestnuts in a pan of water, bring to the boil and simmer for 8–10 minutes until very tender. Drain, place in a bowl and mash roughly.

3 Melt the butter in a small saucepan and cook the garlic for 30 seconds. Drain the parsnips and mash with the garlic butter. Stir in the chestnuts and chopped coriander, then season well.

4 Take about 15ml/1tbsp of the mixture at a time and form into small croquettes, about 7.5cm/3in long. Dip each croquette into the beaten egg and then roll in the breadcrumbs.

5 Heat a little oil in a frying pan and fry the croquettes for 3–4 minutes until golden, turning frequently so they brown evenly. Drain on kitchen paper and then serve at once, garnished with coriander.

PARSNIP, AUBERGINE AND CASHEW BIRYANI

SERVES FOUR TO SIX

INGREDIENTS
1 small aubergine, sliced
275g/10oz basmati rice
3 parsnips
3 onions
2 garlic cloves
2.5cm/1in piece fresh root ginger, peeled
about 60ml/4 tbsp vegetable oil
175g/6oz unsalted cashew nuts
40g/1½oz sultanas
1 red pepper, seeded and sliced
5ml/1 tsp ground cumin
5ml/1 tsp ground coriander
2.5ml/½ tsp chilli powder
120ml/4fl oz/½ cup natural yogurt
300ml/½ pint/1¼ cups vegetable or chicken stock
25g/1oz butter
salt and freshly ground black pepper
sprigs of coriander, to garnish
2 hard-boiled eggs, quartered

1 Sprinkle the aubergine with salt and leave for 30 minutes. Rinse, pat dry and cut into bite-size pieces. Soak the rice in a bowl of cold water for 40 minutes. Peel and core the parsnips. Cut into 1cm/½in pieces. Roughly chop 1 onion and put in a food processor or blender with the garlic and ginger. Add 30–45ml/2–3 tbsp water and process to a paste.

2 Finely slice the remaining onions. Heat 45ml/3 tbsp of the oil in a large flameproof casserole and fry gently for 10–15 minutes until deep golden brown. Remove and drain. Add 40g/1½oz of the cashew nuts to the pan and stir-fry for 2 minutes. Add the sultanas and fry until they swell. Remove and drain.

3 Add the aubergine and pepper to the pan and stir-fry for 4–5 minutes. Drain on kitchen paper. Fry the parsnips for 4–5 minutes. Stir in the remaining cashew nuts and fry for 1 minute. Transfer to the plate with the aubergines.

4 Add the remaining 15ml/1 tbsp of oil to the pan. Add the onion paste. Cook, stirring over a moderate heat for 4–5 minutes until the mixture turns golden. Stir in the cumin, coriander and chilli powder. Cook, stirring, for 1 minute, then reduce the heat and add the yogurt.

5 Bring the mixture slowly to the boil and stir in the stock, parsnips, aubergines and peppers. Season, cover and simmer for 30–40 minutes until the parsnips are tender and then transfer to an ovenproof casserole.

6 Preheat the oven to 150°C/300°F/Gas 2. Drain the rice and add to 300ml/½ pint/1¼ cups of salted boiling water. Cook gently for 5–6 minutes until it is tender but slightly undercooked.

7 Drain the rice and pile it in a mound on top of the parsnips. Make a hole from the top to the base using the handle of a wooden spoon. Scatter the reserved fried onions, cashew nuts and sultanas over the rice and dot with butter. Cover with a double layer of foil and then secure in place with a lid.

8 Cook in the oven for 35–40 minutes. To serve, spoon the mixture on to a warmed serving dish and garnish with coriander sprigs and quartered eggs.

CELERIAC AND BLUE CHEESE ROULADE

CELERIAC ADDS A DELICATE AND SUBTLE FLAVOUR TO THIS ATTRACTIVE DISH. THE SPINACH ROULADE
MAKES AN ATTRACTIVE CONTRAST TO THE CREAMY FILLING BUT YOU COULD USE A PLAIN OR CHEESE
ROULADE BASE INSTEAD. BE SURE TO ROLL UP THE ROULADE WHILE IT IS STILL WARM AND PLIABLE.

SERVES SIX

INGREDIENTS
    15g/½oz butter
    225g/8oz cooked spinach, drained
      and chopped
    150ml/¼ pint/⅔ cup single cream
    4 large eggs, separated
    15g/½oz Parmesan cheese, grated
    pinch of nutmeg
    salt and freshly ground black pepper
For the filling
    225g/8oz celeriac
    lemon juice
    75g/3oz St Agur cheese
    115g/4oz fromage frais
    freshly ground black pepper

1 Preheat the oven to 200°C/400°F/
Gas 6 and line a 34 x 24cm/13 x 9in Swiss
roll tin with non-stick baking paper.

2 Melt the butter in a saucepan and add
the spinach. Cook gently until all the
liquid has evaporated, stirring frequently.
Remove the pan from the heat and stir in
the cream, egg yolks, Parmesan cheese,
nutmeg and seasoning.

4 Bake in the oven for 10–15 minutes
until the roulade is firm to the touch and
lightly golden on top. Carefully turn out
on to a sheet of greaseproof or non-stick
baking paper and peel away the lining
paper. Roll it up with the paper inside
and leave to cool slightly.

5 To make the filling, peel and grate the
celeriac into a bowl and sprinkle well
with lemon juice. Blend the blue cheese
and fromage frais together and mix with
the celeriac and a little black pepper.

6 Unroll the roulade, spread with the
filling and roll up again. Serve at once or
wrap loosely and chill.

3 Whisk the egg whites until stiff, fold
them gently into the spinach mixture and
then spoon into the prepared tin. Spread
the mixture evenly and use a palette
knife to smooth the surface.

BAKED LEEKS WITH CHEESE AND YOGURT TOPPING

Like all vegetables, the fresher leeks are, the better their flavour, and the freshest leeks available should be used for this dish. Small, young leeks are around at the beginning of the season and are perfect to use here.

SERVES FOUR

INGREDIENTS

8 small leeks, about 675g/1½lb
2 small eggs or 1 large one, beaten
150g/5oz fresh goat's cheese
85ml/3fl oz/⅓ cup natural yogurt
50g/2oz Parmesan cheese, grated
25g/1oz fresh white or brown
  breadcrumbs
salt and freshly ground black pepper

1 Preheat the oven to 180°C/350°F/
Gas 4 and butter a shallow ovenproof
dish. Trim the leeks, cut a slit from top to
bottom and rinse well under cold water.

2 Place the leeks in a saucepan of
water, bring to the boil and simmer
gently for 6–8 minutes until just tender.
Remove and drain well using a slotted
spoon, and arrange in the prepared dish.

3 Beat the eggs with the goat's cheese,
yogurt and half the Parmesan cheese,
and season well with salt and pepper.

4 Pour the cheese and yogurt mixture
over the leeks. Mix the breadcrumbs and
remaining Parmesan cheese together
and sprinkle over the sauce. Bake in the
oven for 35–40 minutes until the top is
crisp and golden brown.

CASSAVA AND VEGETABLE KEBABS

THIS IS AN ATTRACTIVE AND DELICIOUS ASSORTMENT OF AFRICAN VEGETABLES, MARINATED IN A SPICY GARLIC SAUCE. IF CASSAVA IS UNAVAILABLE, USE SWEET POTATO OR YAM INSTEAD.

SERVES FOUR

INGREDIENTS
  175g/6oz cassava
  1 onion, cut into wedges
  1 aubergine, cut into bite-size pieces
  1 courgette, sliced
  1 ripe plantain, sliced
  1 red pepper or ½ red pepper, ½ green
    pepper, sliced
  16 cherry tomatoes
For the marinade
  60ml/4 tbsp lemon juice
  60ml/4 tbsp olive oil
  45–60/3–4 tbsp soy sauce
  15ml/1 tbsp tomato purée
  1 green chilli, seeded and finely
    chopped
  ½ onion, grated
  2 garlic cloves, crushed
  5ml/1 tsp mixed spice
  pinch of dried thyme
  rice or couscous, to serve

1 Peel the cassava and cut into bite-size pieces. Place in a bowl, cover with boiling water and leave to blanch for 5 minutes. Drain well.

2 Place all the vegetables, including the cassava, in a large bowl.

3 Blend together all the marinade ingredients and pour over the prepared vegetables. Set aside for 1–2 hours.

4 Preheat the grill and thread all the vegetables and cherry tomatoes on to eight skewers.

5 Grill the vegetables under a low heat for about 15 minutes until tender and browned, turning frequently and basting occasionally with the marinade.

6 Meanwhile, pour the remaining marinade into a small saucepan and simmer for 10 minutes until slightly reduced.

7 Arrange the vegetable kebabs on a serving plate and strain the sauce into a small jug. Serve with rice or couscous.

SALSIFY GRATIN

THE SPINACH IN THIS RECIPE ADDS COLOUR AND MAKES IT GO FURTHER. HOWEVER, IF YOU CAN OBTAIN SALSIFY EASILY AND HAVE THE PATIENCE TO PEEL A LOT OF IT, INCREASE THE QUANTITY AND LEAVE OUT THE SPINACH.

SERVES FOUR

INGREDIENTS
   450g/1lb salsify, cut into
      5cm/2in lengths
   juice of 1½ lemons
   450g/1lb spinach, prepared
   150ml/¼ pint/⅔ cup vegetable stock
   300ml/½ pint/1¼ cups single cream
   salt and freshly ground black pepper

3 Cook the spinach in a large saucepan over a moderate heat for 2–3 minutes until the leaves have wilted, shaking the pan occasionally. Place the stock, cream and seasoning in a small saucepan and heat through very gently, stirring.

4 Arrange the salsify and spinach in layers in the prepared dish. Pour over the stock and cream mixture and bake in the oven for about 1 hour until the top is golden brown and bubbling.

1 Trim away the tops and bottoms of the salsify and peel or scrape away the outer skin. Place each peeled root immediately in water with lemon juice added, to prevent discolouration.

2 Preheat the oven to 160°C/325°F/ Gas 3 and butter an ovenproof dish. Place the salsify in a saucepan of boiling water with the lemon juice. Simmer for about 10 minutes until the salsify is just tender, then drain.

CELERIAC GRATIN

ALTHOUGH CELERIAC HAS A RATHER UNATTRACTIVE APPEARANCE WITH ITS HARD, KNOBBLY SKIN, IT IS A VEGETABLE THAT HAS A VERY DELICIOUS SWEET AND NUTTY FLAVOUR. THIS IS ACCENTUATED IN THIS DISH BY THE ADDITION OF THE SWEET YET NUTTY EMMENTAL CHEESE.

SERVES FOUR

INGREDIENTS
    450g/1lb celeriac
    juice of ½ lemon
    25g/1oz butter
    1 small onion, finely chopped
    30ml/2 tbsp plain flour
    300ml/½ pint/1¼ cups milk
    25g/1oz Emmental cheese, grated
    15ml/1 tbsp capers
    salt and cayenne pepper

1 Preheat the oven to 190°C/375°F/ Gas 5. Peel the celeriac and cut into 5mm/¼in slices, immediately plunging them into a saucepan of cold water acidulated with the lemon juice.

2 Bring the water to the boil and simmer the celeriac for 10–12 minutes until just tender. Drain and arrange the celeriac in a shallow ovenproof dish.

3 Melt the butter in a small saucepan and fry the onion over a gentle heat until soft but not browned. Stir in the flour, cook for 1 minute and then slowly stir in the milk to make a smooth sauce. Stir in the cheese, capers and seasoning to taste and then pour over the celeriac. Cook in the oven for 15–20 minutes until the top is golden brown.

VARIATION
For a less strongly flavoured dish, alternate the layers of celeriac with potato. Slice the potato, cook until almost tender, then drain well before assembling the dish.

BAKED MARROW <u>IN</u> PARSLEY SAUCE

THIS IS A REALLY GLORIOUS WAY WITH A SIMPLE AND MODEST VEGETABLE. TRY TO FIND A SMALL, FIRM AND UNBLEMISHED MARROW FOR THIS RECIPE, AS THE FLAVOUR WILL BE SWEET, FRESH AND DELICATE. YOUNG MARROWS DO NOT NEED PEELING; MORE MATURE ONES DO.

SERVES FOUR

INGREDIENTS

1 small young marrow, about 900g/2lb
30ml/2 tbsp olive oil
15g/½oz butter
1 onion, chopped
15ml/1 tbsp plain flour
300ml/½ pint/1¼ cups milk
  and single cream mixed
30ml/2 tbsp chopped fresh parsley
salt and freshly ground black pepper

1 Preheat the oven to 180°C/350°F/ Gas 4 and cut the marrow into pieces measuring about 5 x 2.5cm/2 x 1in.

2 Heat the oil and butter in a flameproof casserole and fry the onion over a gentle heat until very soft.

3 Add the marrow and sauté for 1–2 minutes and then stir in the flour. Cook for a few minutes, then stir in the milk and cream mixture.

4 Add the parsley and seasoning, stir well and then cover and cook in the oven for 30–35 minutes. If liked, remove the lid for the final 5 minutes of cooking to brown the top. Alternatively, serve the marrow in its rich pale sauce.

COOK'S TIP
Chopped fresh basil or a mixture of basil and chervil also tastes good in this dish.

MARROWS WITH GNOCCHI

A SIMPLE WAY WITH MARROW, THIS DISH MAKES AN EXCELLENT ACCOMPANIMENT TO GRILLED MEAT BUT IT IS ALSO GOOD WITH A VEGETARIAN DISH, OR SIMPLY SERVED WITH GRILLED TOMATOES. GNOCCHI ARE AVAILABLE FROM MOST SUPERMARKETS; ITALIAN DELICATESSENS MAY ALSO SELL FRESH GNOCCHI.

SERVES FOUR

INGREDIENTS

    1 small marrow, cut into
      bite-size chunks
    50g/2oz butter
    400g/14oz packet gnocchi
    ½ garlic clove, crushed
    salt and freshly ground black pepper
    chopped fresh basil, to garnish

1 Preheat the oven to 180°C/350°F/ Gas 4 and butter a large ovenproof dish. Place the marrow, more or less in a single layer, in the dish. Dot all over with the remaining butter.

2 Place a double piece of buttered greaseproof paper over the top. Cover with an ovenproof plate or lid so that it presses the marrow down, and then place a heavy, ovenproof weight on top of that. (Use a couple of old-fashioned scale weights.)

3 Put in the oven to bake for about 15 minutes, by which time the marrow should just be tender.

4 Cook the gnocchi in a large saucepan of boiling salted water for 2–3 minutes, or according to the instructions on the packet. Drain well.

5 Stir the garlic and gnocchi into the marrow. Season and then place the greaseproof paper over the marrow and return to the oven for 5 minutes (the weights are not necessary).

6 Just before serving, sprinkle the top with a little chopped fresh basil.

PASTA WITH SAVOY CABBAGE AND GRUYÈRE

THIS IS AN INEXPENSIVE AND SIMPLE DISH WITH A SURPRISING TEXTURE AND FLAVOUR. THE CABBAGE IS COOKED SO THAT IT HAS PLENTY OF "BITE" TO IT, CONTRASTING WITH THE SOFTNESS OF THE PASTA.

SERVES FOUR

INGREDIENTS

1 small Savoy or green cabbage, thinly sliced
25g/1oz butter
1 small onion, chopped
350g/12oz pasta, e.g. tagliatelle, fettucine or penne
15ml/1 tbsp chopped fresh parsley
150ml/¼ pint/⅔ cup single cream
50g/2oz Gruyère or Cheddar cheese, grated
about 300ml/½ pint/1¼ cups hot vegetable or chicken stock
salt and freshly ground black pepper

1 Preheat the oven to 180°C/350°F/ Gas 4 and butter a large casserole. Place the cabbage in a mixing bowl.

2 Melt the butter in a small frying pan and fry the onion until softened. Stir into the cabbage in the bowl.

3 Cook the pasta according to the instructions, until *al dente*.

4 Drain well and stir into the bowl with the cabbage and onion. Add the parsley and mix well and then pour into the prepared casserole.

5 Beat together the cream and Gruyère or Cheddar cheese and then stir in the hot stock. Season well and pour over the cabbage and pasta, so that it comes about halfway up the casserole. If necessary, add a little more stock.

6 Cover tightly with foil or a lid and cook in the oven for 30–35 minutes, until the cabbage is tender and the stock is bubbling. Remove the lid for the last 5 minutes of the cooking time to brown the top.

SWEETCORN IN A GARLIC BUTTER CRUST

WHETHER YOU ARE SERVING THIS AS A MAIN COURSE OR AS AN ACCOMPANIMENT, IT WILL DISAPPEAR IN A FLASH. EVEN PEOPLE WHO ARE NOT USUALLY KEEN ON CORN ON THE COB HAVE BEEN WON OVER BY THIS RECIPE.

SERVES SIX

INGREDIENTS

6 ripe cobs of corn
salt
225g/8oz/1 cup butter
30ml/2 tbsp olive oil
2 cloves garlic, peeled and crushed
10ml/2 tsp freshly ground black pepper
115g/4oz/1 cup wholemeal
 breadcrumbs
15ml/1 tbsp chopped parsley

1 Boil the corn cobs in salted water until tender, then leave to cool.

2 Melt the butter, and add the oil, garlic and black pepper. Pour the mixture into a shallow dish.

3 Mix the breadcrumbs and parsley in another shallow dish. Roll the corn cobs in the melted butter mixture and then in the breadcrumbs.

4 Grill the cobs under a high grill until the breadcrumbs are golden.

VARIATIONS

• Partially cut through a French loaf at regular intervals. Spread the garlic butter mixture between the slices and bake in a moderate oven for 30 minutes.
• To make garlic croûtons, melt the garlic butter in a pan and add cubes of bread. Toss frequently over a medium heat. When golden brown, add to soups or salads

VEGETABLE AND HERB KEBABS WITH GREEN PEPPERCORN SAUCE

OTHER VEGETABLES CAN BE INCLUDED IN THESE KEBABS, DEPENDING ON WHAT IS AVAILABLE AT THE TIME. THE GREEN PEPPERCORN SAUCE IS ALSO AN EXCELLENT ACCOMPANIMENT TO MANY OTHER DISHES.

SERVES FOUR

INGREDIENTS

   8 bamboo skewers soaked in water for
     1 hour
   24 mushrooms
   16 cherry tomatoes
   16 large basil leaves
   16 thick slices of courgette
   16 large mint leaves
   16 squares of red sweet pepper
To baste
   120ml/4floz/½ cup melted butter
   1 clove garlic, peeled and crushed
   15ml/1 tbsp crushed green peppercorns
   salt
For the green peppercorn sauce
   50g/2oz/¼ cup butter
   45ml/3 tbsp brandy
   250ml/8fl oz/1 cup double cream
   5ml/1 tsp crushed green peppercorns

1 Thread the vegetables on to bamboo skewers. Place the basil leaves immediately next to the tomatoes, and the mint leaves wrapped around the courgette slices.

2 Mix the basting ingredients and baste the kebabs thoroughly. Place the skewers on a barbecue or under the grill, turning and basting regularly until the vegetables are just cooked – about 5–7 minutes.

3 Heat the butter for the sauce in a frying pan, then add the brandy and light it. When the flames have died down, stir in the cream and the peppercorns. Cook for approximately 2 minutes, stirring all the time. Serve the kebabs with the green peppercorn sauce.

SPINACH, WALNUT AND GRUYÈRE LASAGNE WITH BASIL

THIS NUTTY LASAGNE IS A DELICIOUS COMBINATION OF FLAVOURS, WHICH EASILY EQUALS THE TRADITIONAL MEAT AND TOMATO VERSION.

SERVES EIGHT

INGREDIENTS
    350g/12oz/3 cups spinach lasagne
      (quick cooking)
For the walnut and tomato sauce
    45ml/3 tbsp walnut oil
    1 large onion, chopped
    225g/8oz celeriac, finely chopped
    1 x 400g/14oz can chopped tomatoes
    1 large clove garlic, finely chopped
    ½ tsp sugar
    115g/4oz/⅔ cup chopped walnuts
    150ml/¼ pint/⅔ cup Dubonnet
For the spinach and Gruyère sauce
    75g/3oz/generous ⅓ cup butter
    30ml/2 tbsp walnut oil
    1 medium onion, chopped
    75g/3oz/generous ⅓ cup flour
    5ml/1 tsp mustard powder
    1.2 litres/2 pints/5 cups milk
    225g/8oz/2 cups grated Gruyère cheese
    salt and pepper
    ground nutmeg
    500g/1lb frozen spinach, thawed and
      puréed
    30ml/2 tbsp basil, chopped

2 To make the spinach and Gruyère sauce, melt the butter with the walnut oil and add the onion. Cook for 5 minutes, then stir in the flour. Cook for another minute and add the mustard powder and milk, stirring vigorously. When the sauce has come to the boil, take off the heat and add three-quarters of the grated Gruyère. Season to taste with salt, pepper and nutmeg. Finally, add the puréed spinach.

3 Preheat the oven to 180°C/350°F/Gas 4. Layer the lasagne in an ovenproof dish. Start with a layer of the spinach and Gruyère sauce, then add a little walnut and tomato sauce, then a layer of lasagne, and continue until the dish is full, ending with a layer of one of the sauces.

4 Sprinkle the remaining Gruyère over the top of the dish, followed by the basil. Bake for 45 minutes.

1 First make the walnut and tomato sauce. Heat the walnut oil and sauté the onion and celeriac. Cook for about 8–10 minutes. Meanwhile purée the tomatoes in a food processor. Add the garlic to the pan and cook for about 1 minute, then add the sugar, walnuts, tomatoes and Dubonnet. Season to taste. Simmer, uncovered, for 25 minutes.

KOHLRABI STUFFED WITH PEPPERS

IF YOU HAVEN'T SAMPLED KOHLRABI, OR HAVE ONLY EATEN IT IN STEWS WHERE ITS FLAVOUR IS LOST, THIS DISH IS RECOMMENDED. THE SLIGHTLY SHARP FLAVOUR OF THE PEPPERS IS AN EXCELLENT FOIL TO THE MORE EARTHY FLAVOUR OF THE KOHLRABI.

SERVES FOUR

INGREDIENTS
   4 small kohlrabi, about 175–225g/
      6–8oz each
   about 400ml/14fl oz/1⅔ cups hot
      vegetable stock
   15ml/1 tbsp olive or sunflower oil
   1 onion, chopped
   1 small red pepper, seeded and sliced
   1 small green pepper, seeded
      and sliced
   salt and freshly ground black pepper
   flat leaf parsley, to garnish (optional)

1 Preheat the oven to 180°C/350°F/ Gas 4. Trim and top and tail the kohlrabi and arrange in the base of a medium-size ovenproof dish.

2 Pour over the stock to come about halfway up the vegetables. Cover and braise in the oven for about 30 minutes until tender. Transfer to a plate and allow to cool, reserving the stock.

3 Heat the oil in a frying pan and fry the onion for 3–4 minutes over a gentle heat, stirring occasionally. Add the peppers and cook for a further 2–3 minutes, until the onion is lightly browned.

4 Add the reserved vegetable stock and a little seasoning, then simmer, uncovered, over a moderate heat until the stock has almost evaporated.

5 Scoop out the flesh from the kohlrabi and roughly chop. Stir the flesh into the onion and pepper mixture, taste and adjust the seasoning. Arrange the shells in a shallow ovenproof dish.

6 Spoon the filling into the kohlrabi shells. Place in the oven for 5–10 minutes to heat through and then serve, garnished with flat leaf parsley, if liked.

SPINACH ROULADE

A SIMPLE PURÉE OF SPINACH BAKED WITH EGGS ROLLED AROUND A CREAMY RED PEPPER FILLING MAKES AN EXOTIC AND COLOURFUL SUPPER DISH. THIS CAN BE PREPARED IN ADVANCE AND THEN REHEATED WHEN REQUIRED.

SERVES FOUR

INGREDIENTS
    450g/1lb leaf spinach, well washed and
        drained
    fresh nutmeg, grated
    good knob of butter
    45ml/3 tbsp Parmesan cheese, grated
    45ml/3 tbsp double cream
    salt and ground black pepper
    2 eggs, separated
    1 small red pepper, chopped
    1 × 200g/7oz pack soft cheese with
        garlic and herbs

1 Line and then grease a medium-sized Swiss roll tin. Preheat the oven to 190°C/375°F/Gas 5. Cook the spinach with a tiny amount of water then drain well, pressing it through a sieve with the back of a ladle. Chop the spinach finely.

2 Mix the spinach with the nutmeg, butter, Parmesan cheese, cream and seasoning. Cool for 5 minutes then beat in the egg yolks.

3 Whisk the egg whites until they form soft peaks and carefully fold in to the spinach mixture. Spread in to the prepared tin, level and bake for 12–15 minutes until firm.

4 Turn the spinach out upside down on to a clean tea towel and allow it to cool in the tin for half an hour.

5 Meanwhile, simmer the pepper in about 30ml/2 tbsp of water in a covered pan until just soft, then either purée it in a blender or chop it finely. Mix with the soft cheese and season well.

6 When the spinach has cooled, peel off the paper. Trim any hard edges and spread it with the red pepper cream.

7 Carefully roll up the spinach and pepper in the tea towel, leave for 10 minutes to firm up, then serve on a long platter, cut in thick slices.

VARIATION
A thick vegetable purée of any root vegetable also works well as a roulade. Try cooked beetroot or parsnip, flavouring lightly with a mild curry-style spice such as cumin or coriander. The fillings can be varied too, such as finely chopped and sautéd mushroom and onion or grated carrot mixed with fromage frais and chives. Roulades are delicious served warm as well. Sprinkle with cheese and bake in a moderately hot oven for 15 minutes or so.

SALADS
AND
VEGETABLE DISHES

*Brighten your menus with this delightful range of colourful
salads and scrumptious vegetable dishes. Varied
but easy-to-obtain ingredients will guarantee that you
have the perfect recipe and accompaniment for every
style of meal — whatever the season.*

USEFUL DRESSINGS

A GOOD DRESSING CAN MAKE
EVEN THE SIMPLEST COMBINATION
OF FRESH VEGETABLES A
MEMORABLE TREAT. REMEMBER
THAT LIGHTLY COOKED
VEGETABLES ABSORB MORE
FLAVOUR AND ARE LESS GREASY IF
DRESSED WHILE HOT.

HOME-MADE MAYONNAISE

Make this by hand, if possible. If you make it in a food processor, it will be noticeably lighter.

INGREDIENTS

    2 egg yolks
    2.5ml/½ tsp salt
    2.5ml/½ tsp dry mustard
    ground black pepper
    300ml/½ pint/1¼ cups sunflower oil or
      half olive and half sunflower oil
    15ml/1 tbsp wine vinegar
    15ml/1 tbsp hot water

1 Put the yolks into a bowl with the salt and mustard and a grinding of pepper.

2 Stand the bowl on a damp dish cloth. Using a whisk, beat the yolks and the seasoning thoroughly, then beat in a small trickle of oil.

3 Continue trickling in the oil, adding it in very small amounts. The secret of a good, thick mayonnaise is to add the oil very slowly, beating each addition well before you add more. When all the oil is added, mix in the vinegar and hot water.
To make mayonnaise in a blender, use one whole egg and one yolk instead of two yolks. Blend the eggs with the seasonings. Then, with the blades running, trickle in the oil very slowly. Add the vinegar.

ORIGINAL THOUSAND ISLANDS DRESSING

INGREDIENTS

    60ml/4 tbsp sunflower oil
    15ml/1 tbsp fresh orange juice
    15ml/1 tbsp fresh lemon juice
    10ml/2 tsp grated lemon rind
    15ml/1 tbsp finely chopped onion
    5ml/1 tsp paprika
    5ml/1 tsp Worcestershire sauce
    15ml/1 tbsp finely chopped parsley
    salt and ground black pepper

Put all the ingredients into a screw-topped jar, season to taste and shake vigorously. Great with green salads, grated carrot, hot potato, pasta and rice salads.

YOGURT DRESSING

INGREDIENTS

    150g/5oz/⅔ cup natural yogurt
    30ml/2 tbsp mayonnaise
    30ml/2 tbsp milk
    15ml/1 tbsp fresh parsley, chopped
    15ml/1 tbsp fresh chives or spring
      onions, chopped
    salt and ground black pepper

Simply mix all the ingredients together thoroughly in a bowl.

PEPERONATA WITH RAISINS

SLICED ROASTED PEPPERS IN DRESSING WITH VINEGAR-SOAKED RAISINS MAKE A TASTY SIDE SALAD WHICH COMPLEMENTS MANY OTHER DISHES.

SERVES TWO TO FOUR

INGREDIENTS

    90ml/6 tbsp sliced peppers in olive oil,
      drained
    15ml/1 tbsp onion, chopped
    30ml/2 tbsp balsamic vinegar
    45ml/3 tbsp raisins
    30ml/2 tbsp fresh parsley, chopped
    ground black pepper

1 Toss the peppers with the onion and leave to steep for an hour.

2 Put the vinegar and raisins in a small saucepan and heat for a minute, then allow to cool.

3 Mix all the ingredients together thoroughly and spoon into a serving bowl. Serve lightly chilled.

COOK'S TIP
Peperonata is one of the classic Italian antipasto dishes, served at the start of each meal with crusty bread to mop up the delicious juices. Try serving shavings of fresh Parmesan cheese alongside, or buy a good selection of green and black olives to accompany the peperonata. Small baby tomatoes will complete the antipasto.

BEAN SPROUT STIR-FRY

*HOME-GROWN BEAN SPROUTS
TASTE SO GOOD, TOSSED INTO A
TASTY STIR-FRY. THEY HAVE
MORE FLAVOUR AND TEXTURE
THAN SHOP-BOUGHT VARIETIES
AND ARE VERY NUTRITIOUS,
BEING RICH IN VITAMINS AND
HIGH IN FIBRE. OLD DRIED
BEANS WILL NOT SPROUT, SO USE
BEANS THAT ARE WELL WITHIN
THEIR "USE BY" DATE.*

SERVES THREE TO FOUR

INGREDIENTS
    30ml/2 tbsp sunflower or groundnut oil
    225g/8oz mixed sprouted beans
    2 spring onions, chopped
    1 garlic clove, crushed
    30ml/2 tbsp soy sauce
    10ml/2 tsp sesame oil
    15ml/1 tbsp sesame seeds
    30ml/2 tbsp fresh coriander or parsley,
       chopped
    salt and ground black pepper

1 Heat the sunflower or groundnut oil in
a large wok and stir-fry the sprouting
beans, onion and garlic for 3–5 minutes.

2 Add the remaining ingredients, cook for
1–2 minutes more and serve hot.

HOME SPROUTING BEANS
Use any of the following pulses to grow
your own sprouts: green or brown lentils,
aduki beans, mung beans, chick-peas,
flageolet and haricot beans, soya beans or
Indian massor dal.

Cover 45ml/3 tbsp of your chosen
pulses with lukewarm water. Leave to soak
overnight, then drain and rinse.

Place in a large clean jam jar or special
sprouting tray. Cover with muslin and
secure with a rubber band. If using a jam
jar, lay the jar on its side and shake it so
the beans spread along the length. Leave
somewhere warm.

Rinse the beans night and morning with
plenty of cold water, running this through
the muslin, and drain. Carefully place the
jar back on its side.

Repeat this twice a day until roots and
shoots appear. (If nothing happens after
48 hours, the beans are probably too old.)
When the shoots are at least twice the
length of the bean, rinse finally and store
in the fridge for up to two days.

PANZANELLA SALAD

*SLICED JUICY TOMATOES LAYERED
WITH DAY-OLD BREAD SOUNDS
STRANGE FOR A SALAD, BUT IT'S
QUITE DELICIOUS.*

SERVES FOUR TO SIX

INGREDIENTS
    4 thick slices day-old bread, either
    white, brown or rye
    1 small red onion, thinly sliced
    450g/1lb ripe tomatoes, thinly sliced
    115g/4oz Mozzarella cheese
    5ml/1 tbsp fresh basil, shredded, or
       marjoram
    salt and ground black pepper
    120ml/4fl oz/½ cup extra virgin olive oil
    45ml/3 tbsp balsamic vinegar
    juice of 1 small lemon
    stoned and sliced black olives or salted
       capers, to garnish

1 Dip the bread briefly in cold water,
then carefully squeeze out the excess
water. Arrange on the bottom of a shallow
salad bowl.

2 Soak the onion slices in cold water for
about 10 minutes while you prepare the
other ingredients. Drain and reserve.

3 Layer the tomatoes, thinly sliced
cheese, onion, basil or marjoram,
seasoning well in between each layer.
Sprinkle with oil, vinegar and lemon juice.

4 Top with the olives or capers, cover
with cling film and chill in the refrigerator
overnight, if possible.

CAESAR'S SALAD

ON INDEPENDENCE DAY 1924, IN TIJUANA, MEXICO, A RESTAURANTEUR – CAESAR CARDINI – CREATED THIS MASTERPIECE OF NEW AMERICAN CUISINE.

SERVES FOUR

INGREDIENTS
  2 thick slices crustless bread
  2 garlic cloves
  sunflower oil, for frying
  1 cos lettuce, torn in pieces
  50g/2oz/½ cup fresh Parmesan cheese,
    coarsely grated
  2 eggs
For the dressing
  30ml/2 tbsp extra virgin olive oil
  10ml/2 tsp French mustard
  10ml/2 tsp Worcestershire sauce
  30ml/2 tbsp fresh lemon juice

1 Cut the bread into cubes. Heat one of the garlic cloves slowly in about 45ml/ 3 tbsp of the sunflower oil in a saucepan then toss in the bread cubes. Remove the garlic clove.

2 Heat an oven to 190°C/375°F/Gas 5. Spread the garlicky cubes on a baking sheet and bake the bread for about 10–12 minutes until golden and crisp. Remove and allow to cool.

3 Rub the inside of a large salad bowl with the remaining garlic clove and then discard it.

4 Toss in the torn lettuce, sprinkling between the leaves with the cheese. Cover and set the salad aside.

5 Boil a small saucepan of water and cook the eggs for 1 minute only. Remove the eggs, and crack them open into a jug or bowl. The whites should be milky and the yolks raw.

6 Whisk the dressing ingredients into the eggs. When ready to serve pour the dressing over the leaves, toss well together and serve topped with the croûtons.

VARIATION
Why not try a refreshing Italian version of this salad? Use cubed ciabatta bread for the croûtons. Rub the inside of a salad bowl with garlic and spoon in 30–45ml/ 2–3 tbsp of good olive oil. Add a selection of torn salad leaves, including rocket, together with some shaved Parmesan cheese. Do not mix yet. Just before serving, add the croûtons, season well, then toss the leaves with the oil, coating well. Finally squeeze over the juice of a fresh lemon.

CHEF'S SALAD

THIS IS A GOOD OPPORTUNITY
TO USE UP LEFTOVER VEGETABLES
AND PIECES OF CHEESE FROM THE
REFRIGERATOR.

SERVES SIX

INGREDIENTS
   450g/1lb new potatoes, halved if large
   2 carrots, grated coarsely
   ½ small fennel bulb or 2 sticks celery,
      sliced thinly
   50g/2oz sliced button mushrooms
   ¼ cucumber, sliced or chopped
   small green or red pepper, sliced
   60ml/4 tbsp garden peas
   200g/7oz/1 cup cooked pulses (e.g. red
      kidney beans or green lentils)
   1 Ice Gem or baby lettuce or 1 head
      chicory (endive)
   2–3 hard boiled eggs, quartered and/or
      grated cheese, to serve
   ½ carton salad cress, snipped

For the dressing
   60ml/4 tbsp mayonnaise
   45ml/3 tbsp natural yogurt
   30ml/2 tbsp milk
   30ml/2 tbsp chopped fresh chives or
      spring onion tops
   salt and ground black pepper

1 Put all the vegetables and pulses (except the lettuce or chicory) into a large mixing bowl.

2 Line a large platter with the lettuce or chicory leaves – creating a nest for the other salad ingredients. Mix the dressing ingredients together and pour over the salad in the mixing bowl.

3 Toss the salad thoroughly in the dressing, season well then pile into the centre of the lettuce nest.

4 Top the salad with the eggs, cheese or both and sprinkle with the snipped cress. Serve lightly chilled.

POTATO AND RADISH SALAD

*MANY POTATO SALADS ARE
DRESSED IN A THICK SAUCE. THIS
ONE IS QUITE LIGHT, WITH A
FLAVOURSOME YET DELICATE
DRESSING.*

SERVES FOUR TO SIX

INGREDIENTS

450g/1lb new potatoes, scrubbed
45ml/3 tbsp olive oil
15ml/1 tbsp walnut or hazelnut oil
 (optional)
30ml/2 tbsp wine vinegar
10ml/2 tsp coarse grain mustard
5ml/1 tsp clear honey
salt and ground black pepper
about 6–8 radishes, thinly sliced
30ml/2 tbsp fresh chives, chopped

1 Boil the potatoes until just tender.
Drain, return to the pan and cut any large
potatoes in half.

2 Make a dressing with the oils, vinegar,
mustard, honey and seasoning. Mix them
together thoroughly in a bowl.

3 Toss the dressing into the potatoes
while they are still cooling and allow them
to stand for an hour or so.

4 Mix in the radishes and chives, chill
lightly, toss again and serve.

COOK'S TIP

The secret of a good potato salad is to
dress the potatoes while they are still
warm in a vinaigrette-style dressing in
order to let them soak up the flavour as
they cool. You can then mix in an
additional creamy dressing of mayonnaise
and natural yogurt if liked. Sliced celery,
red onion and chopped walnuts would
make a good alternative to the radishes
and, for best effect, serve on a platter
lined with frilly lettuce leaves.

THAI RICE AND SPROUTING BEANS

*THAI RICE HAS A DELICATE
FRAGRANCE AND TEXTURE. THIS
SALAD IS A COLOURFUL
COLLECTION OF THAI FLAVOURS
AND TEXTURES.*

SERVES SIX

INGREDIENTS
    30ml/2 tbsp sesame oil
    30ml/2 tbsp fresh lime juice
    1 small fresh red chilli
    1 garlic clove, crushed
    10ml/2 tsp grated fresh root ginger
    30ml/2 tbsp light soy sauce
    5ml/1 tsp clear honey
    45ml/3 tbsp pineapple juice
    15ml/1 tbsp wine vinegar
    225g/8oz/1¼ cups Thai fragrant rice
    2 spring onions, sliced
    2 rings canned pineapple in natural
      juice, chopped
    150g/5oz/1¼ cups sprouted lentils
    1 small red pepper, sliced
    1 stick celery, sliced
    50g/2oz/½ cup unsalted cashew nuts,
      roughly chopped
    30ml/2 tbsp toasted sesame seeds
    salt and ground black pepper

1 Whisk together the sesame oil, lime juice, the seeded and chopped chilli, garlic, ginger, soy sauce, honey, pineapple juice and vinegar in a large bowl. Stir in the lightly boiled rice.

2 Toss in all the remaining ingredients and mix well. This dish can be served warm or lightly chilled. If the rice grains stick together on cooling, simply stir them with a metal spoon.

CHICORY, CARROT AND ROCKET SALAD

*A BRIGHT AND COLOURFUL
SALAD, WHICH IS IDEAL FOR A
BUFFET OR BARBECUE PARTY.*

SERVES FOUR TO SIX

INGREDIENTS
    3 carrots, coarsely grated
    about 50g/2oz fresh rocket or
      watercress, roughly chopped
    1 large head chicory
For the dressing
    45ml/3 tbsp sunflower oil
    15ml/1 tbsp hazelnut or walnut oil
      (optional)
    30ml/2 tbsp cider or wine vinegar
    10ml/2 tsp clear honey
    5ml/1 tsp grated lemon rind
    15ml/1 tbsp poppy seeds
    salt and ground black pepper

1 Mix the carrot and rocket or watercress together in a large bowl and season well.

2 Shake the dressing ingredients together in a screw-topped jar then pour onto the carrot and greenery. Toss the salad thoroughly.

3 Line a shallow salad bowl with the chicory leaves and spoon the salad into the centre. Serve lightly chilled.

BOUNTIFUL BEAN AND NUT SALAD

THIS IS A GOOD MULTI-PURPOSE DISH. IT CAN BE A COLD MAIN COURSE, A BUFFET PARTY DISH OR A SALAD ON THE SIDE. IT ALSO KEEPS WELL FOR UP TO THREE DAYS IF KEPT IN THE REFRIGERATOR.

SERVES SIX

INGREDIENTS
    75g/3oz/½ cup red kidney, pinto or
        borlotti beans
    75g/3oz/½ cup white cannellini or
        butter beans
    30ml/2 tbsp olive oil
    175g/6oz cut fresh green beans
    3 spring onions, sliced
    1 small yellow or red pepper, sliced
    1 carrot, coarsely grated
    30ml/2 tbsp dried topping onions or
        sun-dried tomatoes, chopped
    50g/2oz/½ cup unsalted cashew nuts or
        almonds, split

For the dressing
    45ml/3 tbsp sunflower oil
    30ml/2 tbsp red wine vinegar
    15ml/1 tbsp coarse grain mustard
    5ml/1 tsp caster sugar
    5ml/1 tsp dried mixed herbs
    salt and ground black pepper

1 Soak the beans, overnight if possible, then drain and rinse well, cover with a lot of cold water and cook according to the instructions on the pack.

2 When cooked, drain and season the beans and toss them in the olive oil. Leave to cool for 30 minutes.

3 In a large bowl, mix in the other vegetables, including the sun-dried tomatoes but not the crisp topping onions, if using, or the nuts.

4 Make up the dressing by shaking all the ingredients together in a screw-topped jar. Toss the dressing into the salad and check the seasoning again. Serve sprinkled with the topping onions, if using, and the split nuts.

GARDEN SALAD AND GARLIC CROSTINI

DRESS A COLOURFUL MIXTURE OF SALAD LEAVES WITH GOOD OLIVE OIL AND FRESH LEMON JUICE.

SERVES FOUR TO SIX

INGREDIENTS

3 thick slices day-old bread
120ml/4fl oz/½ cup extra virgin olive oil
garlic clove, cut
½ small cos or romaine lettuce
½ small oak leaf lettuce
25g/1oz rocket leaves or cress
25g/1oz fresh flat leaf parsley
a few leaves and flowers of nasturtium
flowers of pansy and pot marigold
a handful of young dandelion leaves
sea salt flakes and ground black pepper
juice of 1 fresh lemon

1 Cut the bread into medium size dice about 1cm/½in square.

2 Heat half the oil gently in a frying pan and fry the bread cubes in it, tossing them until they are well coated and lightly browned. Remove and cool.

3 Rub the inside of a large salad bowl with the garlic and discard. Pour the rest of the oil into the bottom of the bowl.

4 Wash, dry and tear the leaves into bite size pieces and pile them into the bowl. Season with salt and pepper. Cover and keep chilled until ready to serve.

5 To serve, toss the leaves in the oil at the bottom of the bowl, then sprinkle with the lemon juice and toss again. Scatter over the crostini and serve immediately.

CALIFORNIAN VIM AND VIT SALAD

FULL OF VITALITY AND VITAMINS, THIS IS A LOVELY LIGHT, HEALTHY SALAD FOR SUNNY DAYS WHEN YOU FEEL FULL OF ENERGY OR WHEN YOU NEED AN EXTRA BOOST.

SERVES FOUR

INGREDIENTS
1 small crisp lettuce, torn in pieces
225g/8oz young spinach leaves
2 carrots, coarsely grated
115g/4oz cherry tomatoes, halved
2 celery sticks, thinly sliced
75g/3oz/½ cup raisins
50g/2oz/½ cup blanched almonds or
  unsalted cashew nuts, halved
30ml/2 tbsp sunflower seeds
30ml/2 tbsp sesame seeds, lightly
  toasted

For the dressing
45ml/3 tbsp extra virgin olive oil
30ml/2 tbsp cider vinegar
10ml/2 tsp clear honey
juice of 1 small orange
salt and ground black pepper

1 Put the salad vegetables, raisins, almonds or cashew nuts and seeds into a large bowl.

2 Put all the dressing ingredients into a screw top jar, shake them up well and pour over the salad.

3 Toss the salad thoroughly and divide it between four small salad bowls. Season and serve lightly chilled.

SCANDINAVIAN CUCUMBER AND DILL

IT'S AMAZING WHAT A TOUCH OF LIGHT SALT CAN DO TO SIMPLE CUCUMBER SLICES. THEY TAKE ON A CONTRADICTORY SOFT YET CRISP TEXTURE AND DEVELOP A GOOD, FULL FLAVOUR. HOWEVER, JUICES CONTINUE TO FORM AFTER SALTING, SO THIS SALAD IS BEST DRESSED JUST BEFORE SERVING.

SERVES FOUR

INGREDIENTS
2 cucumbers
salt
30ml/2 tbsp fresh chives, chopped
30ml/2 tbsp fresh dill, chopped
150ml/¼ pt/⅔ cup soured cream or
  fromage frais
ground black pepper

1 Slice the cucumbers as thinly as possible, preferably in a food processor or a slicer.

2 Place the slices in layers in a colander set over a plate to catch the juices. Sprinke each layer well, but not too heavily, with salt.

3 Leave the cucumber to drain for up to 2 hours, then lay out the slices on a clean tea towel and pat them dry.

4 Mix the cucumber with the herbs, cream or fromage frais and plenty of pepper. Serve as soon as possible.

COOK'S TIP
Lightly salted (or degorged) cucumbers are also delicious as sandwich fillings in wafer thin buttered brown bread. These sandwiches were always served at the traditional British tea-time.

POTATO SALAD <u>WITH</u> CURRY PLANT MAYONNAISE

POTATO SALAD CAN BE MADE WELL IN ADVANCE AND IS THEREFORE A USEFUL BUFFET DISH. ITS POPULARITY MEANS THAT THERE ARE VERY RARELY ANY LEFTOVERS.

SERVES SIX

INGREDIENTS
salt
1kg/2lb new potatoes, in skins
300ml/½ pint/1¼ cups shop-bought
  mayonnaise
6 curry plant leaves, roughly chopped
black pepper
mixed lettuce or other salad greens, to
  serve

1 Place the potatoes in a pan of salted water and boil for 15 minutes or until tender. Drain and place in a large bowl to cool slightly.

2 Mix the mayonnaise with the curry plant leaves and black pepper. Stir these into the potatoes while they are still warm. Leave to cool, then serve on a bed of mixed lettuce or other assorted salad leaves.

TOMATO, SAVORY AND FRENCH BEAN SALAD

SAVORY AND BEANS MUST HAVE BEEN INVENTED FOR EACH OTHER. THIS SALAD MIXES THEM WITH RIPE TOMATOES, MAKING A SUPERB ACCOMPANIMENT FOR ALL VEGETABLE DISHES.

SERVES FOUR

INGREDIENTS
    500g/1lb French beans
    1kg/2lb ripe tomatoes
    3 spring onions, roughly sliced
    15ml/1 tbsp pine nuts
    4 sprigs fresh savory
For the dressing
    30ml/2 tbsp extra virgin olive oil
    juice of 1 lime
    75g/3oz soft blue cheese
    1 clove garlic, peeled and crushed
    salt and pepper

1 Prepare the dressing first so that it can stand a while before using. Place all the dressing ingredients in the bowl of a food processor, season to taste and blend until all the cheese has been finely chopped and you have a smooth dressing. Pour it into a jug.

2 Top and tail the beans, and boil in salted water until they are just cooked. Drain them and run cold water over them until they have completely cooled. Slice the tomatoes, or, if they are fairly small, quarter them.

3 Toss the salad ingredients together, except for the pine nuts and savory. Pour on the salad dressing. Sprinkle the pine nuts over the top, followed by the savory.

French Bean Salad

Although bean salads are delicious served with a simple vinaigrette dressing, this dish is a little more elaborate. It does, however, enhance the fresh flavour of the beans.

SERVES FOUR

INGREDIENTS
    450g/1lb French beans
    15ml/1 tbsp olive oil
    25g/1oz butter
    ½ garlic clove, crushed
    50g/2oz fresh white breadcrumbs
    15ml/1 tbsp chopped fresh
      parsley
    1 egg, hard-boiled and finely chopped
For the dressing
    30ml/2 tbsp olive oil
    30ml/2 tbsp sunflower oil
    10ml/2 tsp white wine vinegar
    ½ garlic clove, crushed
    1.5ml/¼ tsp French mustard
    pinch of sugar
    pinch of salt

1 Trim the French beans and cook in boiling salted water for 5–6 minutes until tender. Drain the beans and refresh them under cold running water and place in a serving bowl.

2 Make the salad dressing by blending the oils, vinegar, garlic, mustard, sugar and salt thoroughly together. Pour over the beans and toss to mix.

COOK'S TIP
For a more substantial salad, boil about 450g/1lb scrubbed new potatoes until tender, cool and then cut them into bite-size chunks. Stir into the French beans and then add the dressing.

3 Heat the oil and butter in a frying pan and fry the garlic for 1 minute. Stir in the breadcrumbs and fry over a moderate heat for about 3–4 minutes until golden brown, stirring frequently.

4 Remove the pan from the heat and stir in the parsley and then the egg. Sprinkle the breadcrumb mixture over the French beans. Serve warm or at room temperature.

RADISH, MANGO AND APPLE SALAD

RADISH IS A YEAR-ROUND VEGETABLE AND THIS SALAD CAN BE SERVED AT ANY TIME OF YEAR, WITH ITS CLEAN, CRISP TASTES AND MELLOW FLAVOURS. SERVE WITH SMOKED FISH, SUCH AS ROLLS OF SMOKED SALMON, OR WITH CONTINENTAL HAM OR SALAMI.

SERVES FOUR

INGREDIENTS
   10–15 radishes
   1 dessert apple, peeled cored and
      thinly sliced
   2 celery stalks, thinly sliced
   1 small ripe mango, peeled and cut
      into small chunks
For the dressing
   120ml/4fl oz/½ cup soured cream
   10ml/2 tsp creamed horseradish
   15ml/1 tbsp chopped fresh dill
   salt and freshly ground black pepper
   sprigs of dill, to garnish

3 Cut through the mango lengthways either side of the stone. Make even criss-cross cuts through each side section. Take each one and bend it back to separate the cubes. Remove the mango cubes with a small knife and add to the bowl. Pour the dressing over the vegetables and fruit and stir gently so that all the ingredients are coated in the dressing. When ready to serve, spoon the salad into an attractive salad bowl and garnish with sprigs of dill.

1 To prepare the dressing, blend together the soured cream, horseradish and dill in a small jug or bowl and season with a little salt and pepper.

2 Top and tail the radishes and then slice them thinly. Add to a bowl together with the thinly sliced apple and celery.

CRUNCHY CABBAGE SALAD WITH PESTO MAYONNAISE

BOTH THE PESTO AND THE MAYONNAISE CAN BE MADE FOR THIS DISH. HOWEVER, IF TIME IS SHORT, YOU CAN BUY THEM BOTH READY-PREPARED AND IT WILL TASTE JUST AS GOOD. ADD THE DRESSING JUST BEFORE SERVING TO PREVENT THE CABBAGE GOING SOGGY.

SERVES FOUR TO SIX

INGREDIENTS
 1 small or ½ medium white cabbage
 3–4 carrots, grated
 4 spring onions, finely sliced
 25–40g/1–1½oz pine nuts
 15ml/1 tbsp chopped fresh mixed
  herbs; parsley, basil and chervil
For the pesto mayonnaise
 1 egg yolk
 about 10ml/2 tsp lemon juice
 200ml/7fl oz/⅞ cup sunflower oil
 10ml/2 tsp pesto
 60ml/4 tbsp natural yogurt
 salt and freshly ground black pepper

1 To make the mayonnaise, place the egg yolk in a blender or food processor and process with the lemon juice. With the machine running, very slowly add the oil, pouring it more quickly as the mayonnaise emulsifies. Season to taste with salt and pepper and a little more lemon juice if necessary. Alternatively, make by hand using a balloon whisk.

2 Spoon 75ml/5 tbsp of mayonnaise into a bowl and stir in the pesto and yogurt, beating well to make a fairly thin dressing. (The remaining mayonnaise will keep for about 3–4 weeks in a screw-top jar in the fridge.)

3 Using a food processor or a sharp knife, thinly slice the cabbage and place in a large salad bowl.

4 Add the carrots and spring onions, together with the pine nuts and herbs, mixing thoroughly with your hands. Stir the pesto dressing into the salad or serve separately in a small dish if preferred.

ROCKET AND GRILLED CHÈVRE SALAD

FOR THIS RECIPE, LOOK OUT FOR CYLINDER-SHAPED GOAT'S CHEESE FROM A DELICATESSEN OR FOR SMALL ROLLS THAT CAN BE CUT INTO HALVES, WEIGHING ABOUT 50G/2OZ. SERVE ONE PER PERSON AS A STARTER OR DOUBLE THE RECIPE AND SERVE TWO EACH FOR A LIGHT LUNCH.

SERVES FOUR

INGREDIENTS

    about 15ml/1 tbsp olive oil
    about 15ml/1 tbsp vegetable oil
    4 slices French bread
    45ml/3 tbsp walnut oil
    15ml/1 tbsp lemon juice
    salt and freshly ground black pepper
    225g/8oz cylinder-shaped goat's
      cheese
    generous handful of rocket leaves
    about 115g/4oz curly endive
For the sauce
    45ml/3 tbsp apricot jam
    60ml/4 tbsp white wine
    5ml/2 tsp Dijon mustard

1 Heat the two oils in a frying pan and fry the slices of French bread on one side only, until lightly golden. Transfer to a plate lined with kitchen paper.

4 Preheat the grill a few minutes before serving the salad. Cut the goat's cheese into 50g/2oz rounds and place each piece on a croûton, untoasted side up. Place under the grill and cook for 3–4 minutes until the cheese melts.

5 Toss the rocket and curly endive in the walnut oil dressing and arrange attractively on four individual serving plates. When the cheese croûtons are ready, arrange on each plate and pour over a little of the apricot sauce.

2 To make the sauce, heat the jam in a small saucepan until warm but not boiling. Push through a sieve, into a clean pan, to remove the pieces of fruit, and then stir in the white wine and mustard. Heat gently and keep warm until ready to serve.

3 Blend the walnut oil and lemon juice and season with a little salt and pepper.

BRAISED CELERY WITH GOAT'S CHEESE

THE SHARP FLAVOUR OF THE CELERY IN THIS DISH IS PERFECTLY COMPLEMENTED BY THE MILD YET TANGY GOAT'S CHEESE. THIS RECIPE IS AN EXAMPLE OF QUICK AND EASY PREPARATION TO MAKE A DELICIOUS ACCOMPANIMENT TO GRILLED MEAT OR STUFFED PANCAKES.

SERVES FOUR

INGREDIENTS
  25g/1oz butter
  1 head of celery, thinly sliced
  175g/6oz mild medium-fat goat's
    cheese
  45–60ml/3–4 tbsp single cream
  salt and freshly ground black pepper

1 Preheat the oven to 180°C/350°F/ Gas 4 and lightly butter a medium-size shallow ovenproof dish.

2 Melt the butter in a heavy-based saucepan and fry the thinly sliced celery for 2–3 minutes, stirring frequently. Add 45–60ml/3–4 tbsp water to the pan, heat gently and then cover and simmer over a gentle heat for 5–6 minutes, until the celery is nearly tender and the water has almost evaporated.

3 Remove the pan from the heat and stir in the goat's cheese and cream. Taste and season with salt and pepper, and then turn into the prepared dish.

4 Cover the dish with buttered grease-proof paper and cook in the oven for 10–12 minutes. Serve at once.

CELERY, AVOCADO AND WALNUT SALAD

THE CRUNCHINESS OF THE CELERY AND WALNUTS CONTRASTS PERFECTLY WITH THE SMOOTH AVOCADO. SERVE IT WITH A SOURED CREAM DRESSING AS SUGGESTED, OR SIMPLY DRESSED WITH A LITTLE OLIVE OIL AND FRESHLY SQUEEZED LEMON JUICE.

SERVES FOUR

INGREDIENTS
  8 tender white celery stalks,
    very thinly sliced
  3 spring onions, finely chopped
  50g/2oz walnut halves
  1 ripe avocado
  lemon juice
For the dressing
  120ml/4fl oz/½ cup soured cream
  15ml/1 tbsp olive oil
  pinch of cayenne pepper

1 Place the celery, spring onions and walnuts in a large salad bowl.

2 Halve the avocado and, using a very sharp knife, cut into thin slices. Peel away the skin from each slice and then sprinkle generously with lemon juice and add to the celery mixture.

3 Lightly beat the soured cream, olive oil and cayenne pepper together in a jug or small bowl. Either fold carefully into the salad or serve separately.

SPINACH AND CANNELLINI BEANS

THIS HEARTY DISH CAN BE MADE WITH ALMOST ANY DRIED BEAN, SUCH AS BLACK-EYED BEANS, HARICOTS OR CHICK-PEAS. IT IS A GOOD DISH TO SERVE ON A COLD EVENING. IF USING CANNED BEANS, DRAIN, THEN RINSE UNDER COLD WATER.

SERVES FOUR

INGREDIENTS

225g/8oz cannellini beans,
 soaked overnight
60ml/4 tbsp olive oil
1 slice white bread
1 onion, chopped
3–4 tomatoes, peeled and chopped
a good pinch of paprika
450g/1lb spinach
1 garlic clove, halved
salt and freshly ground black pepper

1 Drain the beans, place in a saucepan and cover with water. Bring to the boil and boil rapidly for 10 minutes. Cover and simmer for about 1 hour until the beans are tender. Drain.

2 Heat 30ml/2 tbsp of the oil in a frying pan and fry the bread until golden brown. Transfer to a plate.

3 Fry the onion in 15ml/1 tbsp of the oil over a gentle heat until soft but not brown, then add the tomatoes and continue cooking over a gentle heat.

4 Heat the remaining oil in a large pan, stir in the paprika and then add the spinach. Cover and cook for a few minutes until the spinach has wilted.

5 Add the onion and tomato mixture to the spinach, mix well and stir in the cannellini beans. Place the garlic and fried bread in a food processor and process until smooth. Stir into the spinach and bean mixture. Add 150ml/ ¼ pint/⅔ cup cold water and then cover and simmer gently for 20–30 minutes, adding more water if necessary.

SPINACH IN FILO WITH THREE CHEESES

A GOOD CHOICE TO SERVE WHEN VEGETARIANS AND MEAT EATERS ARE GATHERED FOR A MEAL AS, WHATEVER THEIR PREFERENCE, EVERYONE SEEMS PARTIAL TO THIS TASTY DISH.

SERVES FOUR

INGREDIENTS

450g/1lb spinach
15ml/1 tbsp sunflower oil
15g/½oz butter
1 small onion, finely chopped
175g/6oz ricotta cheese
115g/4oz feta cheese, cut into
  small cubes
75g/3oz Gruyère or Emmenthal
  cheese, grated
15ml/1 tbsp fresh chopped chervil
5ml/1 tsp fresh chopped marjoram
salt and freshly ground black pepper
5 large or 10 small sheets filo pastry
40–50g/1½–2oz butter, melted

1 Preheat the oven to 190°C/375°F/ Gas 5. Cook the spinach in a large saucepan over a moderate heat for 3–4 minutes until the leaves have wilted, shaking the saucepan occasionally. Strain and press out the excess liquid.

2 Heat the oil and butter in a saucepan and fry the onion for 3–4 minutes until softened. Remove from the heat and add half of the spinach. Combine, using a metal spoon to break up the spinach.

3 Add the ricotta cheese and stir until evenly combined. Stir in the remaining spinach, again chopping it into the mixture with a metal spoon. Fold in the feta and Gruyère or Emmenthal cheese, chervil, marjoram and seasoning.

4 Lay a sheet of filo pastry measuring about 30cm/12 in square on a work surface. (If you have small filo sheets, lay them side by side, overlapping by about 2.5cm/1in in the middle.) Brush with melted butter and cover with a second sheet; brush this with butter and build up five layers of pastry in this way.

5 Spread the filling over the pastry, leaving a 2.5cm/1in border. Fold the sides inwards and then roll up.

6 Place the roll, seam side down, on a greased baking sheet and brush with the remaining butter. Bake in the oven for about 30 minutes until golden brown.

MIGHTY MUSHROOMS

THERE IS NOW A WIDE VARIETY OF CULTIVATED MUSHROOMS ON SALE IN LARGER SUPERMARKETS AND GREENGROCERS, SO A SIMPLE SIDE DISH OF MUSHROOMS BECOMES QUITE EXCITING.

SERVES FOUR

INGREDIENTS

15g/½oz pack dried porcini mushrooms
    or ceps (optional)
60ml/4 tbsp olive oil
225g/8oz button mushrooms, halved or
    sliced
115g/4oz oyster mushrooms
115g/4oz fresh shiitake mushrooms, or
    25g/1oz dried and soaked mushrooms
2 garlic cloves, crushed
10ml/2 tsp ground coriander
salt and ground black pepper
45ml/3 tbsp fresh parsley, chopped

1 If you are using porcini mushrooms (and they do give a good rich flavour), soak them in a little hot water just to cover for 20 minutes.

2 In a large saucepan, heat the oil and add all the mushrooms, including the soaked porcinis, if using. Stir well, cover and cook gently for 5 minutes.

3 Stir in the garlic, coriander and seasoning. Cook for another 5 minutes until the mushrooms are tender and much of the liquor has been reduced.

4 Mix in the parsley, allow to cool slightly and serve.

TOMATO SAUCE

THIS BASIC SAUCE CAN EITHER BE PART OF A LARGER RECIPE OR SERVED AS A SIDE SAUCE. FOR EXTRA FLAVOUR, ADD RED PEPPER, OR RING THE CHANGES WITH ORANGE RIND AND JUICE AND CHOPPED FRESH HERBS SUCH AS BASIL.

SERVES FOUR TO SIX

INGREDIENTS

1 onion, chopped
2 garlic cloves, crushed
1 small red pepper (optional), chopped
45ml/3 tbsp olive oil
675g/1½lb fresh tomatoes, skinned and
    chopped or 1 × 400g/14oz can
    chopped tomatoes
15g/½oz granulated sugar
salt and ground black pepper
30ml/2 tbsp fresh herbs, chopped (e.g.
    basil, parsley, marjoram; optional)

1 Gently fry the onion, garlic and red pepper, if using, in the oil for 5 minutes until they are soft.

2 Stir in the tomatoes, add the sugar and seasoning to taste, bring to a boil then cover and simmer for 15–20 minutes.

3 The sauce should now be thick and pulpy. If it is a little thin, then boil it – uncovered – so that it reduces down. Stir in the fresh herbs, if using, and then check the seasoning.

COOK'S TIP
Why not make up a large batch of this sauce and freeze it in two-portion sizes?

SPICED OKRA WITH ALMONDS

LONG AND ELEGANTLY SHAPED, IT'S NOT SURPRISING THESE VEGETABLES HAVE THE POPULAR NAME OF "LADY'S FINGERS".

SERVES TWO TO FOUR

INGREDIENTS
225g/8oz okra
50g/2oz/½ cup blanched almonds, chopped
25g/1oz/2 tbsp butter
15ml/1 tbsp sunflower oil
2 garlic cloves, crushed
2.5cm/1in cube fresh root ginger, grated
5ml/1 tsp cumin seeds
5ml/1 tsp ground coriander
5ml/1 tsp paprika

1 Trim just the tops of the okra stems and around the edges of the stalks. They have a sticky liquid which oozes out if prepared too far ahead, so trim them immediately before cooking.

2 In a shallow fireproof dish, fry the almonds in the butter until they are lightly golden, then remove.

3 Add the oil to the pan and fry the okra, stirring constantly, for 2 minutes.

4 Add the garlic and ginger and fry gently for a minute, then add the spices and cook for another minute or so, stirring all the time.

5 Pour in about 300ml/½ pint/1¼ cups water. Season well, cover and simmer for about 5 minutes or so until the okra feel just tender.

6 Finally, mix in the fried almonds and serve piping hot.

VARIATION
Okra are also popular in Louisiana cooking and are an essential ingredient for gumbo, a thick, spicy stew served over hot, steaming rice. Indeed, "gumbo" was the old African word for okra used by the American slaves. You can make a ratatouille-style vegetable stew using okra instead of aubergines, and adding onions, peppers, garlic and tomatoes. Or try them sliced, fried in garlic and spices, then stirred into a pilaff of basmati rice with cauliflower florets and carrots. This makes a colourful and delicious dish – especially when topped with crushed grilled popadums.

STIR-FRY CABBAGE

An often under-rated vegetable, crisp cabbage is wonderful when lightly cooked the Chinese way in a wok. Any cabbage will do, but perhaps the Savoy cabbage is the nicest.

SERVES FOUR

INGREDIENTS
½ small cabbage
30ml/2 tbsp sunflower oil
15ml/1 tbsp light soy sauce
15ml/1 tbsp fresh lemon juice
  (optional)
10ml/2 tsp caraway seeds
ground black pepper

1 Cut the central core from the cabbage, and shred the leaves finely.

2 Heat the oil until quite hot in a wok and then stir-fry the cabbage for about 2 minutes.

3 Toss in the soy sauce, lemon juice, if using, caraway seeds and pepper, to taste.

PERFECT CREAMED POTATOES

If a bowl of plain, mashed potatoes sounds boring, then try serving it this way — in the French style. It is absolutely scrumptious! Use good quality, floury potatoes.

SERVES FOUR

INGREDIENTS
1 kg/2lb potatoes, peeled and diced
45ml/3 tbsp extra virgin olive oil
about 150ml/¼ pint/⅔ cup milk
fresh nutmeg, grated
salt and ground black pepper
a few leaves fresh basil or sprigs fresh
  parsley, chopped

1 Boil the potatoes until just tender and not too mushy. Drain very well. Ideally press the potatoes through a special potato "ricer" (rather like a large garlic press) or mash them well with a potato masher. Do not pass them through a food processor or you will have a gluey mess.

2 Beat the olive oil into the potato and just enough hot milk to make a smooth, thick purée.

3 Flavour to taste with the nutmeg and seasoning then stir in the fresh chopped herbs. Spoon into a warm serving dish and serve as soon as possible.

ROASTED PEPPERS IN OIL

PEPPERS TAKE ON A DELICIOUS, SMOKY FLAVOUR IF ROASTED IN A VERY HOT OVEN. THE SKINS CAN EASILY BE PEELED OFF AND THE FLESH STORED IN OLIVE OIL.

INGREDIENTS
6 large peppers of differing colours
450ml/¾ pint/scant 2 cups olive oil

1 Preheat the oven to the highest temperature, about 230°C/450°F/Gas 8. Lightly grease a large baking sheet.

3 Roast the peppers at the top of the oven until the skins blacken and blister. This will take about 12-15 minutes.

6 Add the oil to the jar to cover the peppers completely, then seal the lid tightly.

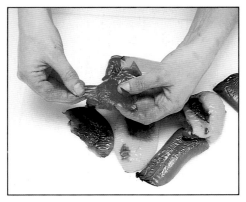

4 Remove the peppers from the oven, cover with a clean tea towel until they are cool, then peel off the skins.

5 Slice the peppers and pack them into a clean preserving jar.

7 Store the peppers in the refrigerator and use them within 2 weeks. Use the oil in dressing or for cooking once the peppers have been eaten.

2 Quarter the peppers, remove the cores and seeds then squash them flat with the back of your hands. Lay the peppers skin side up on the baking tray.

COOK'S TIP
These pepper slices make very attractive presents, especially around Christmas time. You can either buy special preserving jars from kitchen equipment shops or wash out large jam jars, soaking off the labels at the same time. The jars should then be sterilized by placing them upside down in a low oven for about half an hour. Fill the jars while still hot with the sliced peppers, and top up with a good olive oil. Cover immediately with the lid and fix on an attractive label.

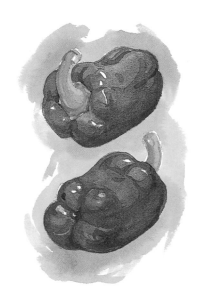

OVEN CHIP ROASTIES

VERY POPULAR WITH CHILDREN AND A MUCH BETTER ALTERNATIVE FOR ALL THE FAMILY TO HIGH-FAT ROAST POTATOES.

SERVES FOUR TO SIX

INGREDIENTS
    4 medium to large baking potatoes
    150ml/¼ pint/⅔ cup olive oil
    5ml/1 tsp dried mixed herbs (optional)
    sea salt flakes

1 Preheat the oven to the highest possible temperature, generally 230°C/450°F/Gas 8. Place a lightly-oiled roasting tin in the oven to get really hot.

2 Cut the potatoes in half lengthways then into long thin wedges. Brush each side lightly with oil.

3 When the oven is really hot, remove the pan carefully and lay the potato slices over it in a single layer.

4 Sprinkle the potatoes with the herbs and salt and place in the oven for about 20 minutes or until they are golden brown, crisp and lightly puffy. Serve immediately.

VARIATION
Parsnips also make fine oven chips. Choose large parsnips which tend to have more flavour. Slice thinly on a diagonal and roast in the same way as above, although you may find they do not take as long to cook. They make great mid-week suppers for kids and grown-ups served with fried eggs, mushrooms and tomatoes.

WARM SPICY DHAL

IF YOU THOUGHT SPLIT YELLOW PEAS WERE ONLY FOR SOUPS, TRY THIS INDIAN INSPIRED DISH. SERVE WITH RICE, CHAPPATIS OR NAAN BREAD AND WHATEVER MAIN DISH YOU LIKE.

SERVES FOUR TO SIX

INGREDIENTS

225g/8oz yellow split peas
2 onions, chopped
1 large bay leaf
600ml/1 pint/2½ cups stock or water
salt and ground black pepper
10ml/2 tsp black mustard seeds
30ml/2 tbsp melted butter
1 garlic clove, crushed
2.5cm/1in cube fresh root ginger
1 small green pepper, sliced
5ml/1 tsp ground turmeric
5ml/1 tsp garam masala
3 tomatoes, skinned and chopped
fresh coriander or parsley, to serve

1 Put the split peas, 1 onion and the bay leaf in the stock or water, in a covered pan. Simmer for 25 minutes, seasoning lightly towards the end.

2 In a separate pan, fry the mustard seeds in the butter for about 30 seconds until they start to pop, then add all the remaining onion, along with the garlic, grated ginger and green pepper.

3 Sauté for about 5 minutes until softened then stir in the remaining spices and fry for a few seconds more.

4 Add the split peas, tomatoes, and a little extra water if it needs it. Cover and simmer for a further 10 minutes, then check the seasoning and serve hot garnished with coriander or parsley.

POTATO LATKES

These little potato pancakes make a pleasant and unusual alternative to chips or roast potatoes.

MAKES ABOUT 24

INGREDIENTS
1 kg/2lb potatoes, peeled and coarsely grated
40g/1½oz/scant ½ cup self-raising flour
2 eggs
15ml/1 tbsp onion, grated
fresh nutmeg, grated
salt and ground black pepper
oil, for shallow frying

1 Soak the grated potato in plenty of cold water for about an hour, then drain well and pat dry with a clean tea towel.

2 Beat together the flour, eggs, onion and nutmeg, then mix in the potato. Season well.

3 Heat a thin layer of oil in a heavy based frying pan and drop about a tablespoon of potato-batter into the pan, squashing it flat, if necessary.

4 Cook the potato until golden brown, then flip over and cook the other side. Drain on kitchen paper towel and keep warm, uncovered in the oven. Repeat with the rest of the mixture.

STIR-FRY AUBERGINE

A speedy side dish with an Oriental touch. The aubergine is stir-fried with red pepper and black beans which gives it an exotic and colourful appearance. Salted black beans are sold either dried or canned. Dried ones will need soaking.

SERVES FOUR

INGREDIENTS
30ml/2 tbsp groundnut oil
1 aubergine, sliced
2 spring onions, sliced diagonally
1 garlic clove, crushed
1 small red pepper, sliced
30ml/2 tbsp oyster sauce
25g/1oz/1 tbsp Chinese salted black beans, soaked if dried
ground black pepper
15ml/1 tbsp fresh coriander or parsley, chopped, to garnish

1 Heat the oil in a wok and stir-fry the aubergine for 2 minutes. Add the onions, garlic and pepper and cook for a further 2 minutes.

2 Add the oyster sauce, black beans and pepper. Cook for a further 1 minute, season with pepper only and serve with fresh coriander or parsley.

PEAS AND LETTUCE

DO NOT DISCARD THE TOUGH,
OUTER LEAVES OF LETTUCE —
THEY ARE DELICIOUS IF THEY ARE
SHREDDED AND COOKED WITH
PEAS.

SERVES FOUR

INGREDIENTS

6 outer leaves of lettuce (e.g. cos,
   Webbs)
1 small onion or shallot, sliced
25g/1oz/2 tbsp butter or sunflower
   margarine (for vegans)
225g/8oz frozen garden peas
fresh nutmeg, grated
salt and ground black pepper

1 Pull off the outer lettuce leaves and
wash them well. Roughly shred the leaves
with your hands.

2 In a saucepan, lightly fry the lettuce
and onion or shallot in the butter or
margarine for 3 minutes.

3 Add the peas, nutmeg to taste and
seasoning. Stir, cover and simmer for
about 5 minutes. This dish can be
drained or served slightly wet.

HASH BROWNS

*A TRADITIONAL AMERICAN
BREAKFAST DISH, HASH BROWNS
CAN BE SERVED ANY TIME OF DAY.*

SERVES FOUR

INGREDIENTS

60ml/4 tbsp sunflower or olive oil
about 450g/1lb diced, cooked potatoes
1 small onion, chopped
salt and ground black pepper

1 Heat the oil in a large, heavy-based
frying pan and when quite hot add the
potatoes in a single layer. Scatter the
onion on top and season well.

2 Cook on a moderate heat until browned
underneath, pressing down on the
potatoes with a spoon or spatula to
squash them together.

3 When the potatoes are nicely browned,
turn them over in sections with a fish slice
and fry on the other side, pressing them
down once again. Serve when heated
through and lightly crispy.

POTATO AND PARSNIP DAUPHINOIS

*LAYERS OF POTATOES AND
PARSNIPS ARE BAKED SLOWLY IN
CREAMY MILK WITH GRATED
CHEESE. THIS IS AN IDEAL
SPECIAL SIDE DISH OR LIGHT
SUPPER DISH.*

SERVES FOUR TO SIX

INGREDIENTS
    1kg/2lb potatoes, thinly sliced
    1 onion, thinly sliced
    450g/1lb parsnips, thinly sliced
    2 garlic cloves, crushed
    50g/2oz/4 tbsp butter
    125g/4oz Gruyère or Cheddar cheese,
      grated
    fresh nutmeg, grated
    salt and ground black pepper
    300ml/½ pint single cream
    300ml/½ pint milk

1 Lightly grease a large shallow ovenproof dish. Preheat the oven to 180°C/350°F/Gas 4.

2 Layer the potatoes with the onion and parsnips. In between each layer, dot the vegetables with garlic and butter, sprinkle over most of the cheese, add the nutmeg and season well.

3 Heat the cream and milk together in a saucepan until it is hot but not boiling. Slowly pour the creamy milk over the vegetables, making sure it seeps underneath them.

4 Scatter the remaining cheese over the vegetables and grate a little more nutmeg on top. Bake for about an hour or so until the potatoes are tender and the cheesy top is bubbling and golden.

ASPARAGUS MIMOSA

*PRETTY SPEARS OF FRESH,
TENDER ASPARAGUS ARE TOSSED
IN A BUTTERY SAUCE AND SERVED
WITH A CHOPPED EGG AND
CHERVIL DRESSING. THIS ALSO
MAKES AN EXCELLENT STARTER.*

SERVES TWO

INGREDIENTS
    225g/8oz fresh asparagus
    salt
    ground black pepper
    50g/2oz/4 tbsp butter, melted
    squeeze of fresh lemon juice
    2 eggs, hard-boiled and chopped
    15ml/1 tbsp fresh chervil, chopped

1 Trim the asparagus stalks and peel off the tough outer layers at the base with a vegetable peeler.

2 Poach the asparagus spears in lightly salted water, until just tender. This will take between 3–6 minutes. (Use a clean, deep frying pan if you don't have an asparagus steamer.)

3 Drain the asparagus well and arrange it on two small plates or one large one. Season well.

4 Mix the melted butter with the lemon juice. Trickle over the spears, sprinkle with the eggs and garnish with the chervil. Serve warm.

STIR-FRIED BRUSSELS SPROUTS

SERVES FOUR

INGREDIENTS
    450g/1lb Brussels sprouts
    15ml/1 tbsp sunflower oil
    6–8 spring onions, cut into 2.5cm/
      1 in lengths
    2 slices fresh root ginger
    40g/1½oz slivered almonds
    150–175ml/5–6fl oz/⅔–¾ cup
      vegetable stock
    salt

1 Remove any large outer leaves and trim the bases of the Brussels sprouts. Cut into slices about 7mm/⅓in thick.

2 Heat the oil in a wok or heavy frying pan and fry the spring onions and the ginger for 2–3 minutes, stirring frequently. Add the almonds and stir-fry over a moderate heat until both the onions and almonds begin to brown.

3 Remove and discard the ginger, reduce the heat and stir in the Brussels sprouts. Stir-fry for a few minutes and then pour in the stock and cook over a gentle heat for 5–6 minutes or until the sprouts are nearly tender.

4 Add a little salt, if necessary, and then increase the heat to boil off the excess liquid. Spoon into a warmed serving dish and serve immediately.

BRUSSELS SPROUTS GRATIN

SERVES FOUR

INGREDIENTS
    150ml/¼ pint/⅔ cup whipping cream
    150ml/¼ pint/⅔ cup milk
    25g/1oz Parmesan cheese, grated
    675g/1½lb Brussels sprouts, thinly
      sliced
    15g/½oz butter
    1 garlic clove, finely chopped
    salt and freshly ground black pepper

1 Preheat the oven to 150°C/300°F/ Gas 2 and butter a shallow ovenproof dish. Blend together the cream, milk, Parmesan cheese and seasoning.

2 Place a layer of Brussels sprouts in the base of the prepared dish, sprinkle with a little garlic and pour over about a quarter of the cream mixture. Add another layer of sprouts and continue building layers in this way, ending with the remaining cream and milk.

3 Cover loosely with greaseproof paper and bake for 1–1¼ hours. Halfway through cooking, remove the paper and press the sprouts under the liquid in the dish. Return to the oven to brown.

PUFFY CREAMED POTATOES

THIS ACCOMPANIMENT CONSISTS OF CREAMED POTATOES INCORPORATED INTO MINI YORKSHIRE PUDDINGS. SERVE THEM WITH ROAST DUCK OR BEEF, OR WITH A VEGETARIAN CASSEROLE. FOR A MEAL ON ITS OWN, SERVE TWO OR THREE PER PERSON AND ACCOMPANY WITH SALADS.

MAKES SIX

INGREDIENTS
    275g/10oz potatoes
    creamy milk and butter for mashing
    5ml/1 tsp chopped fresh parsley
    5ml/1 tsp chopped fresh tarragon
    75g/3oz plain flour
    1 egg
    about 120ml/4fl oz/½ cup milk
    oil or sunflower fat, for baking
    salt and freshly ground black pepper

1 Boil the potatoes until tender and mash with a little milk and butter. Stir in the chopped parsley and tarragon and season well to taste. Preheat the oven to 200°C/400°F/Gas 6.

2 Process the flour, egg, milk and a little salt in a food processor or blender to make a smooth batter.

3 Place about 2.5ml/½ tsp oil or a small knob of sunflower fat in each of six ramekin dishes and place in the oven on a baking tray for 2–3 minutes until the oil or fat is very hot.

4 Working quickly, pour a small amount of batter (about 20ml/4 tsp) into each ramekin dish. Add a heaped tablespoon of mashed potatoes and then pour an equal amount of the remaining batter in each dish. Place in the oven and bake for 15–20 minutes until the puddings are puffy and golden brown.

5 Using a palette knife, carefully ease the puddings out of the ramekin dishes and arrange on a large warm serving dish. Serve at once.

HASSLEBACK POTATOES

THESE RATHER SPLENDID ROAST POTATOES ARE IDEAL TO SERVE FOR DINNER PARTIES OR SPECIAL OCCASIONS SUCH AS CHRISTMAS.

SERVES FOUR

INGREDIENTS

olive or sunflower oil, for roasting
4 medium potatoes, peeled and halved
    lengthways
salt and ground black pepper
15ml/1 tbsp natural colour dried
    breadcrumbs

1 Pour enough oil into a small roasting pan to just cover the base then put into an oven set at 200°C/400°F/Gas 6 to heat.

2 Meanwhile, par boil the potato halves for 5 minutes then drain. Cool slightly and slash about four times from the rounded tops almost down to the flat bottoms.

3 Place the potatoes in the heated roasting pan and spoon over the hot oil. Season well and return the potatoes to the oven for about 20 minutes.

4 Remove the potatoes once more from the oven, prise open the slashes slightly and spoon over the hot oil. Sprinkle the potato tops lightly with breadcrumbs and return to the oven for another 15 minutes or so, until they are golden brown, cooked and crispy.

COOK'S TIP

There are many different ways of roasting potatoes in the oven. First, the choice of potato is important. Choose a variety which holds its shape well and yet is still slightly floury inside. Details on the pack or bag should give you guidance. The oil is important too – choose one which is either flavourless such as sunflower or groundnut oil or one with a lot of good flavour, such as olive oil. Just before serving, try trickling a little sesame seed, walnut or hazelnut oil over the roasted potatoes for a delicious nutty flavour.

BROCCOLI AND CAULIFLOWER WITH A CIDER AND APPLE MINT SAUCE

THE CIDER SAUCE MADE HERE IS IDEAL FOR OTHER VEGETABLES, SUCH AS CELERY OR BEANS. IT IS FLAVOURED WITH TAMARI, A JAPANESE SOY SAUCE AND APPLE MINT.

SERVES FOUR

INGREDIENTS
I large onion, chopped
2 large carrots, chopped
1 large garlic clove
15ml/1 tbsp dill seed
4 large sprigs apple mint
30ml/2 tbsp olive oil
30ml/2 tbsp plain flour
300ml/½ pint/1¼ cups dry cider
500g/1lb broccoli florets
500g/1lb cauliflower florets
30ml/2 tbsp tamari
10ml/2 tsp mint jelly

1 Sauté the onion, carrots, garlic, dill seeds and apple mint leaves in the olive oil until nearly cooked. Stir in the flour and cook for half a minute or so. Pour in the cider and simmer until the sauce looks glossy.

2 Boil the broccoli and cauliflower in separate pans until tender.

3 Pour the sauce into a food processor and add the tamari and the mint jelly. Blend until finely puréed. Pour over the broccoli and cauliflower.

COURGETTE AND CARROT RIBBONS WITH BRIE, BLACK PEPPER AND PARSLEY

THIS RECIPE PRODUCES A DELICIOUS VEGETARIAN MEAL OR SIMPLY A NEW WAY OF PRESENTING COLOURFUL VEGETABLES AS AN ACCOMPANIMENT TO A MAIN COURSE.

SERVES FOUR

INGREDIENTS

1 large green pepper, diced
15ml/1 tbsp sunflower oil
225g/8oz Brie cheese
30ml/2 tbsp crème fraîche
5ml/1 tsp lemon juice
60ml/4 tbsp milk
10ml/2 tsp freshly ground black pepper
30ml/2 tbsp parsley, very finely
  chopped, plus extra to garnish
salt and pepper
6 large courgettes
6 large carrots

1 Sauté the green pepper in the sunflower oil until just tender. Place the remaining ingredients, apart from the carrots and courgettes, in a food processor and blend well. Place the mixture in a saucepan and add the green pepper.

2 Peel the courgettes. Use a potato peeler to slice them into long, thin strips. Do the same with the carrots. Put the courgettes and carrots in separate saucepans, cover with just enough water to cover, then simmer for 3 minutes until barely cooked.

3 Heat the sauce and pour into a shallow vegetable dish. Toss the courgette and carrot strips together and arrange them in the sauce. Garnish with a little finely chopped parsley.

BAKED COURGETTES

WHEN VERY SMALL AND VERY FRESH COURGETTES ARE USED FOR THIS RECIPE IT IS WONDERFUL, BOTH SIMPLE AND DELICIOUS. THE CREAMY YET TANGY GOAT'S CHEESE CONTRASTS WELL WITH THE VERY DELICATE FLAVOUR OF THE YOUNG COURGETTES.

SERVES FOUR

INGREDIENTS
   8 small courgettes, about 450g/1lb
      total weight
   15ml/1 tbsp olive oil, plus extra
      for greasing
   75–115g/3–4oz goat's cheese, cut
      into thin strips
   small bunch fresh mint, finely
      chopped
   freshly ground black pepper

1 Preheat the oven to 180°C/350°F/ Gas 4. Cut out eight rectangles of foil large enough to encase each courgette and brush each with a little oil.

2 Trim the courgettes and cut a thin slit along the length of each.

3 Insert pieces of goat's cheese in the slits. Add a little mint and sprinkle with the olive oil and black pepper.

4 Wrap each courgette in the foil rectangles, place on a baking sheet and bake for about 25 minutes until tender.

COOK'S TIP
Almost any cheese could be used in this recipe. Mild cheeses, however, such as a mild Cheddar or mozzarella, will best allow the flavour of the courgettes to be appreciated.

COURGETTES ITALIAN-STYLE

USE A GOOD QUALITY OLIVE OIL AND SUNFLOWER OIL. THE OLIVE OIL GIVES IT A DELICIOUS FRAGRANCY BUT IS NOT ALLOWED TO OVERPOWER THE COURGETTES.

SERVES FOUR

INGREDIENTS

15ml/1 tbsp olive oil
15ml/1 tbsp sunflower oil
1 large onion, chopped
1 garlic clove, crushed
4–5 medium courgettes, cut into
  1cm/½in slices
150ml/¼ pint/⅔ cup chicken
  or vegetable stock
2.5ml/½ tsp fresh chopped oregano
salt and freshly ground black pepper
chopped fresh parsley, to garnish

3 Stir in the stock, oregano and seasoning and simmer gently for 8–10 minutes, until the liquid has almost evaporated. Spoon the courgettes into a serving dish, sprinkle with parsley and serve.

1 Heat the oils in a large frying pan and fry the onion and garlic over a moderate heat for 5–6 minutes until the onion has softened and is beginning to brown.

2 Add the courgettes and fry for about 4 minutes until they just begin to be flecked with brown. Stir frequently.

PATATAS BRAVAS

THIS IS A CLASSIC SPANISH TAPAS DISH OF DEEP-FRIED CUBES OF POTATO WITH A SPICY TOMATO SAUCE. USE A VERSATILE MAIN CROP VARIETY OF POTATO FOR THIS DISH — MARIS PIPER IS IDEAL — SO THAT THE FLESH RETAINS ITS TEXTURE.

SERVES FOUR

INGREDIENTS
   675g/1½lb potatoes, such as Maris
      Piper or Estima
   oil, for deep frying
For the sauce
   15ml/1 tbsp olive oil
   1 small onion, chopped
   1 garlic clove, crushed
   400g/14oz can tomatoes
   10ml/2 tsp Worcestershire sauce
   5ml/1 tsp wine vinegar
   about 5ml/1 tsp Tabasco sauce

1 Peel and cut the potatoes into small cubes and place in a large bowl of cold water to remove the excess starch.

2 Heat the oil in a medium-sized frying pan and fry the onion and garlic for 3–4 minutes until the onion is soft and just beginning to brown.

3 Pour the tomatoes into a blender or processor, process until smooth and then pour into the pan with the onion. Simmer, uncovered, over a moderate heat for 8–10 minutes until the mixture is thick and reduced, stirring occasionally.

4 Heat the oil in a deep fryer. Drain the potatoes and pat dry with kitchen paper. Fry the potatoes in the hot oil, in batches if necessary, until golden brown. Drain on kitchen paper.

5 Stir the Worcestershire sauce, wine vinegar and Tabasco sauce into the tomato mixture. Add the potatoes, stirring well so that all the potatoes are coated with the sauce. Spoon into individual serving dishes and serve at once.

POTATOES DAUPHINOIS

SERVES FOUR

INGREDIENTS
   675g/1½lb potatoes, peeled and thinly
      sliced
   1 garlic clove
   25g/1oz butter
   300ml/½ pint/1¼ cups single cream
   50ml/2fl oz/¼ cup milk
   salt and white pepper

1 Preheat the oven to 150°C/300°F/ Gas 2. Place the potato slices in a bowl of cold water to remove the excess starch. Drain and pat dry with kitchen paper.

2 Cut the garlic in half and rub the cut side around the inside of a wide shallow ovenproof dish. Butter the dish generously. Blend the cream and milk in a jug.

3 Cover the base of the dish with a layer of potatoes. Dot a little butter over the potato layer, season with salt and pepper and then pour over a little of the cream and milk mixture.

4 Continue making layers, until all the ingredients have been used up, ending with a layer of cream.

5 Bake in the oven for about 1¼ hours. If the dish browns too quickly and seems to be drying out, cover with a lid or with a piece of foil. The potatoes are ready when they are very soft and the top is pale golden brown.

COOK'S TIP
For a slightly speedier version of this recipe, par-boil the potato slices for 3–4 minutes. Drain well and assemble as above. Cook at 160°C/325°F/Gas 3 for 45–50 minutes until the potatoes are completely tender.

GLAZED CARROTS WITH CIDER

THIS RECIPE IS EXTREMELY SIMPLE TO MAKE. THE CARROTS ARE COOKED IN THE MINIMUM OF LIQUID TO BRING OUT THE BEST OF THEIR FLAVOUR, AND THE CIDER ADDS A PLEASANT SHARPNESS.

SERVES FOUR

INGREDIENTS
450g/1lb young carrots
25g/1oz butter
15ml/1 tbsp brown sugar
120ml/4fl oz/½ cup cider
60ml/4 tbsp vegetable stock or water
5ml/1 tsp Dijon mustard
15ml/1 tbsp finely chopped fresh
    parsley

1 Trim the tops and bottoms of the carrots. Peel or scrape them. Using a sharp knife, cut them into julienne strips.

2 Melt the butter in a frying pan, add the carrots and sauté for 4–5 minutes, stirring frequently. Sprinkle over the sugar and cook, stirring for 1 minute or until the sugar has dissolved.

3 Add the cider and stock or water, bring to the boil and stir in the Dijon mustard. Partially cover the pan and simmer for about 10–12 minutes until the carrots are just tender. Remove the lid and continue cooking until the liquid has reduced to a thick sauce.

4 Remove the saucepan from the heat, stir in the parsley and then spoon into a warmed serving dish. Serve as an accompaniment to grilled meat or fish or with a vegetarian dish.

COOK'S TIP
If the carrots are cooked before the liquid in the saucepan has reduced, transfer the carrots to a serving dish and rapidly boil the liquid until thick. Pour over the carrots and sprinkle with parsley.

CARROT, APPLE AND ORANGE COLESLAW

THIS DISH IS AS DELICIOUS AS IT IS EASY TO MAKE. THE GARLIC AND HERB DRESSING ADDS THE NECESSARY CONTRAST TO THE SWEETNESS OF THE SALAD.

SERVES FOUR

INGREDIENTS
350g/12oz young carrots,
    finely grated
2 eating apples
15ml/1 tbsp lemon juice
1 large orange
For the dressing
45ml/3 tbsp olive oil
60ml/4 tbsp sunflower oil
45ml/3 tbsp lemon juice
1 garlic clove, crushed
60ml/4 tbsp natural yogurt
15ml/1 tbsp chopped mixed fresh
    herbs: tarragon, parsley, chives
salt and freshly ground black pepper

1 Place the carrots in a large serving bowl. Quarter the apples, remove the core and then slice thinly. Sprinkle with the lemon juice to prevent them discolouring and then add to the carrots.

2 Using a sharp knife, remove the peel and pith from the orange and then separate into segments.

3 To make the dressing, place all the ingredients in a jar with a tight-fitting lid and shake vigorously to blend.

4 Just before serving, pour the dressing over the salad and toss well together.

RUNNER BEANS WITH GARLIC

DELICATE AND FRESH-TASTING FLAGEOLET BEANS AND GARLIC ADD A DISTINCT FRENCH FLAVOUR TO THIS SIMPLE SIDE DISH. SERVE TO ACCOMPANY ROAST LAMB OR VEAL.

SERVES FOUR

INGREDIENTS
  225g/8oz flageolet beans
  15ml/1 tbsp olive oil
  25g/1oz butter
  1 onion, finely chopped
  1–2 garlic cloves, crushed
  3–4 tomatoes, peeled and chopped
  350g/12oz runner beans, prepared
    and sliced
  150ml/¼ pint/⅔ cup white wine
  150ml/¼ pint/⅔ cup vegetable stock
  30ml/2 tbsp chopped fresh parsley
  salt and freshly ground black pepper

1 Place the flageolet beans in a large saucepan of water, bring to the boil and simmer for ¾–1 hour until tender. Drain.

2 Heat the oil and butter in a large frying pan and sauté the onion and garlic for 3–4 minutes until soft. Add the chopped tomatoes and continue cooking over a gentle heat until they are soft.

3 Stir the flageolet beans into the onion and tomato mixture, then add the runner beans, wine, stock, and a little salt. Stir well. Cover and simmer for 5–10 minutes until the runner beans are tender.

4 Increase the heat to reduce the liquid, then stir in the parsley and season with a little more salt, if necessary, and pepper.

BALTI-STYLE CAULIFLOWER <u>WITH</u> TOMATOES

BALTI IS A TYPE OF MEAT AND VEGETABLE COOKING FROM PAKISTAN AND NORTHERN INDIA. IT CAN REFER BOTH TO THE PAN USED FOR COOKING, WHICH IS LIKE A LITTLE WOK, AND THE SPICES USED. IN THE ABSENCE OF A GENUINE BALTI PAN, USE EITHER A WOK OR A HEAVY FRYING PAN.

SERVES FOUR

INGREDIENTS
   30ml/2 tbsp vegetable oil
   1 onion, chopped
   2 garlic cloves, crushed
   1 cauliflower, broken into florets
   5ml/1 tsp ground coriander
   5ml/1 tsp ground cumin
   5ml/1 tsp ground fennel seeds
   2.5ml/½ tsp garam masala
   pinch of ground ginger
   2.5ml/½ tsp chilli powder
   4 plum tomatoes, peeled, seeded
     and quartered
   175ml/6fl oz/¾ cup water
   175g/6oz fresh spinach, roughly
     chopped
   15–30ml/1–2 tbsp lemon juice
   salt and freshly ground black
     pepper

1 Heat the oil in a balti pan, wok, or large frying pan. Add the onion and garlic and stir-fry for 2–3 minutes over a high heat until the onion begins to brown. Add the cauliflower florets and stir-fry for a further 2–3 minutes until the cauliflower is flecked with brown.

2 Add the coriander, cumin, fennel seeds, garam masala, ginger and chilli powder and cook over a high heat for 1 minute, stirring all the time; then add the tomatoes, water and salt and pepper. Bring to the boil and then reduce the heat, cover and simmer for 5–6 minutes until the cauliflower is just tender.

3 Stir in the chopped spinach, cover and cook for 1 minute until the spinach is tender. Add enough lemon juice to sharpen the flavour and adjust the seasoning to taste.

4 Serve straight from the pan, with an Indian meal or with chicken or meat.

TURNIP TOPS WITH PARMESAN AND GARLIC

TURNIP TOPS HAVE A PRONOUNCED FLAVOUR AND ARE GOOD WHEN COOKED WITH OTHER STRONG-FLAVOURED INGREDIENTS SUCH AS ONIONS, GARLIC AND PARMESAN CHEESE. THEY DO NOT NEED LONG COOKING AS THE LEAVES ARE QUITE TENDER.

SERVES FOUR

INGREDIENTS
   45ml/3 tbsp olive oil
   2 garlic cloves, crushed
   4 spring onions, sliced
   350g/12oz turnip tops, thinly sliced,
     tough stalk removed
   50g/2oz Parmesan cheese, grated
   salt and freshly ground black pepper
   shavings of Parmesan cheese,
     to garnish

1 Heat the olive oil in a large saucepan and fry the garlic for a few seconds. Add the spring onions, stir-fry for 2 minutes and then add the turnip tops.

2 Stir-fry for a few minutes so that the greens are coated in oil, then add about 50ml/2fl oz/¼ cup water. Bring to the boil, cover and simmer until the greens are tender. Stir occasionally and do not allow the pan to boil dry.

3 Bring the liquid to the boil and allow the excess to evaporate and then stir in the Parmesan cheese. Serve at once with extra shavings of cheese, if liked.

INDIAN–STYLE OKRA

WHEN OKRA (BHINDI) IS SERVED IN INDIAN RESTAURANTS IT IS OFTEN FLAT AND SOGGY BECAUSE IT HAS BEEN OVERCOOKED OR LEFT STANDING. HOWEVER, WHEN YOU MAKE THIS DISH YOURSELF, YOU WILL REALIZE HOW DELICIOUS OKRA CAN BE.

SERVES FOUR

INGREDIENTS
    350g/12oz okra
    2 small onions
    2 garlic cloves, crushed
    1cm/½in piece fresh root ginger
    1 green chilli, seeded
    10ml/2 tsp ground cumin
    10ml/2 tsp ground coriander
    30ml/2 tbsp vegetable oil
    juice of 1 lemon

3 Reduce the heat and add the garlic and ginger mixture. Cook for about 2–3 minutes, stirring frequently, and then add the okra, lemon juice and 105ml/7 tbsp water. Stir well, cover tightly and simmer over a low heat for about 10 minutes until tender. Transfer to a serving dish, sprinkle with the fried onion rings and serve at once.

1 Trim the okra and cut into 1cm/½in lengths. Roughly chop one of the onions and place in a food processor or blender with the garlic, ginger, chilli and 90ml/6 tbsp water. Process to a paste. Add the cumin and coriander and blend again.

2 Thinly slice the remaining onion into half rings and fry in the oil for 6–8 minutes until golden brown. Transfer to a plate using a slotted spoon.

STUFFED TOMATOES WITH WILD RICE, CORN AND CORIANDER

THESE TOMATOES COULD BE SERVED AS A LIGHT MEAL WITH CRUSTY BREAD AND A SALAD, OR AS AN ACCOMPANIMENT TO A MAIN COURSE.

<u>SERVES FOUR</u>

INGREDIENTS

8 medium tomatoes
50g/2oz/⅓ cup sweetcorn kernels
30ml/2 tbsp white wine
50g/2oz/¼ cup cooked wild rice
1 clove garlic
50g/2oz/½ cup grated Cheddar cheese
15ml/1 tbsp chopped fresh coriander
salt and pepper
15ml/1 tbsp olive oil

1 Cut the tops off the tomatoes and remove the seeds with a small teaspoon. Scoop out all the flesh and chop finely – also chop the tops.

2 Preheat the oven to 180°C/350°F/Gas 4. Put the chopped tomato in a pan. Add the sweetcorn and the white wine. Cover with a close-fitting lid and simmer until tender. Drain.

3 Mix together all the remaining ingredients except the olive oil, adding salt and pepper to taste. Carefully spoon the mixture into the tomatoes, piling it higher in the centre. Sprinkle the oil over the top, arrange the tomatoes in an ovenproof dish, and bake at 180°C/350°F/Gas 4 for 15–20 minutes until cooked through. Let stand for a few minutes before serving.

PEAS WITH BABY ONIONS AND CREAM

IDEALLY, USE FRESH PEAS AND FRESH BABY ONIONS. FROZEN PEAS ARE AN ACCEPTABLE SUBSTITUTE IF FRESH ONES AREN'T AVAILABLE, BUT FROZEN ONIONS TEND TO BE INSIPID AND ARE NOT WORTH USING. ALTERNATIVELY, USE THE WHITE PARTS OF SPRING ONIONS.

SERVES FOUR

INGREDIENTS
175g/6oz baby onions
15g/½ oz butter
900g/2lb fresh peas (about
  350g/12oz shelled or frozen)
150ml/¼ pint/⅔ cup double cream
15g/½oz plain flour
10ml/2 tsp chopped fresh parsley
15–30ml/1–2 tbsp lemon juice
  (optional)
salt and freshly ground black pepper

1 Peel the onions and halve them if necessary. Melt the butter in a flame-proof casserole and fry the onions for 5–6 minutes over a moderate heat, until they begin to be flecked with brown.

3 Using a small whisk, blend the cream with the flour. Remove the pan from the heat and stir in the combined cream and flour, parsley and seasoning to taste.

4 Cook over a gentle heat for about 3–4 minutes, until the sauce is thick. Taste and adjust the seasoning; add a little lemon juice to sharpen, if liked.

2 Add the peas and stir-fry for a few minutes. Add 120ml/4fl oz/¼ cup water and bring to the boil. Partially cover and simmer for about 10 minutes until both the peas and onions are tender. There should be a thin layer of water on the base of the pan – add a little more water if necessary or, if there is too much liquid, remove the lid and increase the heat until the liquid is reduced.

PAK-CHOI WITH LIME DRESSING

FOR THIS THAI RECIPE, THE COCONUT DRESSING IS TRADITIONALLY MADE USING FISH SAUCE, BUT VEGETARIANS CAN USE MUSHROOM SAUCE INSTEAD. BEWARE, THIS IS A FIERY DISH!

SERVES FOUR

INGREDIENTS
  6 spring onions
  2 pak-choi
  30ml/2 tbsp oil
  3 fresh red chillies, cut in strips
  4 garlic cloves, thinly sliced
  15ml/1 tbsp crushed peanuts
For the dressing
  15–30ml/1–2 tbsp mushroom
    sauce
  30ml/2 tbsp lime juice
  250ml/8fl oz/1 cup coconut milk

1 To make the dressing, blend together the sauce and lime juice, and then stir in the coconut milk.

2 Cut the spring onions diagonally into slices, including all but the very tips of the green parts.

3 Using a large sharp knife, cut the pak-choi into very fine shreds.

4 Heat the oil in a wok and stir-fry the chillies for 2–3 minutes until crisp. Transfer to a plate using a slotted spoon.

5 Stir-fry the garlic for 30–60 seconds until golden brown and transfer to the plate with the chillies.

6 Stir-fry the white parts of the spring onions for about 2–3 minutes and then add the green parts and stir-fry for a further 1 minute. Add to the plate with the chillies and garlic.

7 Bring a large pan of salted water to the boil and add the pak-choi; stir twice and then drain immediately.

8 Place the warmed pak-choi in a large bowl, add the coconut dressing and stir well. Spoon into a large serving bowl and sprinkle with the crushed peanuts and the stir-fried chilli mixture. Serve either warm or cold.

COOK'S TIP
Coconut milk is available in cans from large supermarkets and Chinese stores. Alternatively, creamed coconut is available in packets. To use creamed coconut, place about 115g/4oz in a jug and pour over 250ml/8fl oz/1 cup boiling water. Stir well until dissolved.

SHIITAKE FRIED RICE

SHIITAKE MUSHROOMS HAVE A STRONG MEATY MUSHROOMY AROMA AND FLAVOUR. THIS IS A VERY EASY RECIPE TO MAKE, AND ALTHOUGH IT IS A SIDE DISH IT CAN ALMOST BE A MEAL IN ITSELF.

SERVES FOUR

INGREDIENTS
2 eggs
45ml/3 tbsp vegetable oil
350g/12oz shiitake mushrooms
8 spring onions, sliced diagonally
1 garlic clove, crushed
½ green pepper, chopped
25g/1oz butter
175–225g/6–8oz long grain rice, cooked
15ml/1 tbsp medium dry sherry
30ml/2 tbsp dark soy sauce
15ml/1 tbsp chopped fresh coriander
salt

1 Beat the eggs with 15ml/1 tbsp cold water and season with a little salt.

2 Heat 15ml/1 tbsp of the oil in a wok or large frying pan, pour in the eggs and cook to make a large omelette. Lift the sides of the omelette and tilt the wok so that the uncooked egg can run underneath and be cooked. Roll up the omelette and slice thinly.

3 Remove and discard the mushroom stalks if tough and slice the caps thinly, halving them if they are large.

4 Heat 15ml/1 tbsp of the remaining oil in the wok and stir-fry the spring onions and garlic for 3–4 minutes until softened but not brown. Transfer them to a plate using a slotted spoon.

5 Add the pepper, stir-fry for about 2–3 minutes, then add the butter and the remaining 15ml/1 tbsp of oil. As the butter begins to sizzle, add the mushrooms and stir-fry over a moderate heat for 3–4 minutes until soft.

6 Loosen the rice grains as much as possible. Pour the sherry over the mushrooms and then stir in the rice.

7 Heat the rice over a moderate heat, stirring all the time to prevent it sticking. If the rice seems very dry, add a little more oil. Stir in the reserved onions and omelette slices, the soy sauce and coriander. Cook for a few minutes until heated through and serve.

COOK'S TIP
Unlike risotto, for which rice is cooked along with the other ingredients, Chinese fried rice is always made using cooked rice. If you use 175–225g/6–8oz uncooked long grain, you will get about 450–500g/16–20oz cooked rice, enough for four people.

PARTIES
AND
PICNICS

Whether you are planning an informal open-air meal or elegant, delicious dishes for a special party, you will find a variety of ideas and inspiration to add fun to any occasion.

HERBAL PUNCH

A GOOD PARTY DRINK THAT WILL HAVE PEOPLE COMING BACK FOR MORE, AND A DELIGHTFUL NON-ALCOHOLIC CHOICE FOR DRIVERS.

SERVES THIRTY PLUS

INGREDIENTS
    450ml/¾ pint/2 cups honey
    4 litres/7 pints/8½ US pints water
    450ml/¾ pint/2 cups freshly squeezed
        lemon juice
    3 tbsp fresh rosemary leaves, plus more
        to decorate
    1.5kg/3½ lb/8 cups sliced strawberries
    450ml/¾ pint/2 cups freshly squeezed
        lime juice
    1.75 litres/3 pints/4 US pints sparkling
        mineral water
    ice cubes
    3–4 scented geranium leaves

1 Combine the honey, 1 litre/1¾ pints/4½ cups water, one-eighth of the lemon juice, and the rosemary leaves in a saucepan. Bring to the boil, stirring until all the honey is dissolved. Remove from the heat and allow to stand for about 5 minutes. Strain into a large punch bowl.

2 Press the strawberries through a fine sieve into the punch bowl, add the rest of the water and lemon juice, and the lime juice and sparkling water. Stir gently. Add the ice cubes 5 minutes before serving, and float the geranium and rosemary leaves on the surface.

ANGELICA LIQUEUR

THIS SHOULD BE DRUNK IN TINY GLASSES AFTER A LARGE MEAL. NOT ONLY WILL IT HELP THE DIGESTIVE SYSTEM, IT TASTES SUPERB.

ABOUT 1 LITRE

INGREDIENTS

   1 tsp fennel seeds
   1 tsp aniseed
   20 coriander seeds
   2–3 cloves
   2 tbsp crystallized angelica stems
   225g/8 oz/1 cup caster sugar
   1 bottle vodka

1 Crush the fennel, aniseed and coriander seeds and cloves a little, and chop the crystallized angelica stems.

2 Put the seeds and angelica stems into a large preserving jar.

3 Add the sugar. Pour on the vodka and leave by a sunny window for 2 weeks, swirling the mixture daily.

4 Strain through fine muslin into a sterilized bottle and seal. Leave in a dark cupboard for at least 4 months. Drink in small quantities with a piece of angelica in each glass.

STRAWBERRY AND MINT CHAMPAGNE

THIS IS A SIMPLE CONCOCTION THAT MAKES A BOTTLE OF CHAMPAGNE GO A LOT FURTHER. IT TASTES VERY SPECIAL ON A HOT SUMMER'S EVENING.

SERVES FOUR TO SIX

INGREDIENTS
500g/1lb strawberries
6–8 fresh mint leaves
1 bottle champagne or sparkling
    white wine

2 Strain through a fine sieve into a bowl. Half fill a glass with the mixture and top up with champagne. Decorate with a sprig of mint.

1 Purée the strawberries and mint leaves in a food processor.

MELON, GINGER AND BORAGE CUP

MELON AND GINGER COMPLEMENT EACH OTHER MAGNIFICENTLY. IF YOU PREFER, YOU CAN LEAVE OUT THE POWDERED GINGER — THE RESULT IS MILDER BUT EQUALLY DELICIOUS.

SERVES SIX TO EIGHT

INGREDIENTS
½ large honeydew melon
1 litre/1¾ pints/1 quart ginger beer
1 tsp powdered ginger (or to taste)
borage sprigs with flowers, to
    decorate

2 Pour the purée into a large jug and top up with ginger beer. Add powdered ginger to taste. Pour into glasses and decorate with borage.

1 Discard the seeds from the half melon and scoop the flesh into a food processor. Blend to a thin purée.

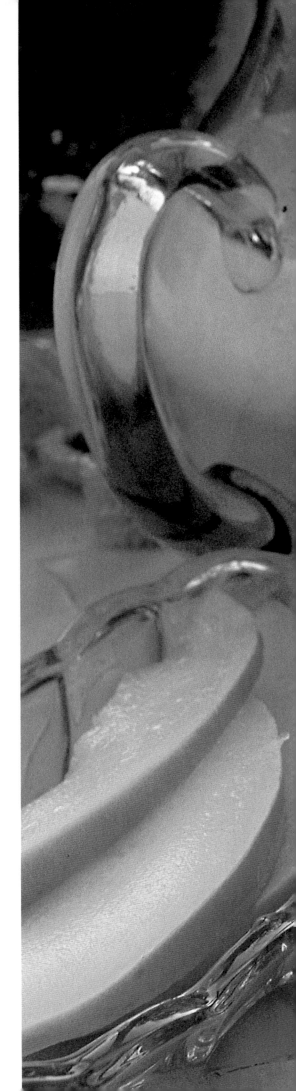

MINT CUP

MINT IS A PERENNIALLY POPULAR FLAVOUR AND THIS DELICATE CUP IS A WONDERFUL MIXTURE WITH AN INTRIGUING TASTE.

SERVES ONE

INGREDIENTS
    4 sprigs fresh mint
    ½ tsp sugar
    crushed ice
    ½ tsp lemon juice
    2 tbsp grapefruit juice
    120ml/4 fl oz/½ cup chilled tonic water
    lemon slices, to decorate

1 Crush two of the sprigs of mint with the sugar and put these into a glass. Fill the glass with crushed ice.

2 Add the lemon juice, grapefruit juice and tonic water. Stir gently and decorate with the remaining mint sprigs and slices of lemon.

ELDERFLOWER SPARKLER

THE FLAVOUR OF ELDERFLOWERS IS BECOMING POPULAR ONCE AGAIN. THIS RECIPE PRODUCES ONE OF THE MOST DELICIOUS DRINKS EVER CONCOCTED.

ABOUT 5 LITRES/8½ PINTS/10 US PINTS

INGREDIENTS
750g/1¾lb/3½ cups caster sugar
475ml/16 fl oz/2 cups hot water
4 large fresh elderflower heads
2 tbsp white wine vinegar
juice and pared rind of 1 lemon
4 litres/7 pints/8½ US pints water

1 Mix the sugar with the hot water. Pour the mixture into a large glass or plastic container. Add all the remaining ingredients. Stir well, cover and leave for about 5 days.

2 Strain off the liquid into sterilized screw-top bottles (glass or plastic). Leave for a further week or so. Serve very cold with slivers of lemon rind.

PASTA AND BEETROOT SALAD

A COLOURFUL, EYE-CATCHING PARTY SALAD.

SERVES EIGHT

INGREDIENTS
2 uncooked beetroots, scrubbed
225g/8 oz pasta shells or twists
45ml/3 tbsp vinaigrette dressing
salt and ground black pepper
2 celery sticks, thinly sliced
3 spring onions, sliced
75g/3 oz/⅔ cup walnuts or hazelnuts,
  roughly chopped
1 dessert apple, cored and sliced

DRESSING
60ml/4 tbsp mayonnaise
45ml/3 tbsp natural yogurt
30 ml/2 tbsp milk
10ml/2 tsp horseradish relish

TO SERVE
curly lettuce leaves
3 eggs, hard boiled and chopped
2 ripe avocados
box of salad cress

1 Boil the beetroots, without peeling them, in lightly salted water until they are just tender. Drain, cool, peel and chop. Set aside.

2 Cook the pasta according to the instructions on the pack, then drain and toss in the vinaigrette and season well. Cool. Mix the pasta with the beetroot, celery, onions, nuts and apple.

3 Stir all the dressing ingredients together and mix into the pasta bowl. Chill well.

4 To serve, line a pretty salad bowl with the lettuce leaves and pile the salad in the centre. When ready to serve, scatter over the chopped egg. Peel and slice the avocados and arrange them on top then sprinkle over the cress.

PICNIC GAZPACHO

THIS VERSION OF THE CLASSIC SPANISH NO-COOK SOUP IS IDEAL FOR TAKING ON PICNICS.

SERVES SIX

INGREDIENTS
1 slice white bread, crusts removed
cold water, to soak
1 garlic clove, crushed
30ml/2 tbsp extra virgin olive oil
30ml/2 tbsp white wine vinegar
6 large ripe tomatoes, skinned and
  finely chopped
1 small onion, finely chopped
2.5ml/½ tsp paprika
good pinch of ground cumin
150ml/¼ pint/⅔ cup tomato juice
salt and ground black pepper

TO GARNISH
1 green pepper, chopped
⅔ cucumber, peeled, seeded and
  chopped

CROUTONS
2 slices bread, cubed and deep fried

1 Soak the bread slice in enough cold water just to cover and leave for about 5 minutes, then mash with a fork.

2 Pound the garlic, oil and vinegar with a pestle and mortar or blend in a food processor. Mix this into the bread.

3 Spoon the mixture into a bowl and stir in the tomatoes, onion, spices and tomato juice. Season well and store in the refrigerator. Prepare the garnishes and store these in separate containers.

4 For a picnic, pour the chilled soup into a flask. Otherwise, pour into a chilled glass salad bowl and hand the garnishes round in smaller bowls.

SPICY POTATO STRUDEL

THIS IS A PERFECT DISH FOR A SPECIAL FAMILY SUPPER.

SERVES FOUR

INGREDIENTS
  1 onion, chopped
  2 carrots, coarsely grated
  1 courgette, chopped
  350g/12oz potatoes, chopped
  65g/2½oz/5 tbsp butter
  10ml/2 tsp mild curry paste
  2.5ml/½ tsp dried thyme
  150ml/¼ pint/⅔ cup water
  salt and ground black pepper
  1 egg, beaten
  30ml/2 tbsp single cream
  50g/2oz Cheddar cheese, grated
  8 sheets filo pastry
  sesame seeds, to sprinkle

1 Fry the onion, carrots, courgette and potatoes in half the butter for 5 minutes until they are soft then add the curry paste and cook for a further minute.

2 Add the thyme, water and seasoning. Continue to cook gently, uncovered for another 10 minutes.

3 Allow the mixture to cool and mix in the egg, cream and cheese. Chill until ready to fill and roll.

4 Melt the remaining butter and lay out four sheets of filo pastry, slightly overlapping them to form a large rectangle. Brush with butter and fit the other sheets on top. Brush again.

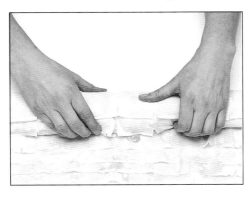

5 Spoon the filling along one long edge, then roll up the pastry. Form it into a circle and brush again with the last of the butter. Sprinkle over the sesame seeds and set on a baking sheet.

6 Heat the oven to 190°C/375°F/Gas 5 then bake the strudel for about 25 minutes until golden and crisp. Allow to stand for 5 minutes or so before cutting.

CORONATION SALAD

THIS FAMOUS SALAD DRESSING WAS CREATED ESPECIALLY FOR THE CORONATION DINNER OF QUEEN ELIZABETH II. IT IS A WONDERFUL ACCOMPANIMENT TO EGGS AND VEGETABLES.

SERVES SIX

INGREDIENTS
    450g/1lb new potatoes
    salt
    45ml/3 tbsp vinaigrette dressing
    3 spring onions, chopped
    ground black pepper
    6 eggs, hard-boiled and halved
    frilly lettuce leaves, to serve
    ¼ cucumber, sliced then cut in shreds
    6 large radishes, sliced
    carton of salad cress
For the dressing
    30ml/2 tbsp olive oil
    1 small onion, chopped
    15ml/1 tbsp mild curry
    powder or korma spice mix
    10ml/2 tsp tomato purée
    30ml/2 tbsp lemon juice
    30ml/2 tbsp sherry
    300ml/½ pint/1¼ cups mayonnaise
    150ml/¼ pint natural yogurt

1 Boil the potatoes in salted water until tender. Drain them and toss in the vinaigrette dressing.

2 Allow the potatoes to cool, stirring in the spring onions and seasoning. Cool the mixture thoroughly.

3 Meanwhile, make the coronation dressing. Heat the oil and fry the onion for 3 minutes until it is soft. Stir in the spice powder and fry for a further minute. Mix in all the other dressing ingredients.

4 Stir the dressing into the potatoes; add the eggs then chill. Line a serving platter with lettuce leaves and pile the salad in the centre. Scatter over the cucumber, radishes and cress.

PASTA AND WILD MUSHROOM MOULD

BAKE PASTA SHAPES IN A GOLDEN CRUMB COATING LAYERED WITH A RICH BECHAMEL SAUCE AND MUSHROOMS. SCRUMPTIOUS!

SERVES FOUR TO SIX

INGREDIENTS
200g/7oz pasta shapes
600ml/1 pint/2½ cups milk
1 bay leaf
small onion stuck with 6 cloves
50g/2oz/4 tbsp butter
45ml/3 tbsp breadcrumbs
10ml/2 tsp dried mixed herbs
40g/1½oz/⅓ cup plain flour
60ml/4 tbsp Parmesan cheese, freshly grated
fresh nutmeg, grated
salt and ground black pepper
2 eggs, beaten
1 × 15g/½oz pack porcini or cepes dried mushrooms
350g/12oz button mushrooms, sliced
2 garlic cloves, crushed
30ml/2 tbsp olive oil
30ml/2 tbsp fresh parsley, chopped

1 Boil the pasta according to the instructions on the pack. Drain and set aside. Heat the milk with the bay leaf and clove-studded onion and stand for 15 minutes. Remove bay leaf and onion.

2 Melt the butter in a saucepan and use a little to brush the inside of a large oval pie dish. Mix the crumbs and mixed herbs together and use them to coat the inside of the dish.

3 Stir the flour into the butter, cook for a minute then slowly add the hot milk to make a smooth sauce. Add the cheese, nutmeg, seasoning and cooked pasta. Cool for 5 minutes then beat in the eggs.

4 Soak the porcini in a little hot water until they are soft. Reserve the liquor and chop the porcinis.

5 Fry the porcini with the sliced mushrooms and garlic in the olive oil for about 3 minutes. Season well, stir in the liquor and reduce down. Add the parsley.

6 Spoon a layer of pasta in the dish. Sprinkle over the mushrooms then more pasta and so on, finishing with pasta. Cover with greased foil. Heat the oven to 190°C/375°F/Gas 5 and bake for about 25–30 minutes. Allow to stand for 5 minutes before turning out to serve.

AUBERGINE BOATS

THESE CAN BE PREPARED AHEAD AND BAKED PRIOR TO EATING. THE HAZELNUT TOPPING CONTRASTS NICELY WITH THE SMOOTH AUBERGINE FILLING.

SERVES FOUR

INGREDIENTS
115g/4oz/⅔ cup brown basmati rice
2 medium size aubergines, halved lengthways
1 onion, chopped
2 garlic cloves, crushed
1 small green pepper, chopped
115g/4oz mushrooms, sliced
45ml/3 tbsp olive oil
75g/3oz Cheddar cheese, grated
1 egg, beaten
2.5ml/½ tsp marjoram
salt and ground black pepper
30ml/2 tbsp hazelnuts, chopped

1 Boil the rice according to the instructions on the pack, drain and then cool. Scoop out the flesh from the aubergines and chop. Blanch the shells in boiling water for 2 minutes then drain upside down.

2 Fry the aubergine flesh, onion, garlic, pepper and mushrooms in the oil, for about 5 minutes.

3 Mix in the rice, cheese, egg, marjoram and seasoning. Arrange the aubergine shells in an ovenproof dish. Spoon the filling inside. Sprinkle over the nuts. Chill until ready to bake.

4 Heat the oven to 190°C/375°F/Gas 5 and bake the aubergines for about 25 minutes until the filling is set and the nuts are golden brown.

CHARGRILLED VEGETABLES WITH SALSA

ENJOY A BARBECUE WITH THESE CHARGRILLED VEGETABLES. SERVE HOT WITH A NO-COOK SALSA.

SERVES FOUR

INGREDIENTS
1 large sweet potato, cut in thick slices
2 courgettes, halved lengthways
salt
2 red peppers, quartered
olive oil, to brush
For the salsa
2 large tomatoes, skinned and finely
   chopped
2 spring onions, finely chopped
1 small green chilli, chopped
juice of 1 small lime
30ml/2 tbsp fresh coriander, chopped
salt and ground black pepper

1 Par-boil the sweet potato for 5 minutes until it is barely tender. Drain and leave to cool.

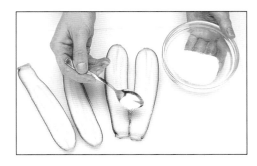

2 Sprinkle the courgettes with a little salt and leave to drain in a colander for 20 minutes, then pat dry.

3 Make the salsa by mixing all the ingredients together, and allow them to stand for about 30 minutes to mellow.

4 Prepare the barbecue until the coals glow or preheat a grill. Brush the potato slices, courgettes and peppers with oil and cook them until they are lightly charred and softened, brushing with oil again and turning at least once. Serve hot accompanied by the salsa.

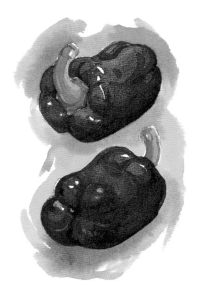

GARDEN VEGETABLE TERRINE

PERFECT FOR A SPECIAL FAMILY PICNIC OR BUFFET, THIS IS A SOFTLY SET, CREAMY TERRINE OF COLOURFUL VEGETABLES WRAPPED IN GLOSSY SPINACH LEAVES. SELECT LARGE SPINACH LEAVES FOR THE BEST RESULTS.

SERVES SIX

INGREDIENTS

225g/8oz fresh leaf spinach
3 carrots, cut in sticks
3–4 long, thin leeks
about 115g/4oz long green beans,
   topped and tailed
1 red pepper, cut in strips
2 courgettes, cut in sticks
115g/4oz broccoli florets
For the sauce
1 egg and 2 yolks
300ml/½ pint/1¼ cups single cream
fresh nutmeg, grated
5ml/1 tsp salt
50g/2oz Cheddar cheese, grated
oil, for greasing
ground black pepper

1 Blanch the spinach quickly in boiling water, then drain, refresh in cold water and drain again. Take care not to break up the leaves, then carefully pat them dry.

2 Grease a 1kg/2lb loaf tin and line the base with a sheet of greaseproof paper. Line the tin with the spinach leaves, trimming any thick stalks. Allow the leaves to overhang the tin.

3 Blanch the rest of the vegetables in boiling, salted water until just tender. Drain and refresh in cold water then, when cool, pat dry with pieces of kitchen paper towel.

4 Place the vegetables into the loaf tin in a colourful mixture, making sure the sticks of vegetables lie lengthways.

5 Beat the sauce ingredients together and slowly pour over the vegetables. Tap the loaf tin to ensure the sauce seeps into the gaps. Fold over the spinach leaves at the top of the terrine.

6 Cover the terrine with a sheet of greased foil then bake in a roasting pan half full of boiling water at 180°C/350°F/ Gas 4 for about 1–1¼ hours until set.

7 Cool the terrine in the tin then chill. To serve, loosen the sides and shake gently out. Serve cut in thick slices.

CRÊPE GALETTE

*A STACK OF LIGHT PANCAKES
LAYERED WITH A LENTIL FILLING
MAKES AN IMPRESSIVE DINNER
PARTY MAIN COURSE.*

SERVES SIX

INGREDIENTS
   115g/4oz/1 cup plain flour
   good pinch of salt
   1 egg
   300ml/½ pint/1¼ cups buttermilk, or
     milk and water, mixed
   oil, for cooking
For the filling
   2 leeks, thinly sliced
   1 small fennel bulb, thinly sliced
   60ml/4 tbsp olive oil
   150g/6oz/¾ cup red lentils
   150ml/¼ pint/⅔ cup dry white wine
   1 × 400g/14oz can chopped tomatoes
   300ml/½ pint/1¼ cups stock
   5ml/1 tsp dried oregano
   salt and ground black pepper
   1 onion, sliced
   225g/8oz mushrooms, sliced
   225g/8oz frozen leaf spinach, thawed
   200g/7oz low-fat cream cheese
   50g/2oz Parmesan cheese

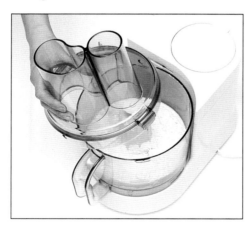

1 Make the pancake batter by mixing the flour, salt, egg and buttermilk or milk and water in a blender until smooth. Set aside while you prepare the filling.

2 Gently fry the leeks and fennel in half the olive oil for 5 minutes, then add the lentils and wine. Cook for a minute until reduced down then stir in the tomatoes and stock.

3 Bring the leek/fennel mixture to a boil, add the oregano and seasoning then simmer for 20 minutes, stirring it occasionally until it thickens.

4 Fry the onion and mushrooms in the remaining olive oil for 5 minutes, stir in the squeezed-dry spinach and heat. Season well then mix in the cream cheese.

5 Make about 12–14 pancakes with the batter in a well heated non-stick frying pan. Lightly grease a 20cm/8in round deep spring-form cake tin and line the base and sides with some of the pancakes, overlapping them as necessary.

6 Layer the remaining pancakes with the two fillings, sprinkling freshly grated Parmesan in between and pressing them down well. Finish with a pancake on top.

7 Cover with foil and set aside to rest. Preheat the oven to 190°C/375°F/Gas 5. Bake for about 40 minutes, then turn out and allow to firm up for 10 minutes before cutting into wedges. Serve with a home-made tomato sauce.

COOK'S TIP
This can be frozen ready made up, but it is probably nicer if frozen in parts – the pancakes interleaved with greaseproof paper and then wrapped in foil and the sauce frozen separately.

POLENTA FINGERS WITH BEANS AND TOMATOES

POLENTA, OR CORNMEAL, IS A POPULAR FAMILY FAVOURITE IN ITALY. IT IS EATEN HOT FROM A BOWL OR ALLOWED TO SET, CUT INTO FINGERS AND GRILLED.

SERVES SIX

INGREDIENTS
  1.75 litres/3 pints/7½ cups milk and water, mixed
  10ml/2 tsp salt
  275g/10oz/1½ cups polenta
  25g/1oz/2 tbsp butter, plus extra for spreading
  50g/2oz Parmesan cheese
  ground black pepper
For the sauce
  1 onion, chopped
  2 garlic cloves, crushed
  30ml/2 tbsp olive oil
  1 × 400g/14oz can chopped tomatoes
  salt and ground black pepper
  good pinch of dried sage
  225g/8oz frozen broad beans

1 In a large saucepan, bring the milk and water to a boil. Stir in the salt. While stirring with a wooden spoon trickle the polenta into the boiling liquid in a steady stream and continue stirring until the mixture has thickened.

2 Lower the heat and simmer for about 20 minutes, stirring frequently. Add the butter, freshly grated cheese and seasoning.

3 Lightly grease a shallow rosting pan and pour in the polenta mixture. Cool, then chill overnight.

4 For the sauce, fry the onion and garlic in the oil for 5 minutes. Add the tomatoes, seasoning and sage and cook for a further 10 minutes. Stir in the broad beans and cook for 5 minutes more.

5 Turn out the polenta and cut into fingers. Grill both sides until brown and crisp. Spread with a little butter and serve accompanied by the tomato and beans.

PAPRIKA AND PARMESAN TARTLETS

PRETTY PINK PASTRY TARTS WITH A TANGY CREAM FILLING ARE IDEAL FOR HANDING ROUND AT DRINKS PARTIES. MAKE THE CASES AHEAD OF THE PARTY AND FILL THEM JUST BEFORE SERVING.

MAKES 18

INGREDIENTS
  225g/8oz/2 cups plain flour
  10ml/2 tsp paprika
  150g/5oz/10 tbsp butter or sunflower margarine
  40g/1½oz/scant ½ cup Parmesan cheese, freshly grated
  cold water, to bind
For the filling
  350g/12oz goats' cheese
  50g/2oz rocket leaves, or watercress, chopped
  30ml/2 tbsp fresh chives, chopped
  salt and ground black pepper
  450g/1lb tomatoes, sliced

1 Sift the flour with the paprika and rub in the butter or margarine. Stir in the Parmesan and mix to a firm dough with cold water.

2 Roll out the pastry and stamp out 18 rounds, large enough to fit into bun tins. Prick the bases well with a fork and chill while you preheat the oven to 190°C/375°F/Gas 5.

3 Bake the tartlets for 15 minutes until crisp. Cool them on a wire rack.

4 Beat the cheese with the rocket or cress, chives and seasoning. Slice the tomatoes, allowing roughly two slices per tart.

5 When ready to serve, spoon the filling into the tarts. Top each one with some tomato and garnish with extra rocket or cress leaves.

TOFU SATAY

GRILL CUBES OF TOFU UNTIL CRISPY THEN SERVE WITH A THAI-STYLE PEANUT SAUCE.

SERVES FOUR TO SIX

INGREDIENTS
    2 × 200g/7oz packs smoked tofu
    45ml/3 tbsp light soy sauce
    10ml/2 tsp sesame oil
    1 garlic clove, crushed
    1 yellow and 1 red pepper
    8–12 fresh bay leaves
    sunflower oil, for grilling
For the sauce
    2 spring onions, finely chopped
    2 garlic cloves, crushed
    good pinch chilli powder
    5ml/1 tsp granulated sugar
    15ml/1 tbsp white vinegar
    30ml/2 tbsp light soy sauce
    45ml/3 tbsp crunchy peanut butter

1 To help them withstand the hot grilling, soak 8–12 wooden satay sticks in water for 20 minutes then drain. Cut the tofu into bite-sized cubes and mix with the soy sauce, sesame oil and garlic. Cover and marinate for 20 minutes.

2 Beat the sauce ingredients together until well blended. Avoid using a food processor for this as the texture should be slightly chunky.

3 Drain the tofu and thread the cubes onto the soaked sticks with the pepper cut into squares and bay leaves. Larger leaves may need to be halved.

4 Heat a grill or barbecue until quite hot. Brush the satays with oil. Grill, turning the sticks occasionally, until the ingredients are browned and crisp. Serve hot with the dipping sauce.

PARTY MOUSSAKA

MOUSSAKA BENEFITS FROM BEING MADE AHEAD OF TIME AND REQUIRING JUST REHEATING ON THE DAY.

SERVES EIGHT

INGREDIENTS

2 large aubergines, thinly sliced
6 courgettes, cut in chunks
150ml/¼ pint/⅔ cup olive oil, plus extra if required
675g/1½ lb potatoes, thinly sliced
2 onions, sliced
3 garlic cloves, crushed
150ml/¼ pint/⅔ cup dry white wine
2 × 400g/14oz cans chopped tomatoes
30ml/2 tbsp tomato purée
1 × 430g/15oz can green lentils
10ml/2 tsp dried oregano
60ml/4 tbsp chopped fresh parsley
225g/8oz/2 cups feta cheese, crumbled
salt and ground black pepper
For the béchamel sauce
40g/1½oz/3 tbsp butter
40g/1½oz/4 tbsp plain flour
600ml/1 pint/2½ cups milk
2 eggs, beaten
115g/4oz grated Parmesan cheese
nutmeg, freshly grated

2 Pour in the wine and cook until reduced, then add the tomatoes, lentils, liquor from the can and the herbs and seasoning. Cover and simmer for 15 minutes.

5 Meanwhile, for the béchamel sauce, put the butter, flour and milk into a saucepan all together and bring slowly to a boil, stirring or whisking constantly. It should thicken and become smooth. Season and add the nutmeg.

1 Lightly salt the aubergines and courgettes in a colander and leave to drain for 30 minutes. Rinse and pat dry. Heat the oil until quite hot in a frying pan and quickly brown the aubergine and courgette slices. Remove them with a slotted spoon and drain on a kitchen paper towel. Brown the potato slices, remove and pat dry. Add the onion and garlic with a little extra oil, if required, and fry for about 5 minutes.

3 In a large ovenproof dish, layer the vegetables, trickling the tomato and lentil sauce in between and scattering over the feta cheese. Finish with a layer of aubergine slices.

4 Cover the vegetables with a sheet of foil and bake at 190°C/375°F/Gas 5 for 25 minutes or until the vegetables are quite soft but not overcooked.

6 Remove the sauce and cool for 5 minutes then beat in the eggs. Pour over the aubergines and sprinkle with the Parmesan. If cooking ahead, cool and chill at this stage.

7 To finish, return to the oven uncovered and bake for a further 25–30 minutes until golden and bubbling hot.

BASMATI AND BLUE LENTIL SALAD

*PUY LENTILS FROM FRANCE
(SOMETIMES KNOWN AS BLUE
LENTILS) ARE SMALL,
DELICIOUSLY NUTTY PULSES,
HIGHLY PRIZED BY GOURMETS.*

SERVES SIX

INGREDIENTS
 115g/4oz/⅔ cup puys de dome lentils,
   soaked
 225g/8oz/1¼ cups basmati rice, rinsed
   well
 2 carrots, coarsely grated
 ⅓ cucumber, halved, seeded and
   coarsely grated
 3 spring onions, sliced
 45ml/3 tbsp fresh parsley, chopped
For the dressing
 30ml/2 tbsp sunflower oil
 30ml/2 tbsp extra virgin olive oil
 30ml/2 tbsp wine vinegar
 30ml/2 tbsp fresh lemon juice
 good pinch of granulated sugar
 salt and ground black pepper

1 Soak the lentils for 30 minutes.
Meanwhile, make the dressing by shaking
all the ingredients together in a screw-
topped jar. Set aside.

2 Boil the lentils in plenty of unsalted
water for 20–25 minutes or until soft.
Drain thoroughly.

3 Boil the basmati rice for 10 minutes,
then drain.

4 Mix together the rice and lentils in the
dressing and season well. Leave to cool.

5 Add the carrots, cucumber, onions and
parsley. Spoon into an attractive serving
bowl and chill before serving.

WILD RICE WITH JULIENNE VEGETABLES

*FOR THE BEST FLAVOUR, BUY A
GOOD QUALITY WILD RICE
(WHICH IS ACTUALLY A CEREAL!)
AND, TO SHORTEN THE COOKING
TIME, SOAK IT OVERNIGHT.*

SERVES FOUR

INGREDIENTS
 115g/4oz/½ cup wild rice
 1 red onion, sliced
 2 carrots, cut in julienne sticks
 2 celery sticks, cut in julienne sticks
 50g/2oz/4 tbsp butter
 150ml/¼ pint/⅔ cup stock or water
 salt and ground black pepper
 2 medium courgettes, cut in thicker
   sticks
 a few toasted almond flakes, to serve

1 Drain the soaked rice, then boil in
plenty of unsalted water for 15–20
minutes, until it is soft and many of the
grains have burst open. Drain.

2 In another saucepan, gently fry the
onion, carrots and celery in the butter for
2 minutes then pour in the stock or water
and season well.

3 Bring to a boil, simmer for 2 minutes
then stir in the courgettes. Cook for 1
more minute then mix in the rice. Reheat
and serve hot sprinkled with the almonds.

OVEN-CRISP ASPARAGUS ROLLS

A LOVELY TREAT IS TO WRAP BLANCHED SPEARS OF FRESH ASPARAGUS IN SLICES OF THIN BREAD AND BAKE IN A BUTTERY GLAZE UNTIL CRISP.

SERVES EIGHT

INGREDIENTS

8 thick spears of fresh asparagus
salt
115g/4oz/8½ tbsp butter, softened
15ml/1 tbsp coarse grained mustard
grated rind of 1 lemon
ground black pepper
8 slices thin white bread, crusts
  removed

1 Trim the asparagus stalks, peeling the tough woody skin at the base. Blanch until just tender in a shallow pan of boiling, salted water. Drain and refresh in cold water. Pat dry.

2 Blend two-thirds of the butter with the mustard, lemon rind and seasoning. Spread over the slices of bread.

3 Lay an asparagus spear on the edge of each bread slice and roll it up tightly. Place the rolls join side down on a lightly greased baking sheet.

4 Melt the remaining butter and brush over the rolls. Heat the oven to 190°C/375°F/Gas 5 and bake for 12–15 minutes until golden and crisp. Cool slightly before serving.

VARIATION

Asparagus has always been something of a luxury as its season is so short, but imports from across the world mean that it is available almost all year round, albeit at a price!

Thin baby asparagus, known as sprue, can be eaten raw in salads or stir-fried quickly. Opinions differ about the merits of green or white asparagus spears. The latter are forced in the dark (hence their white colour), but some people copnsider them to have a better flavour and texture.

RATATOUILLE TART

A DELICIOUS MEDITERRANEAN FILLING ON A PASTRY BASE.

SERVES SIX

INGREDIENTS

115g/4oz/1 cup plain flour
115g/4oz/¾ cup wholemeal flour
5ml/1 tsp dried mixed herbs
salt and ground black pepper
115g/4oz/½ cup sunflower margarine
45–60ml/3–4 tbsp cold water
For the filling
1 small aubergine, thickly sliced
salt
45ml/3 tbsp olive oil
1 onion, sliced
1 red or yellow pepper, sliced
2 garlic cloves, crushed
2 courgettes, thickly sliced
2 tomatoes, skinned and sliced
ground black pepper
30ml/2 tbsp fresh basil, chopped
150g/5oz Mozzarella cheese, sliced
30ml/2 tbsp pine nuts

1 Mix the two flours with the herbs and seasoning then rub in the margarine until it resembles fine crumbs. Mix to a firm dough with water.

2 Roll out the pastry and line a 23cm/ 9in round flan tin. Prick the base, line with foil and baking beans then allow to rest in the fridge.

3 Meanwhile, sprinkle the aubergine lightly with salt and leave to drain for 30 minutes in a colander. Rinse and pat dry.

4 Heat the oil in a frying pan and fry the onion and pepper for 5 minutes, then add the garlic, courgettes and aubergines. Fry for a further 10 minutes, stirring the mixture occasionally.

5 Stir in the tomatoes and seasoning, cook for a further 3 minutes, add the basil then remove the pan from the heat and allow to cool.

6 Heat the oven to 200°C/400°F/Gas 6. Place the flan on a baking sheet and bake for 25 minutes, removing the foil and baking beans for the last 5 minutes. Cool and then, if possible, remove the case from the tin.

7 When ready to serve, spoon the vegetables into the case using a slotted spoon so any juices drain off and don't soak into the pastry. Top with the cheese slices and pine nuts. Toast under a preheated grill until golden and bubbling. Serve warm.

FILO BASKETS WITH GINGER DILL VEGETABLES

MAKE UP SOME ELEGANT FILO BASKETS, THEN FILL WITH SOME CRISPLY STEAMED VEGETABLES TOSSED IN A TASTY SAUCE.

SERVES FOUR

INGREDIENTS
    4 sheets of filo pastry
    40g/1½oz/3 tbsp butter, melted
For the filling
    30ml/2 tbsp olive oil
    15ml/1 tbsp fresh root ginger, grated
    2 garlic cloves, crushed
    3 shallots, sliced
    225g/8oz mushrooms, chestnut or
      brown, sliced
    115g/4oz oyster mushrooms, sliced
    1 courgette, sliced
    200g/7oz crème fraîche
    30ml/2 tbsp fresh dill, chopped
    salt and ground black pepper
    dill and parsley sprigs, to serve

1 Cut the filo sheets into four. Line four Yorkshire pudding tins, angling the layers so that the corners form a pretty star shape. Brush between each layer with butter. Set aside.

2 Heat the oven to 190°C/375°F/Gas 5. Bake the cases for about 10 minutes until golden brown and crisp. Remove and cool.

3 For the filling, heat the oil and sauté the ginger, garlic and shallots for 2 minutes, then add the mushrooms and courgette. Cook for another 3 minutes.

4 Mix in the crème fraîche, chopped dill and seasoning. Heat until just bubbling then spoon into the filo cases. Garnish with the dill and parsley and serve.

GADO GADO SALAD WITH PEANUT SAMBAL

INDONESIANS ENJOY A SALAD OF LIGHTLY-STEAMED VEGETABLES TOPPED WITH A PEANUT SAUCE.

SERVES SIX

INGREDIENTS
    225g/8oz new potatoes, halved
    2 carrots, cut in sticks
    115g/4oz green beans
    ½ small cauliflower, broken into florets
    ¼ firm white cabbage, shredded
    200g/7oz bean or lentil sprouts
    4 eggs, hard-boiled and quartered
    bunch watercress, trimmed
For the sauce
    90ml/6 tbsp crunchy peanut butter
    300ml/½ pint/1¼ cups cold water
    1 garlic clove, crushed
    30ml/2 tbsp dark soy sauce
    15ml/1 tbsp dry sherry
    10ml/2 tsp caster sugar
    15ml/1 tbsp fresh lemon juice
    5ml/1 tsp anchovy essence

1 Fit a steamer or metal colander over a pan of gently boiling water. Cook the potatoes for 10 minutes.

2 Add the rest of the vegetables and sprouts and steam for a further 10 minutes until tender. Cool and serve on a platter with the egg quarters surrounded by the watercress.

3 Beat all the sauce ingredients together until smooth. Put the sauce in a small bowl and drizzle over each individual serving of salad.

PERSIAN RICE AND LENTILS WITH A TAHDEEG

PERSIAN OR IRAN CUISINE IS AN EXOTIC, DELICIOUS ONE. FLAVOURS ARE INTENSE AND SOMEHOW MORE SOPHISTICATED THAN OTHER EASTERN STYLES. A TAHDEEG IS THE GOLDEN RICE CRUST THAT FORMS AT THE BOTTOM OF THE SAUCEPAN.

SERVES EIGHT

INGREDIENTS

450g/1lb basmati rice, rinsed
   thoroughly and soaked
2 onions, 1 chopped, 1 thinly sliced
2 garlic cloves, crushed
150ml/¼ pint/⅔ cup sunflower oil
150g/5oz/1 cup green lentils, soaked
600ml/1 pint/2½ cups stock
50g/2oz/⅓ cup raisins
10ml/2 tsp ground coriander
45ml/3 tbsp tomato purée
salt and ground black pepper
few strands of saffron
1 egg yolk, beaten
10ml/2 tsp natural yogurt
75g/3oz/6 tbsp butter, melted and
   strained
extra oil, for frying

1 Boil the rinsed and drained rice in plenty of well salted water for 3 minutes only. Drain again.

2 Meanwhile, fry the chopped onion and garlic in 30ml/2 tbsp of oil for 5 minutes then add the lentils, stock, raisins, coriander, tomato purée and seasoning. Bring to a boil, then cover and simmer for 20 minutes. Set aside.

3 Soak the saffron strands in a little hot water. Remove about 120ml/8 tbsps of the rice and mix with the egg yolk and yogurt. Season well.

4 In a large saucepan, heat about two-thirds of the remaining oil and scatter the egg and yogurt rice evenly over the base.

5 Scatter the remaining rice into the pan, alternating it with the lentils. Build up in a pyramid shape away from the sides of the pan, finish with plain rice on top.

6 With a long wooden spoon handle, make three holes down to the bottom of the pan and drizzle over the butter. Bring to a high heat, then wrap the pan lid in a clean, wet tea towel and place firmly on top. When a good head of steam appears, turn the heat down to low. Cook for about 30 minutes.

7 Meanwhile, fry the sliced onion in the remaining oil until browned and crisp. Drain well and set aside.

8 Remove the rice pan from the heat, still covered and stand it briefly in a sink of cold water for a minute or two to loosen the base. Remove the lid and mix a few spoons of the white rice with the saffron water prepared in Step 3.

9 Toss the rice and lentils together in the pan and spoon out onto a serving dish in a mound. Scatter the saffron rice on top. Break up the rice crust on the bottom (the prized tahdeeg) and place around the mound. Scatter the onions on top of the saffron rice and serve.

VEGETABLE FRITTERS WITH TZATZIKI

Spicy deep-fried aubergine and courgette slices served with a creamy Greek yogurt and dill dip make a good, simple party starter.

SERVES FOUR TO SIX

INGREDIENTS

½ cucumber, coarsely grated
225g/8oz Greek natural yogurt
15ml/1 tbsp extra virgin olive oil
10ml/2 tsp fresh lemon juice
30ml/2 tbspfresh dill, chopped
15ml/1 tbsp fresh mint, chopped
1 garlic clove, crushed
salt and ground black pepper
1 large aubergine, thickly sliced
2 courgettes, thickly sliced
1 egg white, beaten
40g/1½oz/4 tbsp plain flour
10ml/2 tsp ground coriander
ground cumin

1 For the dip, mix the cucumber, yogurt, oil, lemon juice, dill, mint, garlic and seasoning. Spoon into a bowl then set aside.

2 Layer the aubergine and courgettes in a colander and sprinkle them with salt. Leave for 30 minutes. Rinse in cold water then pat dry.

3 Put the egg white into a bowl. Mix the flour, coriander and cumin with seasoning and put into another bowl.

4 Dip the vegetables first into the egg white then into the seasoned flour and set aside.

5 Heat about 2.5cm/1in of oil in a deep frying pan until quite hot, then fry the vegetables a few at a time until they are golden and crisp.

6 Drain and keep warm while you fry the remainder. Serve warm on a platter with a bowl of the tzatziki dip lightly sprinkled with paprika.

MUSHROOM SAUCERS

RECIPES THAT ARE ALREADY PORTIONED ARE A GREAT BOON FOR THE HOST, AS GUESTS CAN HELP THEMSELVES WITHOUT FEELING THEY ARE TAKING MORE THAN THEIR FAIR SHARE.

SERVES EIGHT

INGREDIENTS

8 large, flat mushrooms with stalks
  removed, wiped clean and chopped
45ml/3 tbsp olive oil
salt and ground black pepper
1 onion, sliced
5ml/1 tsp cumin seeds
450g/1 lb leaf spinach, stalks trimmed,
  and shredded
1 × 425g/15oz can red kidney beans,
  drained
1 × 200g/7oz pack soft cheese with
  garlic and herbs
2 medium tomatoes, halved, seeded
  and sliced in strips

1 Heat the oven to 190°C/375°F/Gas 5. Lightly grease a large, shallow ovenproof dish. Brush the mushrooms with some oil, place them in the dish and season well. Cover with foil and bake for 15–20 minutes. Uncover, drain and reserve the juices.

2 Fry the onion and chopped mushroom stalks in the remaining oil for 5 minutes until soft. Then add the cumin seeds and mushroom juices and cook for a minute longer until reduced down.

3 Stir in the spinach and fry until the leaves begin to wilt, then mix in the beans and heat well. Add the cheese, stirring until melted and season again.

4 Divide the mixture between the mushroom cups and return to the oven to heat through. Serve garnished with tomato slices.

BIRDS' NESTS

A RECIPE FROM AN OLD HAND-WRITTEN COOKERY BOOK DATED 1887. THESE ARE ALSO KNOWN AS WELSH EGGS BECAUSE THEY RESEMBLE SCOTCH EGGS BUT THEY HAVE LEEKS IN THE FILLING.

SERVES SIX

INGREDIENTS

6 eggs, hard-boiled
flour, seasoned with salt and paprika
1 leek, chopped
10ml/2 tsp sunflower oil
115g/4oz/2 cups fresh white
   breadcrumbs
grated rind and juice of 1 lemon
50g/2oz/½ cup vegetarian shredded
   suet
60ml/4 tbsp fresh parsley, chopped
5ml/1 tsp dried thyme
salt and ground black pepper
1 egg, beaten
75g/3oz/½ cup dried breadcrumbs
oil, for deep fat frying
lettuce and tomato slices, to garnish

1 Peel the hard-boiled eggs and toss in the seasoned flour.

2 Fry the leeks in the oil for about 3 minutes. Remove, cool, then mix with the fresh breadcrumbs, lemon rind and juice, suet, herbs and seasoning. If the mixture is a little dry add a little water.

3 Shape the mixture around the eggs, then toss first into the beaten egg, then the dried breadcrumbs. Set aside on a plate to chill for 30 minutes.

4 Pour enough oil to fill one-third of a deep fat fryer and heat to a temperature of 190°C/375°F/Gas 5, and fry the eggs, three at a time, for about 3 minutes. Remove and drain on kitchen paper towel.

5 Serve, cool, cut in half on a platter lined with lettuce and garnished with tomato slices.

PORTABLE SALADS

A CLEVER VICTORIAN NOTION FOR TRANSPORTING SAUCY SALADS NEATLY TO A PICNIC SITE.

<u>SERVES SIX</u>

INGREDIENTS

   1 large, deep crusty loaf
   softened butter or margarine, for
     spreading
   few leaves of crisp lettuce
   4 eggs, hard-boiled and chopped
   350g/12oz new potatoes, boiled and
     sliced
   1 green pepper, thinly sliced
   2 carrots, coarsely grated
   3 spring onions, chopped
   115g/4oz Gouda cheese, grated
   salt and ground black pepper
For the dressing
   30ml/2 tbsp mayonnaise
   30ml/2 tbsp natural yogurt
   30ml/2 tbsp milk
   1 garlic clove, crushed (optional)
   15ml/1 tbsp fresh dill, chopped

1 Cut the top from the loaf and scoop out the bread inside. Use this for making fresh breadcrumbs and freeze for later.

2 Spread the inside of the loaf lightly with the softened butter or margarine, then line with the lettuce leaves.

3 Mix the eggs with the vegetables and cheese. Season well. Beat the dressing ingredients together and mix into the egg and vegetables.

4 Spoon the dressed salad into the hollow and lined loaf, replace the lid and wrap in cling film. Chill until ready to transport. To serve, spoon the salad onto plates and cut the crust into chunks.

GUACAMOLE, BASIL AND TOMATO PITTA BREADS

THIS IS A FAVOURITE FAMILY RECIPE — THE FRESH BASIL AND TOMATOES ARE PERFECT PARTNERS FOR EACH OTHER AND FOR THE SPICY GUACAMOLE.

SERVES SIX

INGREDIENTS
    6 large pitta breads
    1–2 large beef tomatoes, sliced
    12 basil leaves
    2 large ripe avocados
    1 tomato
    ½ red onion
    1 clove garlic, peeled and crushed
    15ml/1 tbsp lime juice
    2ml/¼ tsp chilli powder
    30ml/2 tbsp chopped fresh dill

1 Open the ends of the pitta breads to make pockets and place a couple of slices of tomato and two basil leaves in each one.

2 Roughly chop the avocados, the remaining tomato and the red onion. Mix all the remaining ingredients briefly in a food processor.

3 Add the mixture from the food processor to the roughly chopped avocado, tomato and onions, and stir gently. Fill the pockets with the avocado mixture and serve immediately.

BRIE AND GRAPE SANDWICHES WITH MINT

A SLIGHTLY UNUSUAL SANDWICH COMBINATION, WHICH WORKS WELL, JUDGING BY THE SPEED WITH WHICH THE SANDWICHES DISAPPEAR AT FAMILY PICNICS OR SUMMER TEA PARTIES.

SERVES FOUR

INGREDIENTS
    8 slices wholewheat bread
    butter for spreading
    350g/12oz ripe Brie cheese
    30–40 large grapes
    16 fresh mint leaves

OTHER SANDWICH IDEAS
• Feta cheese, black olives, lettuce, tomato and freshly chopped mint in pitta bread.
• Italian salami, cream cheese, tomato and fresh basil on ciabatta bread.
• Sliced chicken breast, mayonnaise and dill sprigs on wholewheat bread.
• Grilled mozzarella and sun-dried tomato focaccia bread sandwich, with black olives, fresh rocket and basil leaves.
• Hummus, lettuce and freshly chopped coriander on French bread.
• Parma ham, green olives and rocket leaves on poppy-seeded white bread.

1 Butter the bread. Cut the Brie into thick slices, to be divided among the sandwiches.

2 Place the Brie slices on four slices of bread. Peel, halve and seed the grapes and put on top of the Brie. Chop the mint finely by hand or in a food processor, and sprinkle the mint over the Brie and grapes. Place the other four slices of bread over the top and cut each sandwich in half.

MARBLED QUAILS' EGGS

HARD-BOILED QUAILS' EGGS RE-BOILED IN SMOKY CHINA TEA ASSUME A PRETTY MARBLED SKIN. IT'S QUITE A TREAT TO DIP THEM INTO A FRAGRANT SPICY SALT AND HAND THEM ROUND WITH DRINKS. SZECHUAN PEPPERCORNS CAN BE BOUGHT FROM ORIENTAL FOOD STORES.

SERVES FOUR TO SIX

INGREDIENTS
    12 quails' eggs
    600ml/1 pint/2½ cups strong lapsang
        souchong tea
    15ml/1 tbsp dark soy sauce
    15ml/1 tbsp dry sherry
    2 star anise pods
    lettuce leaves, to serve
    milled Szechuan red peppercorns
    sea salt, to mix

1 Place the quails' eggs in cold water and bring to the boil. Time them for 2 minutes from when the water boils.

2 Remove the eggs from the pan and run them under cold water to cool. Tap the shells all over so they are crazed, but do not peel yet.

3 In a saucepan, bring the tea to the boil and add the soy sauce, sherry and star anise. Re-boil the eggs for about 15 minutes, partially covered, so the liquid does not boil dry.

4 Cool the eggs, then peel them and arrange on a small platter lined with lettuce leaves. Mix the ground red peppercorns with equal quantities of salt and place in a small side dish.

BEETROOT ROULADE

THIS ROULADE IS SIMPLE TO MAKE, YET WILL CREATE A STUNNING IMPRESSION. PREPARE IT IN THE AUTUMN WHEN BEETROOTS ARE AT THEIR BEST.

SERVES SIX

INGREDIENTS
    225g/8oz fresh beetroot, cooked and
        peeled
    2.5ml/½ tsp ground cumin
    25g/1oz/2 tbsp butter
    10ml/2 tsp grated onion
    4 eggs, separated
    salt and ground black pepper
For the filling
    150ml/¼ pint/⅔ cup crème fraîche or
        double cream
    10ml/2 tsp white wine vinegar
    good pinch dry mustard powder
    5ml/1 tsp sugar
    45ml/3 tbsp fresh parsley, chopped
    30ml/2 tbsp fresh dill, chopped
    45ml/3 tbsp horseradish relish

1 Line and grease a Swiss roll tin, and then preheat the oven to 190°C/375°F/Gas 5.

2 Roughly chop the beetroot then blend to a purée in a food processor and beat in the cumin, butter, onion, egg yolks and seasoning. Turn the beetroot purée into a large bowl.

3 In another bowl, that is spotlessly clean, whisk the egg whites until they form soft peaks. Fold them into the beetroot mixture carefully.

4 Spoon the mixture into the Swiss roll tin, level and bake for about 15 minutes until just firm to touch.

5 Have ready a clean tea towel laid over a wire rack. Turn the beetroot out onto the towel, and remove the paper carefully in strips.

6 Beat the crème fraîche or cream until lightly stiff, then fold in the remaining ingredients. Spread this mixture onto the beetroot. Roll up the roulade in the towel and allow it to cool.

PAN BAGNA

You need three elements for this French picnic classic: a really fresh French baguette, ripe juicy tomatoes and good, extra virgin olive oil.

SERVES THREE TO FOUR

INGREDIENTS
1 long French stick, split in half
1 garlic clove, halved
60–90ml/4–6 tbsp extra virgin olive oil
3-4 ripe tomatoes, thinly sliced
salt and ground black pepper
1 small green pepper, thinly sliced
50g/2oz Gruyère cheese, thinly sliced
a few stoned black olives, sliced
6 fresh basil leaves

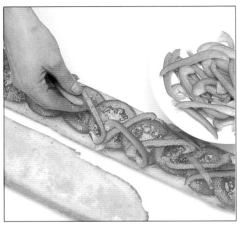

1 Rub the cut surface of the bread with the garlic and discard the clove. Brush over half of the olive oil on both halves of the bread.

2 Lay the tomato slices on top, season well and top with the pepper. Drizzle over the remaining oil.

3 Top the tomatoes with the cheese slices, olives and basil leaves. Sandwich the loaf together firmly and wrap it in cling film for an hour or more. Serve cut diagonally in thick slices.

SANDWICHES, ROLLS AND FILLINGS

THERE IS AN INCREASING VARIETY OF WONDERFUL BREADS AND ROLLS NOW FOR THE PICNIC PACKER TO CHOOSE FROM — NOT ONLY VARIATIONS ON WHITE AND WHOLEMEAL BREADS, BUT ALSO FLAVOURED BREADS SUCH AS ONION, WALNUT, TOMATO SWIRL AND BLACK OLIVE. MAKE SURE THE BREADS ARE FRESH AND SPREAD THEM RIGHT UP TO THE EDGES. ONCE FILLED, WRAP IN CLING FILM AND CHILL UNTIL REQUIRED. REMEMBER TO ALLOW TO RETURN TO ROOM TEMPERATURE BEFORE EATING.

FILLING IDEAS
Unless specified, keep the fillings in separate layers rather than mixing the ingredients together.

• De-rinded Brie or Camembert, mixed with chopped walnuts or pecans and served with frisee or curly endive lettuce.

• Yeast extract, scrambled egg (made without milk) and bean sprout (especially good with alfalfa sprouts). Spread the bread or roll with yeast extract rather than mixing it into the egg.

• Fry onions in olive oil until crisp and brown. Cool. Layer with shredded, young raw spinach leaves and grated cheese mixed with a little mayonnaise.

• Real English cucumber sandwiches. Peel strips from a whole cucumber to leave it stripey, then slice thinly on a mandoline or a food processor slicer. Sprinkle lightly with salt and leave to drain for 30 minutes in a colander. Pat dry. Sprinkle lightly with a little vinegar and black pepper. Sandwich in very fresh bread and cut off the crusts.

SESAME EGG ROLL

A JAPANESE-INSPIRED IDEA. AN EGG PANCAKE IS ROLLED UP WITH A CREAMY WATERCRESS FILLING AND SERVED IN THICK SLICES.

SERVES THREE TO FOUR

INGREDIENTS
   3 eggs
   15ml/1 tbsp soy sauce
   15ml/1 tbsp sesame seeds
   5ml/1 tsp sesame seed oil
   salt and ground black pepper
   15ml/1 tbsp sunflower oil
   75g/3oz cream cheese with garlic
   1 bunch watercress, chopped

1 Beat the eggs with the soy sauce, sesame seeds, sesame seed oil and seasoning.

2 Heat the sunflower oil in a large frying pan until quite hot, then pour in the egg mixture, tilting the pan so it covers the whole base. Cook until firm.

3 Allow the pancake to stand in the pan for a few minutes then turn out onto a chopping board and cool completely.

4 Beat the cream cheese until soft, season well then mix in the chopped watercress. Spread this over the egg pancake then roll it up quite firmly. Wrap in cling film and chill.

HOME-MADE COLESLAW

FORGET SHOP-BOUGHT COLESLAW! MAKING YOUR OWN AT HOME IS QUITE QUICK AND EASY TO DO — AND IT TASTES FRESH, CRUNCHY AND WONDERFUL!

SERVES FOUR TO SIX

INGREDIENTS
   ¼ firm white cabbage
   1 small onion, finly chopped
   2 celery sticks, thinly sliced
   2 carrots, coarsely grated
   5–10ml/1–2 tsp caraway seeds
    (optional)
   1 dessert apple, cored and chopped
    (optional)
   50g/2oz/½ cup walnuts, chopped
    (optional)
   salt and ground black pepper
For the dressing
   45ml/3 tbsp mayonnaise
   30ml/2 tbsp single cream or natural
    yogurt
   5ml/1 tsp lemon rind, grated
   salt and ground black pepper

1 Cut and discard the core from the cabbage quarter then shred the leaves finely. Place this in a large bowl.

2 Into the cabbage, toss the onion, celery and carrot, plus the caraway seeds, apple and walnuts, if using. Season well.

3 Mix the dressing ingredients together and stir into the vegetables. Cover and allow to stand for 2 hours, stirring occasionally, then chill the coleslaw lightly before serving.

MALFATTI WITH RED SAUCE

IF YOU EVER FELT DUMPLINGS WERE A LITTLE HEAVY, TRY MAKING THESE LIGHT ITALIAN SPINACH AND RICOTTA MALFATTI INSTEAD. SERVE THEM WITH A SIMPLE TOMATO AND RED PEPPER SAUCE.

SERVES FOUR TO SIX

INGREDIENTS
    450g/1lb fresh leaf spinach, stalks
      trimmed
    1 small onion, chopped
    1 garlic clove, crushed
    15ml/1 tbsp olive oil
    400g/14oz ricotta cheese
    75g/3oz/⅔ cup dried breadcrumbs
    50g/2oz/½ cup plain flour
    5ml/1 tsp salt
    50g/2oz Parmesan cheese, freshly
      grated
    fresh nutmeg, grated, to taste
    3 eggs, beaten
    25g/1oz/2 tbsp butter, melted
For the sauce
    1 large, red pepper, chopped
    1 small red onion, chopped
    30ml/2 tbsp olive oil
    1 × 400g/14oz can chopped tomatoes
    150ml/¼ pint/⅔ cup water
    good pinch dried oregano
    salt and ground black pepper
    30ml/2 tbsp single cream

1 Blanch the spinach in the tiniest amount of water until it is limp, then drain well, pressing it through a sieve with the back of a ladle or spoon. Chop very finely.

COOK'S TIP
Quenelles are oval-shaped dumplings. To shape the Malfatti into quenelles you need two dessertspoons. Scoop up the mixture with one spoon, making sure it is mounded up, then, using the other spoon, scoop the mixture off the first spoon, twisting the top spoon into the bowl of the second.
    Repeat this action two or three times until the quenelle is smooth, and then gently knock it off onto a plate ready to cook.

2 Lightly fry the onion and garlic in the oil for 5 minutes then mix with the spinach together with the ricotta, breadcrumbs, flour, salt, most of the Parmesan and nutmeg.

3 Allow the mixture to cool, add the eggs and melted butter, then mould into 12 small 'sausage' shapes using two dessertspoons.

4 Meanwhile, make the sauce by lightly sautéeing the pepper and onion in the oil for 5 minutes. Add the tomatoes, water, oregano and seasoning. Bring to a boil, then simmer for 5 minutes.

5 When cooked, remove from the heat and blend to a puree in a food processor. Return to the pan, then stir in the cream. Check the seasoning.

6 Bring a shallow pan of salted water to a gentle boil and drop the malfatti into it a few at a time and poach them for about 5 minutes. Drain them well and keep them warm.

7 Arrange the malfatti on warm plates and drizzle over the sauce. Serve topped with the remaining Parmesan.

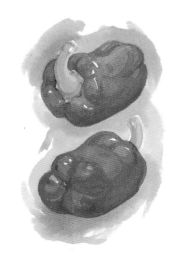

CURRIED MANGO CHUTNEY DIP

A QUICKLY MADE, TANGY AND
SPICY DIP OR DRESSING, IDEAL AS
A DIP FOR STRIPS OF PITTA
BREAD, GRISSINI OR STICKS OF
FRESH CHOPPED VEGETABLES.

SERVES FOUR TO SIX

INGREDIENTS
  1 onion, chopped
  1 garlic clove, crushed
  30ml/2 tbsp sunflower oil
  10ml/2 tsp mild curry powder
  225g/8oz natural Greek style yogurt
  30ml/2 tbsp mango chutney
  salt and ground black pepper
  30ml/2 tbsp fresh parsley, chopped

1 Gently fry the onion and garlic in the oil for 5 minutes until they are soft. Add the curry powder and cook for a further minute then allow the mixture to cool.

2 Spoon into a food processor with the yogurt, chutney and seasoning and blend until smooth.

3 Stir in the parsley and chill before serving with a variety of vegetable crudites and strips of bread.

NUTTY MUSHROOM PÂTÉ

SPREAD THIS DELICIOUS,
MEDIUM-TEXTURE PÂTÉ ON
CHUNKS OF CRUSTY FRENCH
BREAD AND EAT WITH CRISP
LEAVES OF LETTUCE AND SWEET
LITTLE, CHERRY TOMATOES.

SERVES FOUR TO SIX

INGREDIENTS
  1 onion, chopped
  1 garlic clove, crushed
  15ml/1 tbsp sunflower oil
  30ml/2 tbsp water
  15ml/1 tbsp dry sherry
  225g/8oz button mushrooms, chopped
  salt and ground black pepper
  75g/3oz/¾ cup cashew nuts or walnuts,
    chopped
  150g/5oz low-fat soft cheese
  15ml/1 tbsp soy sauce
  few dashes Worcestershire sauce
  fresh parsley, chopped, and a little
    paprika, to serve

1 Gently fry the onion and garlic in the oil for 3 minutes then add the water, sherry and mushrooms. Cook, stirring for about 5 minutes. Season to taste and allow to cool a little.

2 Put the mixture into a food processor with the nuts, cheese and sauces. Blend to a rough purée – do not allow it to become too smooth.

3 Check the seasoning then spoon into a serving dish. Swirl the top and serve lightly chilled sprinkled with parsley and paprika.

ANTIPASTI WITH AIOLI

*FOR A SIMPLE STARTER, MAKE A BOWL OF THE CLASSIC AIOLI AND SERVE IT WITH A SELECTION OF
VEGETABLES AND BREADS.*

SERVES FOUR TO SIX

INGREDIENTS
  4 garlic cloves
  2 egg yolks
  2.5ml/½ tsp salt
  ground black pepper
  300ml/½ pint/1¼ cups extra virgin
  olive oil
To serve
  red or yellow pepper, cut in thick strips
  fennel, cut in slivers
  radishes, halved if large
  button mushrooms
  broccoli florets
  grissini sticks
  French bread, thinly sliced

1 Crush the garlic into a bowl then beat
in the egg yolks, salt and some ground
black pepper.

2 Stand the bowl on a damp cloth and
slowly trickle in the oil, drip by drip;
whisking with a balloon whisk until you
have a thick, creamy sauce. As the sauce
thickens, you can add the oil in slightly
larger amounts.

3 Spoon the aioli into a bowl. Arrange the
dipping food around the bowl and serve
lightly chilled.

CAMEMBERT FRITTERS

*THESE DEEP-FRIED CHEESES ARE QUITE SIMPLE TO DO. THEY ARE SERVED WITH A RED ONION MARMALADE,
WHICH CAN BE MADE IN ADVANCE AND STORED IN THE REFRIGERATOR.*

SERVES FOUR

INGREDIENTS
For the marmalade
  1 kg/2 lb red onions, sliced
  45ml/3 tbsp sunflower oil
  45ml/3 tbsp olive oil
  15ml/1 tbsp coriander berries, crushed
  2 large bay leaves
  45ml/3 tbsp granulated sugar
  90ml/6 tbsp red wine vinegar
  10ml/2 tsp salt
For the cheese
  8 individual portions of Camembert
  1 egg, beaten
  125g/4oz/1 cup natural colour dried
    breadcrumbs, to coat
  oil, for deep fat frying

1 Make the marmalade first. In a large
saucepan, gently fry the onions in the oil,
covered, for 20 minutes or so or until they
are soft.

2 Add the remaining marmalade
ingredients, stir well and cook uncovered
for a further 10–15 minutes until most of
the liquid has been absorbed. Cool and
then set aside.

3 Prepare the cheese by first scratching
the mould rind lightly with a fork. Dip first
in egg then in breadcrumbs to coat well.
Dip and coat a second time if necessary.
Store on a plate.

4 Pour oil into a deep fat fryer to one-
third full; heat to 190°C/375°F/Gas 5.

5 Carefully lower the coated cheeses into
the hot oil three or four at a time and fry
until golden and crisp, about 2 minutes or
less.

6 Drain well on kitchen paper and fry the
rest, reheating the oil in between. Serve
hot with some of the marmalade.

VIRTUALLY
VEGETARIAN

For all those demi-vegetarians who just cannot resist fish and seafood, here is a selection of starter and main-course dishes that combine unusual vegetables and herbs with fish in nutritious and flavoursome ways.

WATERCRESS SOUP

SERVES FOUR

INGREDIENTS
  15ml/1 tbsp sunflower oil
  15g/½oz butter
  1 medium onion, finely chopped
  1 medium potato, diced
  about 175g/6oz watercress
  400ml/14fl oz/1⅔ cups fish
    or vegetable stock
  400ml/14fl oz/1⅔ cups milk
  lemon juice
  salt and freshly ground black pepper
  soured cream, to serve (optional)

1 Heat the oil and butter in a large saucepan and fry the onion over a gentle heat until soft but not browned. Add the potato, fry gently for 2–3 minutes and then cover and sweat for 5 minutes over a gentle heat, stirring occasionally.

2 Strip the watercress leaves from the stalks and roughly chop the stalks.

3 Add the stock and milk to the pan, stir in the chopped stalks and season with salt and pepper. Bring to the boil and then simmer gently, partially covered, for 10–12 minutes until the potatoes are tender. Add all but a few of the watercress leaves and simmer for 2 minutes.

4 Process the soup in a food processor or blender, and then pour into a clean saucepan and heat gently with the reserved watercress leaves. Taste when hot and add a little lemon juice and adjust the seasoning.

5 Pour the soup into warmed soup dishes and swirl in a little soured cream, if using, just before serving.

COOK'S TIP
Provided you leave out the cream, this is a low calorie but nutritious soup, which, served with crusty bread, makes a satisfying meal.

WATERCRESS AND TWO-FISH TERRINE

THIS IS A PRETTY, DELICATE DISH, IDEAL FOR A SUMMER BUFFET PARTY OR PICNIC. SERVE WITH LEMON MAYONNAISE OR SOURED CREAM, AND A WATERCRESS AND GREEN SALAD.

SERVES SIX TO EIGHT

INGREDIENTS
  350g/12oz monkfish, filleted
  175g/6oz lemon sole, filleted
  salt and freshly ground black pepper
  1 egg and 1 egg white
  45–60ml/3–4 tbsp lemon juice
  40–50g/1½–2oz fresh white
    breadcrumbs
  300ml/½ pint/1¼ cups whipping
    cream
  75g/3oz smoked salmon
  175g/6oz watercress, roughly chopped

1 Preheat the oven to 180°C/350°F/ Gas 4 and line a 1.5 litre/2½ pint/6¼ cup loaf tin with non-stick baking paper.

2 Cut the fish into rough chunks, discarding the skin and bones. Put the fish into a food processor with a little seasoning.

3 Process briefly and add the egg and egg white, lemon juice, breadcrumbs and cream. Process to a paste. Put the mixture into a bowl. Take 75ml/5 tbsp of the mixture and process with the smoked salmon. Transfer to a separate bowl. Take 75ml/5 tbsp of the white fish mixture and process with the watercress.

4 Spoon half of the white fish mixture into the base of the prepared loaf tin and smooth the surface with a palette knife.

5 Spread over the watercress mixture, then the smoked salmon mixture and finally spread over the remaining white fish mixture and smooth the top.

6 Lay a piece of buttered non-stick baking paper on top of the mixture and then cover with foil. Place the loaf tin in a roasting tin, half-filled with boiling water and cook in the oven for 1¼–1½ hours. Towards the end of the cooking time the terrine will begin to rise, which indicates that it is ready.

7 Allow to cool in the tin and then turn on to a serving plate and peel away the baking paper. Chill for 1–2 hours.

GNOCCHI WITH OYSTER MUSHROOMS

Gnocchi make an unusual and pleasant alternative to pasta. They are bland on their own but bring out the flavour of the oyster mushrooms in this dish while their soft texture contrasts with the firmness of the mushrooms.

SERVES FOUR

INGREDIENTS
    225g/8oz oyster mushrooms
    15ml/1 tbsp olive oil
    1 medium onion, finely chopped
    1 garlic clove, crushed
    4 plum tomatoes, peeled and chopped
    45–60ml/3–4 tbsp vegetable stock
      or water
    salt and freshly ground black pepper
    2 x 300g/11oz packet plain potato
      gnocchi
    a good knob of butter
    10ml/2 tsp chopped fresh parsley
    Parmesan cheese, cut in shavings,
      to serve

1 Trim the mushrooms and tear into smaller pieces, if they are large. Heat the oil in a large frying pan and fry the onion and garlic over a low heat for about 4–5 minutes until softened but not browned.

2 Increase the heat, add the mushrooms to the pan and sauté for about 3–4 minutes, stirring constantly.

3 Stir in the chopped tomatoes, stock or water and seasoning and then cover and simmer for about 8 minutes until the tomatoes are very soft and reduced to a pulp. Stir occasionally.

4 Cook the gnocchi in a large pan of salted boiling water for 2–3 minutes (or according to the instructions on the packet) then drain well and toss with the butter. Place in a large warmed serving bowl and stir in the chopped parsley.

5 Pour the mushroom and tomato mixture over the top, stir briefly and sprinkle with the Parmesan cheese.

COOK'S TIP
If the mushrooms are very large, the stalks are likely to be tough, therefore they should be discarded. Always tear rather than cut oyster mushrooms.

SEAFOOD AND OYSTER MUSHROOM STARTER

This dish is remarkably quick to prepare. It can be made into a more substantial dish by stirring 275–350g/10–12oz cooked pasta shells into the sauce at the end.

SERVES FOUR

INGREDIENTS
    15ml/1 tbsp olive oil
    15g/½oz butter
    1 garlic clove, crushed
    175g/6oz oyster mushrooms,
      halved or quartered
    115–175g/4–6oz peeled prawns
    115g/4oz cooked mussels, optional
    juice of ½ lemon
    15ml/1 tbsp medium dry sherry
    150ml/¼ pint/⅔ cup double cream
    salt and freshly ground black pepper

1 Heat the oil and butter in a frying pan and sauté the garlic for a few minutes, then add the mushrooms. Cook over a moderate heat for 4–5 minutes until soft, stirring from time to time.

2 Reduce the heat and stir in the prawns, mussels and lemon juice. Cook for 1 minute, stirring continuously. Stir in the sherry and cook for 1 minute.

3 Add the cream and cook gently until heated through but not boiling. Taste and adjust the seasoning and then spoon into warmed serving dishes. Serve immediately with chunks of Italian bread.

CUCUMBER AND TROUT MOUSSE

THIS IS A VERY LIGHT, REFRESHING MOUSSE, MAKING THE MOST OF THE CLEAN TASTE OF CUCUMBER.
SERVE IT AS A STARTER OR FOR A LIGHT LUNCH WITH A GREEN SALAD. YOU COULD ALSO ARRANGE A
CHERVIL LEAF OR A VERY THIN LEMON SLICE ON THE TOPPING.

SERVES SIX

INGREDIENTS

1 small cucumber
2–3 smoked trout fillets, about
    175g/6oz total weight
115g/4oz fromage frais
15ml/1 tbsp powdered gelatine
150ml/¼ pint/⅔ cup vegetable stock
12–14 pimiento stuffed olives, sliced
30ml/2 tbsp lemon juice
5ml/1 tsp finely chopped fresh
    tarragon
150ml/¼ pint/⅔ cup whipping cream
2 egg whites
salt and freshly ground black pepper
peeled prawns and lemon wedges,
    to garnish
For the topping
    15ml/1 tbsp powdered gelatine
    90ml/6 tbsp vegetable stock

1 Lightly oil six ramekin dishes. To prepare the topping, take one quarter of the cucumber and slice thinly. Sprinkle the gelatine over the stock, leave to soak for a few minutes and then place over a saucepan of simmering water and stir until completely dissolved.

2 Spoon a little of the gelatine mixture into each dish and arrange two or three cucumber slices on top. Put in the fridge to set. Pour over the remaining gelatine mixture and return to the fridge to set.

3 To make the mousse, peel and very finely dice the remaining cucumber and put in a bowl. Flake the fish, discarding the skin and any bones and add to the cucumber. Beat in the fromage frais.

4 Sprinkle the gelatine over 30ml/2 tbsp of water in a bowl and leave to soak for a few minutes. Place over a saucepan of simmering water and stir until dissolved.

5 Heat the stock. Stir in the dissolved gelatine and leave until cool but not set. Pour over the trout and stir in the olives, lemon juice, tarragon and seasoning.

6 Lightly whip the cream and whisk the egg whites until stiff. Fold the cream into the trout mixture, followed by the egg whites. Spoon the mousse into the ramekin dishes, levelling the surface. Cover and chill for 1–2 hours and then unmould on to serving plates.

7 Garnish with any remaining cucumber slices together with a few peeled prawns and some lemon wedges.

SWEETCORN AND SCALLOP CHOWDER

FRESH HOME-GROWN SWEETCORN IS IDEAL FOR THIS CHOWDER, ALTHOUGH CANNED OR FROZEN SWEET-CORN ALSO WORKS WELL. THIS SOUP IS ALMOST A MEAL IN ITSELF AND MAKES A PERFECT LUNCH DISH.

SERVES FOUR TO SIX

INGREDIENTS

2 ears of corn or 200g/7oz frozen or canned sweetcorn
600ml/1 pint/2½ cups milk
15g/½oz butter or margarine
1 small leek or onion, chopped
40g/1½oz smoked streaky bacon, finely chopped
1 small garlic clove, crushed
1 small green pepper, seeded and diced
1 celery stalk, chopped
1 medium potato, diced
15ml/1 tbsp plain flour
300ml/½ pint/1¼ cups vegetable stock
4 scallops
115g/4oz cooked fresh mussels
pinch of paprika
150ml/¼ pint/⅔ cup single cream (optional)
salt and freshly ground black pepper

1 Using a sharp knife, slice down the ears of the corn to remove the kernels. Place half of the kernels in a food processor or blender and process with a little of the milk.

2 Melt the butter or margarine in a large saucepan and gently fry the leek or onion, bacon and garlic for 4–5 minutes until the leek is soft but not browned. Add the green pepper, celery and potato and sweat over a gentle heat for a further 3–4 minutes, stirring frequently.

3 Stir in the flour and cook for about 1–2 minutes until the mixture is golden and frothy. Gradually stir in the milk and corn mixture, stock, the remaining milk and corn kernels and seasoning.

4 Bring to the boil and then reduce the heat to a gentle simmer, and cook, partially covered, for 15–20 minutes until the vegetables are tender.

5 Pull the corals away from the scallops and slice the white flesh into 5mm/¼in slices. Stir the scallops into the soup, cook for 4 minutes and then stir in the corals, mussels and paprika. Allow to heat through for a few minutes and then stir in the cream, if using. Adjust the seasoning to taste and serve.

CHINESE LEAVES AND MOOLI WITH SCALLOPS

A SPEEDY STIR-FRY MADE USING CHINESE CABBAGE, MOOLI AND SCALLOPS. BOTH THE MOOLI AND CHINESE LEAVES HAVE A PLEASANT CRUNCHY "BITE". YOU NEED TO WORK QUICKLY, SO HAVE EVERYTHING PREPARED BEFORE YOU START COOKING.

SERVES FOUR

INGREDIENTS
   10 prepared scallops
   60–75ml/4–5 tbsp vegetable oil
   3 garlic cloves, finely chopped
   1cm/½in piece fresh root ginger,
      finely sliced
   4–5 spring onions, cut lengthways
      into 2.5cm/1in pieces
   30ml/2 tbsp medium dry sherry
   ½ mooli (daikon radish), cut into
      1cm/½in slices
   1 Chinese cabbage, chopped
      lengthways into thin strips
For the marinade
   5ml/1 tsp cornflour
   1 egg white, lightly beaten
   pinch of white pepper
For the sauce
   5ml/1 tsp cornflour
   45ml/3 tbsp oyster sauce

6 Heat another 30ml/2 tbsp of oil in the wok, add the remaining garlic, ginger and spring onions and stir-fry for 1 minute. Add the corals, stir-fry briefly and transfer to a dish.

1 Rinse the scallops and separate the corals from the white meat. Cut each scallop into 2–3 pieces and slice the corals. Place them on two dishes.

2 For the marinade, blend together the cornflour, egg white and white pepper. Pour half over the scallops and the rest over the corals. Leave for 10 minutes.

3 To make the sauce, blend the cornflour with 60ml/4 tbsp water and the oyster sauce and set aside.

4 Heat about 30ml/2 tbsp of the oil in a wok, add half of the garlic and let it sizzle, and then add half the ginger and half of the spring onions. Stir-fry for about 30 seconds and then stir in the scallops (not the corals).

5 Stir-fry for ½–1 minute until the scallops start to become opaque and then reduce the heat and add 15ml/ 1 tbsp of the sherry. Cook briefly and then spoon the scallops and the cooking liquid into a bowl and set aside.

7 Heat the remaining oil and add the mooli. Stir-fry for about 30 seconds and then stir in the cabbage. Stir-fry for about 30 seconds and then add the oyster sauce mixture and about 60ml/4 tbsp water. Allow the cabbage to simmer briefly and then stir in the scallops and corals, together with all their liquid and cook briefly to heat through.

CAULIFLOWER, PRAWN AND BROCCOLI TEMPURA

ALL SORTS OF VEGETABLES ARE DELICIOUS DEEP FRIED JAPANESE-STYLE (TEMPURA). FIRM VEGETABLES, SUCH AS CAULIFLOWER AND BROCCOLI, ARE BEST BLANCHED BEFORE FRYING BUT MANGE-TOUT, RED AND GREEN PEPPER SLICES, AND MUSHROOMS CAN SIMPLY BE DIPPED IN THE BATTER AND FRIED.

SERVES FOUR

INGREDIENTS
- ½ cauliflower
- 275g/10oz broccoli
- 8 raw prawns
- 8 button mushrooms (optional)
- sunflower or vegetable oil, for deep frying
- lemon wedges and sprigs of coriander, to garnish (optional)
- soy sauce, to serve

For the batter
- 115g/4oz plain flour
- pinch of salt
- 2 eggs, separated
- 175ml/6fl oz/¾ cup iced water
- 30ml/2 tbsp sunflower or vegetable oil

1 Cut the cauliflower and broccoli into medium-size florets. Blanch all the florets for 1–2 minutes. Drain. Refresh under cold running water. Put to one side. Peel the prawns but leave their tails intact. Put to one side.

2 To make the batter, place the flour and salt in a bowl, blend together the egg yolks and water and stir into the flour, beating well to make a smooth batter.

3 Beat in the oil and then whisk the egg whites until stiff and fold into the batter. Heat the oil for deep frying to 190°C/ 375°F. Coat a few of the vegetables and prawns in the batter.

4 Fry for 2–3 minutes until lightly golden and puffy. Transfer to a plate lined with kitchen paper and keep warm while frying the remaining tempura.

5 Arrange on individual plates, garnish with lemon and coriander, if liked, and serve with little bowls of soy sauce.

COOK'S TIP
Try cooking other vegetables in this way, such as aubergines and courgettes, or even the delicate young leaves of cauliflower, or celery leaves.

FENNEL AND MUSSEL PROVENÇAL

SERVES FOUR

INGREDIENTS

2 large fennel bulbs
1.75kg/4–4½lb fresh mussels in
   their shells, well scrubbed under
   cold water and beards removed
175ml/6fl oz/¾ cup water
sprig of thyme
25g/1oz butter
4 shallots, finely chopped
1 garlic clove, crushed
250ml/8fl oz/1 cup white wine
10ml/2 tsp plain flour
175ml/6fl oz/¾ cup single cream
15ml/1 tbsp chopped fresh parsley
salt and freshly ground black pepper
sprig of dill, to garnish

1 Trim the fennel, cut into slices 5mm/¼in thick and then cut into 1cm/½in sticks. Cook in a little salted water until just tender and drain.

2 Discard any mussels that are damaged or do not close. Put in a large saucepan, add the water and thyme, cover tightly, bring to the boil and cook for about 5 minutes until the mussels open, shaking occasionally.

3 Transfer the mussels to a plate and discard any that are unopened. When cool enough to handle, remove them from their shells, reserving a few in their shells for a garnish.

4 Melt the butter in a saucepan and fry the shallots and garlic for 3–4 minutes until softened but not browned. Add the fennel, fry briefly for 30–60 seconds and then stir in the wine and simmer gently until the liquid is reduced by half.

5 Blend the flour with a little extra wine or water. Add the cream, parsley and seasoning to the saucepan and heat gently. Stir in the blended flour and the mussels. Cook over a low heat until the sauce thickens. Season to taste and pour into a warmed serving dish. Garnish with dill and reserved mussels in their shells.

BRAISED FENNEL WITH TOMATOES

SERVES FOUR

INGREDIENTS

3 small fennel bulbs
30–45ml/2–3 tbsp olive oil
5–6 shallots, sliced
2 garlic cloves, crushed
4 tomatoes, peeled and chopped
about 175ml/6fl oz/¾ cup dry
   white wine
15ml/1 tbsp chopped fresh basil or
   2.5ml/½ tsp dried
40–50g/1½–2oz fresh white breadcrumbs
salt and freshly ground black pepper

1 Preheat the oven to 150°C/300°F/ Gas 2. Trim the fennel bulbs and cut into slices about 1cm/½ in thick.

2 Heat the olive oil in a large saucepan and fry the shallots and garlic for about 4–5 minutes over a moderate heat until the shallots are slightly softened. Add the tomatoes, stir-fry briefly and then stir in 150ml/¼ pint/⅔ cup of the wine, the basil and seasoning. Bring to the boil, add the fennel, then cover and cook for 5 minutes.

3 Arrange the fennel in layers in an ovenproof dish. Pour over the tomato mixture and sprinkle the top with half the breadcrumbs. Bake in the oven for about 1 hour. From time to time, press down on the breadcrumb crust with the back of a spoon and sprinkle over another layer of breadcrumbs and a little more of the wine. The crust slowly becomes golden brown and very crunchy.

SWEET PEPPER CHOUX <u>WITH</u> ANCHOVIES

THE RATATOUILLE VEGETABLES IN THIS DISH ARE ROASTED INSTEAD OF STEWED, AND HAVE A WONDERFUL AROMATIC FLAVOUR. ANY COMBINATION OF RED, GREEN OR YELLOW PEPPERS CAN BE USED. FOR VEGETARIANS, OMIT THE ANCHOVIES.

SERVES SIX

INGREDIENTS
    300ml/½ pint/1¼ cups water
    115g/4oz butter or margarine
    150g/5oz plain flour
    4 eggs
    115g/4oz Gruyère or Cheddar cheese,
      finely diced
    5ml/1 tsp Dijon mustard
    salt
For the filling
    3 peppers; red, yellow and green
    1 large onion, cut into eighths
      or sixteenths
    3 tomatoes, peeled and quartered
    1 courgette, sliced
    6 basil leaves, torn in strips
    1 garlic clove, crushed
    30ml/2 tbsp olive oil
    about 18 black olives, pitted
    45ml/3 tbsp red wine
    175ml/6fl oz/¾ cup passata or puréed
      canned tomatoes
    50g/2oz can anchovy fillets, drained
    salt and freshly ground black pepper

1 Preheat the oven to 240°C/475°F/ Gas 9 and grease six individual oven-proof dishes. To prepare the filling, halve the peppers, discard the seeds and core and cut into 2.5cm/1in chunks.

2 Place the peppers, onion, tomatoes and courgette in a roasting tin. Add the basil, garlic and olive oil, stirring so the vegetables are well coated. Sprinkle with salt and pepper and then roast for about 25–30 minutes until the vegetables are just beginning to blacken at the edges.

3 Reduce the oven temperature to 200°C/400°F/Gas 6. To make the choux pastry, put the water and butter together in a large saucepan, heat until the butter melts. Remove from the heat and add all the flour at once. Beat well with a wooden spoon for about 30 seconds until smooth. Allow to cool slightly.

4 Beat in the eggs, one at a time, and then continue beating until the mixture is thick and glossy. Stir in the cheese and mustard, then season with salt and pepper. Spoon the mixture around the sides of the prepared dishes.

5 Spoon the vegetables into a large mixing bowl, together with any juices or scrapings from the base of the pan. Add the olives and stir in the wine and passata or puréed tomatoes. (Or you can stir these into the roasting tin but allow the tin to cool slightly otherwise the liquid will boil and evaporate.)

6 Divide the pepper mixture between the six dishes and arrange the drained anchovy fillets on top. Bake in the oven for about 25–35 minutes until the choux pastry is puffy and golden. Serve hot with a fresh green salad.

SAMPHIRE <u>WITH</u> CHILLED FISH CURRY

EVEN IF YOU'RE A BIG CURRY FAN, DON'T BE TEMPTED TO ADD TOO MUCH CURRY PASTE TO THIS DISH.
YOU NEED ONLY THE MEREST HINT OF MILD CURRY PASTE SO THAT THE FLAVOUR OF THE SAMPHIRE AND
FISH CAN STILL BE APPRECIATED.

<u>SERVES FOUR</u>

INGREDIENTS
175g/6oz samphire
350g/12oz fresh salmon steak or fillet
350g/12oz lemon sole fillets
fish stock or water
115g/4oz large peeled prawns
25g/1oz butter
1 small onion, very finely chopped
10ml/2 tsp mild curry paste
5–10ml/1–2 tsp apricot jam
150ml/¼ pint/⅔ cup soured cream
sprig of mint, to garnish (optional)

1 Trim the samphire and blanch in boiling water for about 5 minutes until tender. Drain and set aside.

2 Place the salmon and lemon sole in a large frying pan, cover with fish stock or water and bring to the boil. Reduce the heat, cover and cook for 6–8 minutes until the fish is tender.

COOK'S TIP
As the samphire has a fresh salty tang of the sea, there is not really any need to add extra salt to this recipe.

3 Transfer the fish to a plate and when cool enough to handle, break the salmon and sole into bite-size pieces, removing any skin and bones. Place in a mixing bowl with the prawns.

4 Melt the butter in a saucepan and gently fry the onion for 3–4 minutes until soft but not brown. Add the curry paste, cook for 30 seconds, then remove from the heat. Stir in the jam. Allow to cool and then stir in the soured cream.

5 Pour the curry cream over the fish. Arrange the samphire around the edge of a serving plate and spoon the fish into the centre. Garnish with a sprig of mint.

LEEK AND MONKFISH WITH THYME SAUCE

MONKFISH IS A WELL-KNOWN FISH NOW, THANKS TO ITS EXCELLENT FLAVOUR AND FIRM TEXTURE.

SERVES FOUR

INGREDIENTS

1kg/2lb monkfish, cubed
salt and pepper
75g/3oz/generous ⅓ cup butter
4 leeks, sliced
15ml/1 tbsp flour
150ml/¼ pint/⅔ cup fish or vegetable
  stock
10ml/2 tsp finely chopped fresh thyme,
  plus more to garnish
juice of 1 lemon
150ml/¼ pint/⅔ cup single cream
radicchio, to garnish

1 Season the fish to taste. Melt about a third of the butter in a pan, and fry the fish for a short time. Put to one side. Fry the leeks in the pan with another third of the butter until they have softened. Put these to one side with the fish.

2 In a saucepan, melt the rest of the butter, add the remaining butter from the pan, stir in the flour, and add the stock. As the sauce begins to thicken, add the thyme and lemon juice.

3 Return the leeks and monkfish to the pan and cook gently for a few minutes. Add the cream and season to taste. Do not let the mixture boil again, or the cream will separate. Serve immediately garnished with thyme and radicchio leaves.

FISH STEW <u>WITH</u> CALVADOS, PARSLEY <u>AND</u> DILL

THIS RUSTIC STEW HARBOURS ALL SORTS OF INTERESTING FLAVOURS AND WILL PLEASE AND INTRIGUE.
MANY VARIETIES OF FISH CAN BE USED, JUST CHOOSE THE FRESHEST AND BEST.

SERVES FOUR

INGREDIENTS
1kg/2lb assorted white fish
15ml/1 tbsp chopped parsley, plus a
  few leaves to garnish
225g/8oz mushrooms
1 x 225g/8oz can of tomatoes
salt and pepper
10ml/2 tsp flour
15g/½oz/1 tbsp butter
450ml/¾ pint/1⅞ cups cider
45ml/3 tbsp Calvados
1 large bunch fresh dill sprigs,
  reserving 4 fronds to garnish

1 Chop the fish roughly and place it in a casserole or stewing pot with the parsley, mushrooms, tomatoes and salt and pepper to taste.

2 Preheat the oven to 180°C/350°F/Gas 4. Work the flour into the butter. Heat the cider and stir in the flour and butter mixture a little at a time. Cook, stirring, until it has thickened slightly.

3 Add the cider mixture and the remaining ingredients to the fish and mix gently. Cover and bake for about 30 minutes. Serve with garnish.

SMOKED SALMON, LEMON AND DILL PASTA

THIS HAS BEEN TRIED AND TESTED AS BOTH A MAIN-DISH SALAD AND A STARTER, AND THE ONLY PREFERENCE STATED WAS THAT AS A MAIN DISH YOU GOT A LARGER PORTION, SO THAT MADE IT BETTER.

SERVES TWO (as a main course)

INGREDIENTS
  salt
  350g/12oz/3 cups pasta twists
  6 large sprigs fresh dill, chopped, plus
    more sprigs to garnish
  juice of one half lemon
  30ml/2 tbsp extra virgin olive oil
  15ml/1 tbsp white wine vinegar
  300ml/½ pint/1¼ cups double cream
  pepper
  170g/6oz smoked salmon

1 Boil the pasta in salted water until it is just cooked. Drain and run under the cold tap until completely cooled.

2 Make the dressing by combining all the remaining ingredients, apart from the smoked salmon and reserved dill, in the bowl of a food processor and blend well. Season to taste.

3 Slice the salmon into small strips. Placed the cooled pasta and the smoked salmon in a mixing bowl. Pour on the dressing and toss carefully. Transfer to a serving bowl and garnish with the dill sprigs.

AVOCADO AND PASTA SALAD WITH CORIANDER

SERVED AS ONE OF A VARIETY OF SALADS OR ALONE, THIS TASTY COMBINATION IS SURE TO PLEASE. THE DRESSING IS FAIRLY SHARP, YET TASTES WONDERFULLY FRESH.

SERVES FOUR

INGREDIENTS
  115g/4oz/1¼ cups pasta shells or bows
  900ml/1½ pints/3¾ cups stock
  4 sticks celery, finely chopped
  2 avocados, chopped
  1 clove garlic, peeled and chopped
  15ml/1 tbsp finely chopped fresh
    coriander, plus some whole leaves to
    garnish
  115g/4oz/1 cup mature Cheddar
    cheese, grated
For the dressing
  150ml/¼ pint/⅔ cup extra virgin olive
    oil
  15ml/1 tbsp cider vinegar
  30ml/2 tbsp lemon juice
  grated rind of 1 lemon
  5ml/1 tsp French mustard
  15ml/1 tbsp fresh coriander, chopped
  salt and pepper

1 Bring the stock to the boil, add the pasta, and simmer for about 10 minutes until just cooked. Drain and cool under cold running water.

2 Mix the celery, avocados, garlic and chopped coriander in a bowl and add the cooled pasta. Sprinkle with the grated Cheddar.

3 To make the dressing, place all the ingredients in a food processor and process until the coriander is finely chopped. Serve separately, or pour over the salad and toss before serving. Garnish with coriander leaves.

DESSERTS AND BAKES

Here is a variety of mouthwatering sweet treats, from delicate summer fools to Christmas pudding — as well as great recipes for freshly baked cakes and crusty home-made breads.

SUMMER FRUIT GÂTEAU <u>WITH</u> HEARTSEASE

NO ONE COULD RESIST THE APPEAL OF LITTLE HEARTSEASE PANSIES. THIS CAKE WOULD BE LOVELY FOR A SENTIMENTAL SUMMER OCCASION IN THE GARDEN.

SERVES SIX TO EIGHT

INGREDIENTS
   100g/3¾oz/scant ½ cup soft margarine,
      plus more to grease mould
   100g/3¾oz/scant ½ cup sugar
   10ml/2 tsp clear honey
   150g/5oz/1¼ cups self-raising flour
   3ml/½ tsp baking powder
   30ml/2 tbsp milk
   2 eggs, plus white of one more for
      crystallizing
   15ml/1 tbsp rosewater
   15ml/1 tbsp Cointreau
   16 heartsease pansy flowers
   caster sugar, as required, to crystallize
   icing sugar, to decorate
   500g/1lb strawberries
   strawberry leaves, to decorate

1 Crystallize the heartsease pansies by painting them with lightly beaten egg white and sprinkling with caster sugar. Leave to dry.

2 Preheat the oven to 190°C/375°F/ Gas 5. Grease and lightly flour a ring mould. Take a large mixing bowl and add the soft margarine, sugar, honey, flour, baking powder, milk and 2 eggs to the mixing bowl and beat well for 1 minute. Add the rosewater and the Cointreau and mix well.

3 Pour the mixture into the pan and bake for 40 minutes. Allow to stand for a few minutes and then turn out on to the plate that you wish to serve it on.

4 Sift icing sugar over the cake. Fill the centre of the ring with strawberries. Decorate with crystallized heartsease flowers and some strawberry leaves.

BORAGE, MINT <u>AND</u> LEMON BALM SORBET

BORAGE HAS SUCH A PRETTY FLOWER HEAD THAT IT IS WORTH GROWING JUST TO MAKE THIS RECIPE, AND TO FLOAT THE FLOWERS IN SUMMER DRINKS. THE SORBET ITSELF HAS A VERY REFRESHING, DELICATE TASTE, PERFECT FOR A HOT AFTERNOON.

SERVES SIX TO EIGHT

INGREDIENTS
   500g/1lb/2⅛ cups sugar
   500ml/17fl oz/2⅛ cups water
   6 sprigs mint, plus more to decorate
   6 lemon balm leaves
   250ml/8fl oz/1 cup white wine
   30ml/2 tbsp lemon juice
   borage sprigs, to decorate

1 Place the sugar and water in a saucepan with the washed herbs. Bring to a boil. Remove from the heat and add the wine. Cover and cool. Chill for several hours, then add the lemon juice. Freeze in a suitable container. As soon as the mixture begins to freeze, stir it briskly and replace in the freezer. Repeat every 15 minutes for at least 3 hours or until ready to serve.

2 To make the small ice bowls, pour about 1cm/½in cold, boiled water into small freezer-proof bowls about 600ml/1 pint/1¼ US pints in capacity, and arrange some herbs in the water. Place in the freezer. Once this has frozen add a little more water to cover the herbs and freeze.

3 Place a small freezer-proof bowl inside each larger bowl and put a heavy weight inside such as a metal weight from some scales. Fill with more cooled boiled water, float more herbs in this and freeze.

4 To release the ice bowls, warm the inner bowl with a small amount of very hot water and twist it out. Warm the outer bowl by standing it in very hot water for a few seconds, then tip out the ice bowl. Spoon the sorbet into the ice bowls, decorate with sprigs of mint and borage and serve.

LEMON MERINGUE BOMBE WITH MINT CHOCOLATE

THIS EASY ICE CREAM WILL CAUSE A SENSATION AT A DINNER PARTY — IT IS UNUSUAL BUT QUITE THE MOST DELICIOUS COMBINATION OF TASTES THAT YOU CAN IMAGINE.

SERVES SIX TO EIGHT

INGREDIENTS

2 large lemons
150g/5oz/⅔ cup granulated sugar
3 small sprigs fresh mint
150ml/¼ pint/⅔ cup whipping cream
600ml/1 pint/2½ cups Greek natural
   yogurt
2 large meringues
225g/8oz good-quality mint chocolate,
   grated

1 Slice the rind off the lemons with a potato peeler, then squeeze them for juice. Place the lemon rind and sugar in a food processor and blend finely. Add the cream, yogurt and lemon juice and process thoroughly. Pour the mixture into a mixing bowl and add the meringues, roughly crushed.

2 Reserve one of the mint sprigs and chop the rest finely. Add to the mixture. Pour into a 1.2 litre/2 pint/1¼ US pint glass bowl and freeze for 4 hours.

3 When the ice cream has frozen, scoop out the middle and pour in the grated mint chocolate. Replace the ice cream to cover the chocolate and refreeze.

4 To turn out, dip the basin in very hot water for a few seconds to loosen the ice cream, then turn the basin upside down over the serving plate. Decorate with grated chocolate and a sprig of mint.

APPLE MINT AND PINK GRAPEFRUIT FOOL

*APPLE MINT CAN EASILY RUN RIOT IN THE HERB GARDEN; THIS IS AN EXCELLENT WAY OF USING UP AN
ABUNDANT CROP.*

SERVES FOUR TO SIX

INGREDIENTS

  500g/1lb tart apples, peeled, cored and
    sliced
  225g/8oz pink grapefruit segments
  45ml/3 tbsp clear honey
  30ml/2 tbsp water
  6 large sprigs apple mint, plus more to
    garnish
  150ml/¼ pint/⅔ cup double cream
  300ml/½ pint/1¼ cups custard

1 Place the apples, grapefruit, honey, water and apple mint in a pan, cover and simmer for 10 minutes until soft. Leave in the pan to cool, then discard the apple mint. Purée the mixture in a food processor.

2 Whip the double cream until it forms soft peaks, and fold into the custard, keeping 2 tablespoonfuls to decorate. Carefully fold the cream into the apple and grapefruit mixture. Serve in individual glasses, chilled and decorated with swirls of cream and small sprigs of apple mint.

PASSION FRUIT AND ANGELICA SYLLABUB

PASSION FRUIT'S UNIQUE FRAGRANCE AND FLAVOUR MAKE THIS SYLLABUB QUITE IRRESISTIBLE.

SERVES SIX

INGREDIENTS
  6 passion fruit
  15ml/1 tbsp chopped crystallized
    angelica, plus more to decorate
  grated rind and juice of 2 limes
  120ml/4fl oz/½ cup white wine
  50g/2oz/⅓ cup icing sugar
  300ml/½ pint/1¼ cups double cream
  150ml/¼ pint/⅔ cup Greek natural
    yogurt

1 Scoop out the flesh, seeds and juice of the passion fruit and divide between 6 serving dishes. Place the crystallized angelica in a food processor with the lime rind and juice, and blend to a purée.

2 In a large bowl, mix the lime purée with the wine and sugar. Stir until the sugar is dissolved.

3 Whip the cream until it begins to form soft peaks and then gradually beat in the wine mixture – the cream should thicken slightly. Whisk in the yogurt.

4 Spoon the cream mixture over the passion fruit, and refrigerate until ready to serve. Decorate with more crystallized angelica before serving.

JAPANESE FRUIT SALAD WITH MINT AND COFFEE

THIS DESSERT WAS SERVED IN A JAPANESE DEPARTMENT STORE. ALTHOUGH IT SOUNDS A LITTLE STRANGE, IT WORKS VERY WELL — THE COFFEE FLAVOR IS EXCELLENT WITH THE FRUIT.

SERVES SIX

INGREDIENTS
- 12 canned lychees and the juice from the can
- 1 small fresh pineapple
- 2 large ripe pears
- 2 fresh peaches
- 12 strawberries
- 6 small sprigs mint plus 12 extra sprigs to decorate
- 1 tablespoon instant coffee granules
- 2 tablespoons boiling water
- ⅔ cup heavy cream

1 Peel the fruit as necessary and chop into equal-sized pieces. Place all the fruit in a large glass bowl and pour the lychee juice on top.

2 Put the mint, coffee granules and boiling water in a food processor. Blend until smooth. Add the cream and process again briefly.

3 Serve the fruit salad drained and chilled, with sprigs of mint, and the coffee sauce on the side.

CLEMENTINES IN BEAUMES DE VENISE WITH GERANIUM

THE FANTASTIC BONUS OF USING THIS RECIPE IS THAT YOU HAVE HALF A BOTTLE OF BEAUMES DE VENISE LEFT OVER, WHICH SIMPLY HAS TO BE DRUNK AS A DIGESTIF.

SERVES SIX

INGREDIENTS
    10 whole clementines
    ½ bottle Muscat de Beaumes de Venise
      or other dessert wine
    12 scented geranium leaves
    orange leaves, to decorate

1 Peel the clementines and remove the pith. Place the clementines in a glass dish and pour over the wine.

2 Add the scented geranium leaves and refrigerate overnight. Discard the leaves, then serve chilled and decorated with orange leaves. Any juice left over can be served as a digestif.

CHOCOLATE MINT TRUFFLE FILO PARCELS

THESE EXQUISITE LITTLE PARCELS ARE UTTERLY IRRESISTIBLE. THERE WILL BE NO LEFTOVERS.

EIGHTEEN PARCELS

INGREDIENTS

15ml/1 tbsp very finely chopped mint
75g/3oz/¾ cup ground almonds
50g/2oz plain chocolate, grated
2 dessert apples, peeled and grated
115g/4oz crème fraîche or fromage
  frais
9 large sheets filo pastry
75g/3oz/⅓ cup butter, melted
15ml/1 tbsp icing sugar
15ml/1 tbsp cocoa powder, to dust

1 Preheat the oven to 190°C/375°F/ Gas 5. Mix the mint, almonds, chocolate, crème fraîche or fromage frais and grated apple in a bowl. Cut the filo pastry sheets into 7.5cm/3in squares, and cover with a cloth to stop them drying out.

2 Brush a square of filo pastry with melted butter, lay on a second sheet, brush again, and place a spoonful of filling in the middle of the top sheet. Bring in all four corners and twist to form a purse shape. Repeat to make 18 parcels.

3 Place the filo parcels on a baking sheet, well brushed with melted butter. Bake for approximately 10 minutes. Leave to cool and then dust with the icing sugar, and then with the cocoa powder.

YOGURT WITH APRICOTS AND PISTACHIOS

IF YOU ALLOW A THICK YOGURT TO DRAIN OVERNIGHT, IT BECOMES EVEN THICKER AND MORE LUSCIOUS. ADD HONEYED APRICOTS AND NUTS AND YOU HAVE AN EXOTIC YET SIMPLE DESSERT.

SERVES FOUR

INGREDIENTS
    450g/1lb Greek-style yogurt
    175g/6oz/⅔ cup no-need-to-soak, dried
      apricots, snipped
    15ml/1 tbsp clear honey
    orange rind, grated
    30ml/2 tbsp unsalted pistachios,
      roughly chopped
    ground cinnamon

VARIATION
For a simple dessert, strain the fruit, cover with yogurt and sprinkle with demerara sugar and a little mixed spice or cinnamon.

1 Place the yogurt in a fine sieve and allow it to drain overnight in the fridge over a bowl.

2 Discard the whey from the yogurt. Place the apricots in a saucepan, barely cover with water and simmer for just 3 minutes, to soften. Drain and cool, then mix with the honey.

3 Mix the yogurt with the apricots, orange rind and nuts. Spoon into sundae dishes, sprinkle over a little cinnamon and chill.

FRESH PINEAPPLE SALAD

THIS REFRESHING SALAD CAN BE PREPARED AHEAD. ORANGE FLOWER WATER IS AVAILABLE FROM MIDDLE EASTERN FOOD STORES OR GOOD DELICATESSENS.

SERVES FOUR

INGREDIENTS
    1 small ripe pineapple
    icing sugar, to taste
    15ml/1 tbsp orange flower water, or
      more if liked
    115g/4oz/good ½ cup fresh dates,
      stoned and quartered
    225g/8oz fresh strawberries, sliced
    few fresh mint sprigs, to serve

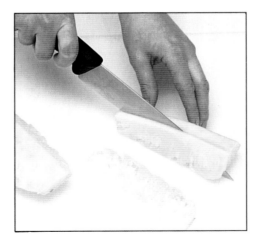

1 Cut the skin from the pineapple and, using the tip of a vegetable peeler, remove as many brown 'eyes' as possible. Quarter lengthways, remove the core then slice.

2 Lay the pineapple in a shallow, pretty, glass bowl. Sprinkle with sugar and orange flower water.

3 Add the dates and strawberries to the pineapple, cover and chill for a good 2 hours, stirring once or twice. Serve lightly chilled decorated with a few mint sprigs.

WALNUT AND RASPBERRY MERINGUE

MAKE SURE YOU BEAT THE EGG WHITES STIFFLY TO FORM A GOOD FIRM FOAM FOR THE MERINGUE. WHEN YOU FOLD IN THE NUTS, THE MERINGUE WILL HOLD ITS SHAPE.

SERVES FOUR TO SIX

INGREDIENTS

3 egg whites
few drops of fresh lemon juice
175g/6oz/1 cup caster sugar
75g/3oz/¾ cup walnuts, finely chopped
450g/1lb fresh raspberries
200ml/7fl oz/¾ cup crème fraîche or
  double cream
few drops vanilla essence
icing sugar, to taste

2 Whisk the egg whites in a spotlessly clean and grease-free bowl with the few drops of lemon juice. (This gives a more stable foam).

3 When the whites are softly stiff, gradually whisk in the sugar until thick and glossy. Quickly and carefully fold in the nuts.

6 Whip the creme fraîche or cream with the vanilla and sugar, until the mixture is quite stiff.

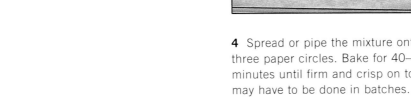

1 Preheat the oven to 160°C/325°F/ Gas 3. Draw three 20cm/8in circles on non-stick baking parchment. Place the circles on baking sheets.

4 Spread or pipe the mixture onto the three paper circles. Bake for 40–50 minutes until firm and crisp on top. This may have to be done in batches.

5 Cool on a wire rack and peel off the paper. Store in an airtight container until ready to serve.

7 Reserve a few raspberries for decoration, crush those remaining and mix into the creamy cheese mixture.

8 Spread the fruit cream on the three meringues. Sandwich them together and decorate the top layer with the reserved raspberries.

COOK'S TIP

Don't waste egg whites if you have recipes which call for yolks only. They do freeze very well and can be stored in batches of 3 or 4 whites at a time. In fact, when thawed, frozen egg whites make a much better foam.

RUM-BAKED BANANAS

THIS IS A QUICK, HOT PUDDING WHICH BAKES IN JUST MINUTES. WHEN COOKED, BANANAS HAVE A REALLY FULL FLAVOUR THAT IS ENHANCED BY RUM AND ORANGE. SERVE THIS DISH WITH A TRICKLE OF CREAM.

SERVES FOUR

INGREDIENTS
    4 bananas
    grated rind and juice of 1 orange
    30ml/2 tbsp dark rum
    50g/2oz/¼ cup soft brown sugar
      (optional)
    good pinch of ground ginger
    fresh nutmeg, grated
    40g/1½oz/3 tbsp butter

1 Preheat the oven to 180°C/350°F/ Gas 4. Peel the bananas and slice them into four large ramekins.

2 Spoon the orange juice and rum over the sliced bananas. Sprinkle over the sugar, if using, orange rind and spices. Dot with butter.

3 Cover the ramekins with small pieces of foil or buttered papers and bake for 15 minutes. Allow to cool slightly before serving with cream or natural yogurt.

MUESLI BARS

INSTEAD OF BUYING EXPENSIVE CRUNCHY OAT BARS, BAKE YOUR OWN. THEY ARE MUCH NICER AND REALLY QUITE EASY TO MAKE. USE MUESLI WITH NO ADDED SUGAR FOR THE BARS.

MAKES 12-16

INGREDIENTS
    350g/12oz/4 cups muesli
    75ml/5 tbsp sunflower oil
    75ml/5 tbsp clear honey
    5ml/1 tsp mixed spice
    1 egg, beaten
    30g/1½oz molasses sugar (optional)

COOK'S TIP
To make your own muesli, buy bags of flaked grains and oats from your local health food store. As these will make a large amount, you need to make sure you eat a lot of muesli! Choose jumbo porridge oats, barley flakes and wheat flakes, then add seeds, dried fruits and nuts of your choice.

1 Preheat the oven to 160°C/325°F/ Gas 3. Grease and line a shallow baking tin about 18 × 28cm/7 × 11in.

2 Mix all the ingredients together and spoon into the tin, patting the mixture until it is level.

3 Bake for 30–35 minutes until light brown round the edges. Remove, cool slightly then mark into 12–16 pieces.

4 Cool completely, turn out onto a wire rack and break into the marked pieces. Store in an airtight tin.

CHOCOLATE AND LEMON FROMAGE FRAIS

WHAT BETTER EXCUSE TO ADD IRON TO YOUR DIET THAN BY EATING SOME GOOD DARK (PLAIN) CHOCOLATE? CONTRAST THE RICHNESS WITH TANGY FROMAGE FRAIS.

SERVES FOUR

INGREDIENTS
    120g/5oz dark (plain) chocolate
    45ml/3 tbsp water
    15ml/1 tbsp rum, brandy or whisky
      (optional)
    grated rind of 1 lemon
    200g/7oz low-fat fromage frais
To decorate
    kumquats, sliced
    sprigs of mint

1 Break up the chocolate into a heatproof bowl. Add the water and either melt very slowly over a pan of gently simmering water or in a microwave on full power for 2–2½ minutes.

2 Stir well until smooth and allow the chocolate to cool for 10 minutes. Stir in the spirit, if using, and the lemon rind and fromage frais.

3 Spoon into four elegant wine glasses and chill until set. Decorate with kumquats and sprigs of mint.

ORANGE, HONEY AND MINT TERRINE

VERY REFRESHING AND EASY TO MAKE, THIS IS AN IDEAL DESSERT TO SERVE AFTER A RICH MEAL AS IT IS A GOOD PALATE CLEANSER.

SERVES SIX

INGREDIENTS
    8–10 oranges
    600ml/1 pint/2½ cups fresh orange
      juice
    30ml/2 tbsp clear honey
    20ml/4 tsp agar-agar
    45ml/3 tbsp fresh mint, chopped
    mint leaves to decorate (optional)

1 Grate the rind from two oranges and put this aside. Cut the peel and membrane from all the oranges, then slice each one thinly, removing any pips, and saving any juice.

2 Heat the orange juice (plus any saved) with the honey, reserved rind and agar-agar. Stir the mixture until it dissolves.

3 Pack the orange slices into a 1kg/2lb loaf tin, sprinkling the mint in between. Slowly pour over the hot orange juice. Tap the tin lightly so that all the juice settles.

4 Chill the terrine overnight, if possible, until it is quite firm. When ready to serve, dip the tin briefly into very hot water and turn the terrine out to a wet platter. Decorate with more mint leaves, if you wish. Serve cut into thick slices.

HALVA

THE GREEKS LOVE HOME-MADE HALVA WHICH THEY COOK IN A SAUCEPAN WITH SEMOLINA, OLIVE OIL, SUGAR, HONEY AND ALMONDS. YOU CAN EITHER EAT IT WARM, OR ALLOW IT TO SET AND CUT IT INTO SLICES OR SQUARES.

MAKES 12-16 PIECES

INGREDIENTS
  400g/14oz/2 cups granulated sugar
  1 litre/1¾ pints/4½ cups water
  2 cinnamon sticks
  225ml/8fl oz/1 cup olive oil
  350g/12oz/3 cups semolina
  75g/3oz/¾ cup blanched almonds, 6–8
    halved, the rest chopped
  120ml/4fl oz/½ cup clear honey
  ground cinnamon, to serve

1 Reserve 60ml/4 tbsp sugar and dissolve the rest in the water over a gentle heat, stirring from time to time.

2 Add the cinnamon sticks, bring to a boil then simmer for 5 minutes. Cool and remove the cinnamon sticks.

3 Heat the olive oil in a large heavy-based saucepan and, when it is quite hot, stir in the semolina. Cook, stirring occasionally, until it turns a golden brown, then add the chopped almonds and cook for a further minute or so.

4 Keep the heat low and stir in the syrup, taking care as the semolina may spit. Bring the mixture to a boil, stirring it constantly. When it is just smooth, remove the pan from the heat and stir in the honey.

5 Cool slightly and mix in the reserved sugar. Pour the halva into a greased and lined shallow tin, pat it down and mark into squares.

6 Sprinkle the halva lightly with ground cinnamon and fix one almond half on each square. When set, cut up and serve.

RICE CONDÉ SUNDAE

COOK A RICE PUDDING ON TOP OF THE STOVE FOR A LIGHT CREAMY TEXTURE, WHICH IS PARTICULARLY GOOD SERVED COLD AND TOPPED WITH FRUITS.

SERVES FOUR

INGREDIENTS

50g/2oz/⅓ cup pudding rice
600ml/1 pint/2½ cups milk
5ml/1 tsp vanilla essence
2.5ml/½ tsp ground cinnamon
30g/1½oz granulated sugar
To serve
Choose from: strawberries, raspberries or blueberries
chocolate sauce
flaked toasted almonds

1 Put the rice, milk, vanilla essence, cinnamon and sugar into a medium-sized saucepan. Bring to the boil, stirring constantly, and then turn down the heat to a gentle simmer.

2 Cook the rice for about 30–40 minutes, stirring occasionally. Add extra milk if it reduces down too quickly.

3 Make sure the grains are soft, then remove the pan from the heat and allow the rice to cool, stirring it occasionally. When cold, chill the rice in the refrigerator.

4 Just before serving, stir the rice and spoon into four sundae dishes. Top with fruits, chocolate sauce and almonds.

VARIATION

Milk puddings are at last enjoying something of a comeback in popularity. Instead of simple pudding rice try using a thai fragrant or Jasmine rice for a delicious natural flavour. For a firmer texture, an Italian Arborio rice makes a good pudding too.

There's no need to use a lot of high-fat milk or cream either. A pudding made with semi-skimmed or even fat-free milk can be just as nice and is much more healthy.

PEAR <u>and</u> HAZELNUT FLAN

IF YOU HAVE DIFFICULTY
FINDING GROUND HAZELNUTS,
GRIND YOUR OWN OR USE
GROUND ALMONDS.

<u>SERVES SIX TO EIGHT</u>

INGREDIENTS
115g/4oz/1 cup plain flour
115g/4oz/¾ cup wholemeal flour
about 45ml/3 tbsp cold water
115g/4oz/8 tbsp sunflower margarine
For the filling
50g/2oz/½ cup self-raising flour
115g/4oz/1 cup ground hazelnuts
5ml/1 tsp vanilla essence
50g/2oz caster sugar
50g/2oz/4 tbsp butter, softened
2 eggs, beaten
45ml/3 tbsp raspberry jam
1 × 400g/14oz can pears in juice
a few chopped hazelnuts, to decorate

1 Stir the flours together in a large mixing bowl, then rub in the margarine until it resembles fine crumbs. Mix to a firm dough with the water.

2 Roll out the pastry and use it to line a 23–25cm/9–10in flan tin, pressing it firmly up the sides after trimming, so the pastry sits above the tin a little. Prick the base, line with greaseproof paper and fill with baking beans. Chill for 30 minutes.

3 Preheat the oven to 200°C/400°F/ Gas 6. Place the flan tin on a baking sheet and bake blind for 20 minutes, removing the paper and beans for the last 5 minutes.

4 Meanwhile, beat all the filling ingredients together except for the jam and pears. If the mixture is a little thick, stir in some of the pear juice.

5 Reduce the oven temperature to 180°C/350°F/Gas 4. Spread the jam on the pastry base and spoon over the filling.

6 Drain the pears well and arrange them cut side down in the filling. Scatter over the nuts and bake for 30 minutes until risen, firm and golden brown.

VARIATION
This is also good made with ground almonds and canned apricots or pineapple pieces. For chocoholics, add 30ml/2 tbsp of cocoa powder to the flour in the filling (and even to the pastry if you want to make it richer and more chocolatey) plus a little grated lemon rind. Instead of raspberry jam, you could use chocolate spread and top with the pears in the recipe.

THREE-FRUITS COMPOTE

MIXING DRIED FRUITS WITH FRESH ONES MAKES A GOOD COMBINATION, ESPECIALLY IF FLAVOURED DELICATELY WITH A LITTLE ORANGE FLOWER WATER.

SERVES SIX

INGREDIENTS

175g/6oz/1 cup dried, no-need-to-soak
   apricots
1 small ripe pineapple
1 small ripe melon
15ml/1 tbsp orange flower water

1 Put the apricots into a saucepan with 300ml/½ pint water. Bring to the boil, then simmer for 5 minutes. Leave to cool.

2 Peel and quarter the pineapple then cut the core from each quarter and discard. Cut the flesh into chunks.

3 Seed the melon and scoop balls from the flesh. Save any juices which fall from the fruits and tip them into the apricots.

4 Stir in the orange flower water and mix all the fruits together. Pour into an attractive serving dish and chill lightly before serving.

VARIATION

A good fruit salad needn't be a boring mixture of multi-coloured fruits swimming in sweet syrup. Instead of the usual apple, orange and grape type of salad, give it a theme, such as red berry fruits or a variety of sliced green fruits – even a dish of just one fruit nicely prepared and sprinkled lightly with some sugar and fresh lemon juice can look beautiful and tastes delicious. Do not use more than three fruits in a salad so that the flavours remain distinct.

RED BERRY TART WITH LEMON CREAM FILLING

JUST RIGHT FOR WARM SUMMER DAYS, THIS FLAN IS BEST FILLED JUST BEFORE SERVING SO THE PASTRY REMAINS MOUTH-WATERINGLY CRISP. SELECT RED BERRY FRUITS SUCH AS STRAWBERRIES, RASPBERRIES OR REDCURRANTS.

SERVES SIX TO EIGHT

INGREDIENTS

150g/5oz/1¼ cups plain flour
25g/1oz/¼ cup cornflour
30g/1½oz icing sugar
100g/3½oz/8 tbsp butter
5ml/1 tsp vanilla essence
2 egg yolks, beaten

For the filling

200g/7oz cream cheese, softened
45ml/3 tbsp lemon curd
grated rind and juice of 1 lemon
icing sugar, to sweeten (optional)
225g/8oz mixed red berry fruits
45ml/3 tbsp redcurrant jelly

1 Sift the flour, cornflour and icing sugar together, then rub in the butter until the mixture resembles breadcrumbs. This can be done in a food processor.

2 Beat the vanilla into the egg yolks, then mix into the crumbs to make a firm dough, adding cold water if necessary.

3 Roll out and line a 23cm/9in round flan tin, pressing the dough well up the sides after trimming. Prick the base of the flan with a fork and allow it to rest in the refrigerator for 30 minutes.

4 Preheat the oven to 200°C/400°F/Gas 6. Line the flan with greaseproof paper and baking beans. Place the tin on a baking sheet and bake for 20 minutes, removing the paper and beans for the last 5 minutes. When cooked, cool and remove the pastry case from the flan tin.

VARIATION

There are all sorts of delightful variations to this recipe. For instance, leave out the redcurrant jelly and sprinkle lightly with icing sugar or decorate with fresh mint leaves. Alternatively, top with sliced kiwi fruits or bananas.

5 Cream the cheese, lemon curd and lemon rind and juice, adding icing sugar to sweeten, if you wish. Spread the mixture into the base of the flan.

6 Top the flan with the fruits. Warm the redcurrant jelly and trickle it over the fruits just before serving.

AVOCADO AND LIME ICE CREAM

THEIR RICH TEXTURE MAKES AVOCADOS PERFECT FOR A SMOOTH, CREAMY AND DELICIOUS ICE CREAM.

SERVES FOUR TO SIX

INGREDIENTS
  4 egg yolks
  300ml/½ pint/1¼ cups whipping cream
  115g/4oz/½ cup granulated sugar
  2 ripe avocadoes
  grated rind of 2 limes
  juice of 1 lime
  2 egg whites
  few unsalted pistachio nuts, to serve

2 As the cream rises to the top of the pan at the point of boiling, remove it from the heat.

4 Peel and mash the avocados until they are smooth then beat them into the custard with the lime rind and juice. Check for sweetness. Ice cream should be quite sweet before freezing as it loses flavour when ice cold. Add extra sugar now if you think it is needed.

5 Pour the mixture into a shallow container and freeze it until it is slushy. Beat it well once or twice as it freezes to stop large ice crystals forming.

1 Beat the yolks in a heatproof bowl. In a saucepan, heat the cream with the sugar, stirring it well until it dissolves.

3 Gently pour the beaten egg yolks into the scalded cream, adding them in small amounts from a height above the saucepan. This stops the mixture from curdling. Allow the mixture to cool, stirring it occasionally, then chill.

COOK'S TIP
If you have an ice cream machine, then simply pour the mixture into the basin and switch on. There is no need to add the egg whites as air is already beaten in with the paddle.

6 Whisk the egg whites until softly stiff and fold into the ice cream. Return the mixture to the freezer and freeze until firm. Cover and label. Use within four weeks, decorated with pistachio nuts.

HONEY AND LEMON SPICY MINCEMEAT

LIKE CHRISTMAS PUDDING, MINCEMEAT IS BEST MADE A FEW WEEKS AHEAD TO ALLOW THE FLAVOURS TO MATURE. THIS MIXTURE IS LIGHTER THAN MOST TRADITIONAL RECIPES.

MAKES 1.5KG/3LB

INGREDIENTS
150g/6oz/1 cup vegetarian shredded
  suet
225g/8oz/1½ cups currants
1 large cooking apple, coarsely grated
grated rind of 2 lemons
grated rind and juice of 1 orange
115g/4oz/¾ cup no-need-to-soak
  prunes, chopped
115g/4oz/¾ cup stoned dates, chopped
150g/6oz/1 cup raisins
225g/8oz/1¼ cups sultanas
115g/4oz/1 cup flaked almonds
90ml/6 tbsp clear honey
60ml/4 tbsp brandy or rum
5ml/1 tsp mixed spice
2.5ml/½ tsp ground cloves or all-spice

1 Mix all the ingredients together well in a large mixing bowl. Cover and store in a cool place for two days, stirring the mixture occasionally.

2 Sterilize clean jam jars by placing them in a warm oven for 30 minutes. Cool, then fill with mincemeat, and seal with wax discs and screw tops. Label and store until required.

CINNAMON AND TREACLE COOKIES

THE SMELL OF HOME-MADE COOKIES BAKING IS BETTERED ONLY BY THEIR WONDERFUL TASTE! THESE COOKIES ARE SLIGHTLY STICKY, SPICY AND NUTTY.

MAKES 24

INGREDIENTS
30ml/2 tbsp black treacle
50g/2oz/4 tbsp butter or margarine
115g/4oz/1 cup plain flour
1.5ml/¼ tsp bicarbonate of soda
2.5ml/½ tsp ground ginger
5ml/1 tsp ground cinnamon
40g/1½oz/¼ cup soft brown sugar
15ml/1 tbsp ground almonds or
  hazelnuts
1 egg yolk
115g/4oz/1 cup icing sugar, sifted

1 Heat the treacle and butter or margarine until it just begins to melt.

2 Sift the flour into a large bowl with the bicarbonate of soda and spices, then stir in the sugar and almonds or nuts.

3 Beat the treacle mixture briskly into the bowl together with the egg yolk and draw the ingredients together to form a firm but soft dough.

4 Roll the dough out on a lightly floured surface to a 5 mm/¼ in thickness and stamp out shapes, such as stars, hearts or circles. Re-roll the trimmings for more shapes. Place on a very lightly greased baking sheet and chill for 15 minutes.

5 Meanwhile, preheat the oven to 190°C/375°F/Gas 5. Prick the cookies lightly with a fork and bake them for 12–15 minutes until just firm. Cool on wire trays to crisp up.

6 To decorate, mix the icing sugar with a little lukewarm water to make it slightly runny, then drizzle it over the biscuits on the wire tray.

MINCE PIES WITH ORANGE CINNAMON PASTRY

MAKES 18

INGREDIENTS

225g/8oz/2 cups plain flour
30g/1½oz icing sugar
10ml/2 tsp ground cinnamon
120g/5oz/10 tbsp butter
grated rind of 1 orange
about 60ml/4 tbsp ice cold water
225g/8oz/1½ cups vegetarian
   mincemeat
1 beaten egg, to glaze
icing sugar, to dust

1 Sift together the flour, icing sugar and cinnamon then rub in the butter until it forms crumbs. (This can be done in a food processor.) Stir in the grated orange rind.

2 Mix to a firm dough with the ice cold water. Knead lightly, then roll out to a 6mm/¼in thickness.

3 Using a 7cm/2½in round cutter, cut out 18 circles, re-rolling as necessary. Then cut out 18 smaller 5cm/2in circles. If liked, cut out little shapes from the centre of the smaller circles.

4 Line two bun tins with the 18 larger circles – they will fill one and a half tins. Spoon a small spoonful of mincemeat into each pastry case and top with the smaller pastry circles, pressing the edges lightly together to seal.

5 Glaze the tops of the pies with egg and leave to rest in the refrigerator for 30 minutes. Preheat the oven to 200°C/400°F/Gas 6.

6 Bake the pies for 15–20 minutes until they are golden brown. Remove them to wire racks to cool. Serve just warm and dusted with icing sugar.

CHRISTMAS PUDDING

DRIED PRUNES AND APRICOTS ADD AN UNUSUAL TEXTURE AND DELICIOUS FLAVOUR TO THIS RECIPE.

MAKES TWO 1 LITRE/2 PINT PUDDINGS

INGREDIENTS

275g/10oz/5 cups fresh white
  breadcrumbs
225g/8oz/2 cups shredded vegetarian
  suet or ice cold butter, coarsely grated
115g/4oz/1 cup plain flour
225g/8oz/1 cup soft brown sugar
10ml/2 tsp ground mixed spice
350g/12oz/2½ cups currants
350g/12oz/2½ cups raisins
225g/8oz/1¾ cups sultanas
150g/6oz/1 cup stoned no-need-to-soak
  prunes, chopped
115g/4oz/¾ cup no-need-to-soak dried
  apricots, chopped
115g/4oz/¾ cup candied peel, chopped
115g/4oz/¾ cup glacé cherries, washed
  and chopped
rind of 1 large lemon, grated
4 eggs, beaten
30ml/2 tbsp black treacle
150ml/¼ pint/⅔ cup beer or milk
60ml/4 tbsp brandy or rum

1 Grease two 1 litre/2 pint pudding basins and line the bases with small discs of greaseproof paper.

2 Mix all the ingredients together well. If you intend to put lucky coins or tokens in the mixture, boil them first to ensure they are clean and wrap them in foil.

3 Pack the mixture into the two basins, pushing it down lightly.

4 Cover each pudding with greased greaseproof paper and a double thickness of foil. Secure it round the rim with kitchen string.

5 Place two old china saucers in the base of two large saucepans. Stand the basins on the saucers, pour boiling water to come two-thirds of the way up and boil gently for about 6 hours, checking the water level regularly and topping it up with more boiling water.

6 When cooked, cool the puddings, remove the foil and paper then re-cover to store. On Christmas Day, re-boil for about 2 hours and serve with brandy butter and cream or custard.

RICH CHOCOLATE CAKE

A SIMPLE ALL-IN-ONE CAKE SANDWICHED WITH A SIMPLE BUT VERY DELICIOUS GANACHE ICING.

SERVES SIX TO EIGHT

INGREDIENTS
    115g/4oz/1 cup self-raising flour
    25g/1oz/3 tbsp cocoa
    5ml/1 tsp baking powder
    120g/5oz/10 tbsp butter, softened, or
       sunflower margarine
    120g/5oz/¾ cup caster sugar
    3 eggs, beaten
    30ml/2 tbsp water
For the icing
    150g/5oz bar of dark (plain) chocolate
    150ml/¼ pint/⅔ cup double cream
    5ml/1 tsp vanilla essence
    30ml/2 tbsp apricot or raspberry jam

1 Grease and line a deep 20cm/8in round cake tin. Preheat the oven to 160°C/325°F/Gas 3.

2 Put all the cake ingredients into a large bowl or food processor. Beat very well with a wooden spoon or blend in a food processor until the mixture is smooth and creamy.

3 Spoon into the cake tin and bake for about 40–45 minutes or until risen and springy to the touch. Cool upside down on a wire rack for 15 minutes, then turn out and set aside to cool completely.

4 To make the icing, break the chocolate into a heatproof bowl and pour in the cream and vanilla essence. Melt in a microwave on full power for 2–3 minutes, or over a pan of simmering water.

5 Cool the icing, stirring it occasionally and chill lightly until it thickens. Split the cake in half. Spread the jam on one half and half the icing on top of that.

6 Sandwich the two halves together and spread the rest of the icing on top, swirling it attractively or marking it with the tip of a table knife. Decorate as desired with candies or candles – even edible flowers can add a nice touch.

PASSION CAKE

THIS CAKE IS ASSOCIATED WITH PASSION SUNDAY. THE CARROT AND BANANA GIVE A RICH, MOIST TEXTURE.

SERVES SIX TO EIGHT

INGREDIENTS
200g/7oz/1¾ cups self-raising flour
10ml/2 tsp baking powder
5ml/1 tsp cinnamon
2.5ml/½ tsp fresh nutmeg, grated
120g/5oz/10 tbsp butter, softened, or
   sunflower margarine
120g/5oz/¾ cup soft brown sugar
grated rind of 1 lemon
2 eggs, beaten
2 carrots, coarsely grated
1 ripe banana, mashed
115g/4oz/¾ cup raisins
50g/2oz/½ cup walnuts or pecans,
   chopped
30ml/2 tbsp milk
For the frosting
200g/7oz cream cheese, softened
30g/1½oz icing sugar
juice of 1 lemon
grated rind of 1 orange
6–8 walnuts, halved
coffee crystal sugar, to sprinkle

1 Line and grease a deep 20cm/8in cake tin. Preheat the oven to 180°C/350°F/Gas 4. Sift the flour, baking powder and spices into a bowl.

2 Using an electric mixer, cream the butter and sugar with the lemon rind until it is light and fluffy, then beat in the eggs. Fold in the flour mixture, then the carrots, banana, raisins, nuts and milk.

3 Spoon the mixture into the prepared tin, level the top and bake for about 1 hour until it is risen and the top is springy to touch. Turn the tin upside down and allow the cake to cool in the tin for 30 minutes. Turn onto a wire rack.

4 When cold, split the cake in half. Cream the cheese with the icing sugar, lemon juice and orange rind, then sandwich the two halves together with half of the frosting.

5 Spread the rest of the frosting on top, swirling it attractively. Decorate with the walnut halves and sprinkle with the coffee crystal sugar.

CHOCOLATE AND MINT FUDGE CAKE

CHOCOLATE AND MINT ARE POPULAR PARTNERS AND THEY BLEND WELL IN THIS UNUSUAL RECIPE. THE FRENCH HAVE BEEN USING POTATO FLOUR IN CAKES FOR YEARS. MASHED POTATO WORKS JUST AS WELL.

ONE CAKE

INGREDIENTS
    6–10 fresh mint leaves
    170g/6oz/¾ cup caster sugar
    115g/4oz/½ cup butter, plus extra to
        grease pan
    75g/3oz/½ cup freshly made mashed
        potato
    50g/2oz semi-sweet chocolate, melted
    170g/6oz1½ cups self-raising flour
    pinch of salt
    2 eggs, beaten
For the filling
    4 fresh mint leaves
    115g/4oz/½ cup butter
    115g/4oz/⅞ cup icing sugar
    30ml/2 tbsp chocolate and mint liqueur
For the topping
    225g/8oz/1 cup butter
    50g/2oz/¼ cup granulated sugar
    30ml/2 tbsp chocolate mint liqueur
    30ml/2 tbsp water
    170g/6oz/1½ cups icing sugar
    25g/1oz/¼ cup cocoa powder
    pecan halves, to decorate

1 Tear the mint leaves into small pieces and mix with the caster sugar. Leave overnight. When you use the flavoured sugar, remove the leaves and discard them.

2 Preheat the oven to 200°C/400°F/ Gas 6. Cream the butter and sugar with the mashed potato, then add the melted chocolate. Sift in half the flour with a pinch of salt and add half of the beaten eggs. Mix well, then add the remaining flour and the eggs.

3 Grease and line a 20cm/8in pan and pile in the mixture. Bake for 25–30 minutes or until a skewer or pointed knife stuck into the centre comes away clean.

4 Turn out on to a wire rack to cool. When cool, split into two layers.

5 Chop the mint leaves in a food processor, then add the butter and sugar. Once the cake is cool, sprinkle the chocolate mint liqueur over both halves and sandwich together with the filling.

6 Put the butter, granulated sugar, liqueur and water into a small pan. Melt the butter and sugar, then boil for 5 minutes. Sieve the icing sugar and cocoa and add the butter and liqueur mixture. Beat until cool and thick. Cover the cake with this mixture, and decorate with the pecan halves.

STRAWBERRY MINT SPONGE

THIS COMBINATION OF FRUIT, MINT AND ICE CREAM IS A REAL WINNER.

ONE CAKE

INGREDIENTS
    6–10 fresh mint leaves, plus more to
        decorate
    170g/6oz/¾ cup caster sugar
    170g/6oz/¾ cup butter, plus extra to
        grease pan
    3 eggs
    170g/6oz/1½ cups self-raising flour
    1.2 litres/2 pints/2½ US pints
        strawberry ice cream
    600ml/1 pint/2½ cups double cream
    30ml/2 tbsp mint liqueur
    350g/12oz/2 cups fresh strawberries

1 Tear the mint into pieces and mix with the caster sugar. Leave overnight.

2 Grease and line a deep springform cake pan. Preheat the oven to 190°C/ 375°F/Gas 5. Remove the mint from the sugar. Mix the butter and sugar together and add the flour, then the eggs. Pile the mixture into the tin.

3 Bake for 20–25 minutes, or until a skewer or pointed knife inserted in the middle comes away clean. Turn out on to a wire rack to cool. When cool, carefully split horizontally into two equal halves.

4 Clean the cake tin and line it with clear non-PVC film. Put the bottom half of the cake back in the pan. Spread on the ice cream mixture and level the top. Put on the top half of the cake and freeze for 3–4 hours.

5 Whip the cream with the mint liqueur. Remove the cake from the freezer and quickly spread a layer of whipped cream all over it, leaving a rough finish. Put the cake back into the freezer until about 10 minutes before serving. Decorate the cake with the strawberries and place fresh mint leaves on the plate around the cake.

CARROT CAKE AND GERANIUM CHEESE

AT A PINCH YOU CAN JUSTIFY CARROT CAKE AS BEING GOOD FOR YOU — AT LEAST THIS IS AN EXCUSE
FOR TAKING A GOOD MANY CALORIES ON BOARD. BUT THE FLAVOUR IS DEFINITELY WORTH IT.

ONE CAKE

INGREDIENTS
    2–3 scented geranium leaves
        (preferably with a lemon scent)
    225g/8oz/2 cups icing sugar
    115g/4oz/1 cup self-raising flour
    3ml/½ tsp ground cinnamon
    3ml/½ tsp ground cloves
    200g/7oz/1 cup soft brown sugar
    225g/8oz/1½ cups grated carrot
    150g/5oz/½ cup sultanas
    150g/5oz/½ cup finely chopped
        preserved ginger
    150g/5oz/½ cup pecan nuts
    150ml/¼ pint/⅔ cup sunflower oil
    2 eggs, lightly beaten
    butter to grease tin
For the cream cheese topping
    60g/2¼oz/generous ¼ cup cream cheese
    30g/1¼oz/2 tbsp softened butter
    5ml/1 tsp grated lemon rind

1 Put the geranium leaves, torn into
small- to medium-sized pieces, in a small
bowl and mix with the icing sugar. Leave
in a warm place overnight for the sugar to
take up the scent of the leaves.

2 Sift the flour and spices together.
Add the soft brown sugar, carrots,
sultanas, ginger and pecans. Stir well,
then add the oil and beaten eggs. Mix
with an electric beater for about 5
minutes, or 10–15 minutes longer by
hand.

3 Preheat the oven to 180°C/350°F/
Gas 4. Grease a 13 x 23cm/5 x 9in loaf
tin, line the base with greaseproof paper,
and then grease the paper. Pour the
mixture into the tin and bake for about 1
hour. Remove the cake from the oven,
leave to stand for a few minutes, and then
turn it out on to a wire rack until cool.

4 While the cake is cooling, make the
cream cheese topping. Remove the pieces
of geranium leaf from the icing sugar and
discard them. Place the cream cheese,
butter and lemon rind in a bowl. Using an
electric beater or a wire whisk, gradually
add the icing sugar, beating well until
smooth.

5 Once the cake has cooled, cover the
top with the cream cheese mixture.

LAVENDER COOKIES

INSTEAD OF LAVENDER YOU CAN USE ANY OTHER FLAVOURING, SUCH AS CINNAMON, LEMON, ORANGE OR
MINT.

ABOUT THIRTY

INGREDIENTS
    150g/5oz/⅝ cup butter, plus more to
        grease baking sheets
    115g/4oz/½ cup granulated sugar
    1 egg, beaten
    15ml/1 tbsp dried lavender flowers
    170g/6oz/1½ cups self-raising flour
    assorted leaves and flowers to decorate

1 Preheat the oven to 180°C/350°F/
Gas 4. Cream the butter and sugar
together, then stir in the egg. Mix in the
lavender flowers and the flour.

2 Grease two baking sheets and drop
spoonfuls of the mixture on them. Bake
for about 15–20 minutes, until the
biscuits are golden.

OATMEAL AND DATE BROWNIES

THESE BROWNIES ARE MARVELLOUS FOR SPECIAL BRUNCHES OR AS A TEA TIME TREAT. THE SECRET OF CHEWY, MOIST BROWNIES IS NOT TO OVERCOOK THEM.

MAKES 16

INGREDIENTS
   150g/5oz dark (plain) chocolate
   50g/2oz/4 tbsp butter
   75g/3oz/¾ cup quick-cook porridge oats
   25g/1oz/3 tbsp wheat germ
   25g/1oz/⅓ cup milk powder
   2.5ml/½ tsp baking powder
   2.5ml/½ tsp salt
   50g/2oz/½ cup walnuts, chopped
   50g/2oz/⅓ cup dates, chopped
   50g/2oz/¼ cup molasses sugar
   5ml/1 tsp vanilla essence
   2 eggs, beaten

1 Break the chocolate into a heatproof bowl and add the butter. Melt it either in a microwave on full power for 2 minutes, stirring once, or in a pan over very gently simmering water.

2 Cool the chocolate, stirring it occasionally. Grease and line a 20cm/8 in square cake tin. Preheat the oven to 180°C/350°F/Gas 4.

3 Combine all the dry ingredients together in a bowl then beat in the melted chocolate, vanilla and eggs.

4 Pour the mixture into the prepared cake tin, level the top and bake for about 20–25 minutes until it is firm around the edges yet still soft in the centre.

5 Cool the brownies in the tin then chill them. When they are more solid, turn them out of the tin and cut into 16 squares. Store in an airtight tin.

COOK'S TIP
These make a marvellous lunch box or picnic snack, and if you store them for a day or two before eating they will become more moist and even more chewy.

CUT-AND-COME-AGAIN FRUIT CAKE

A RICH FRUIT CAKE KEEPS WELL FOR QUITE SOME TIME SO IT IS IDEAL TO HAVE ONE ON HAND FOR WHEN YOU FEEL LIKE A SLICE OF SOMETHING SWEET OR WHEN GUESTS DROP IN.

SERVES EIGHT TO TEN

INGREDIENTS

225g/8oz/1 cup butter, softened, or
   sunflower margarine
225g/8oz/1 cup soft brown sugar
4 eggs, beaten
15ml/1 tbsp black treacle
350g/12oz/3 cups plain flour
5ml/1 tsp mixed spice
45ml/3 tbsp milk
900g/2lb dried mixed fruit (e.g. raisins,
   currants, cherries)
50g/2oz/½ cup flaked almonds
grated rind of 1 lemon
a few blanched almond halves
   (optional)
a little milk, to glaze (optional)
30ml/2 tbsp brandy or rum (optional)

1 Preheat the oven to 140°C/275°F/ Gas 1. Grease and line a deep 20cm/8in cake tin with doubled greaseproof paper.

2 Cream the butter or margarine and sugar until light and fluffy. Beat in the eggs with the treacle and stir into the creamed mixture.

3 Sift the flour and spice and fold this into the mixture, alternating it with the milk. Stir in the dried fruit, almonds and lemon rind. Spoon the mixture into the prepared tin. Dip the almond halves in a little milk and arrange them on top.

4 Bake on a shelf one position below the centre of the oven for about 3 hours. When cooked, the top of the cake will feel quite firm and a skewer inserted into the centre will come out clean.

5 Allow the cake to cool for 10 minutes then, if using the brandy or rum, make small holes in the top of the cake with a thin skewer. Slowly pour the spirit over the cake.

6 Allow the cake to cool completely in the tin, then turn it out and remove the paper. Wrap it in clean greaseproof paper and foil or store in an airtight container for one week before cutting.

THAI RICE CAKE

*A CELEBRATION GÂTEAU MADE
FROM FRAGRANT THAI RICE
COVERED WITH A TANGY CREAM
ICING.*

SERVES EIGHT TO TEN

INGREDIENTS
    225g/8oz/1¼ cups Thai fragrant or
        Jasmine rice
    1 litre/1¾ pints/4½ cups milk
    115g/4oz/¾ cup caster sugar
    6 cardamom pods, crushed open
    2 bay leaves
    300ml/½ pint/1¼ cups whipping cream
    6 eggs, separated
For the topping
    300ml/½ pint/1¼ cups double cream
    200g/7oz quark
    5ml/1 tsp vanilla essence
    grated rind of 1 lemon
    30g/1½oz caster sugar
    soft berry fruits and sliced star or kiwi
        fruits, to decorate

2 Return the rice to the pan with the milk, sugar, cardamom and bay leaves. Bring to the boil, then lower the heat and simmer the mixture for 20 minutes, stirring it occasionally.

3 Allow the mixture to cool, then remove the bay leaves and any cardamom husks. Turn into a bowl. Beat in the cream and then the egg yolks. Preheat the oven to 180°C/350°F/Gas 4.

4 Whisk the egg whites until they are softly stiff and fold into the rice mixture. Spoon into the prepared tin and bake for 45–50 minutes until risen and golden brown. The centre should be slightly wobbly – it will firm up as it cools.

5 Chill overnight in the tin. Turn out on to a large serving plate. Whip the double cream until stiff then mix in the quark, vanilla essence, lemon rind and sugar.

1 Grease and line a deep 25cm/10in round cake tin. Boil the rice in unsalted water for 3 minutes then drain.

COOK'S TIP
This is a good cake to serve to those with a gluten allergy as it is flour free.

6 Cover the top and sides of the cake with the cream, swirling it attractively. Decorate with soft berry fruits and sliced star or kiwi fruits.

APPLE AND APRICOT CRUMBLE

LIGHTLY COOK THE FRUIT BASE FIRST FOR THE BEST RESULTS.

SERVES FOUR TO SIX

INGREDIENTS
   1 × 425g/15oz can apricot halves in
      natural juice
   450g/1lb cooking apples, peeled and
      sliced
   granulated sugar, to taste (optional)
   grated rind of 1 orange
   fresh nutmeg, grated
For the topping
   200g/7oz/1¾ cups plain flour
   50g/2oz/½ cup rolled porridge oats
   120g/5oz/10 tbsp butter or sunflower
      margarine
   50g/2oz/¼ cup soft brown sugar
   demerara sugar, to sprinkle

1 Preheat the oven to 190°C/375°F/
Gas 5. Drain the apricots, reserving a little
of the juice.

2 Put the apples into a saucepan with a
little of the reserved apricot juice and
sugar to taste. Simmer for just 5 minutes
to cook the fruit lightly.

3 Transfer the apples into an ovenproof
pie dish and stir in the apricots, orange
rind and nutmeg to taste.

4 Rub the flour, oats and butter or
margarine together until they form fine
crumbs. (You can use a food processor if
you prefer.) Mix in the soft brown sugar.

5 Scatter the crumble over the fruit,
spreading it evenly. Sprinkle with a little
demerara sugar. Bake for about
30 minutes until golden and crisp on top.
Allow to cool slightly before serving.

FRENCH APPLE CAKE

WITH ITS MOIST TEXTURE AND
FRUITY FLAVOUR, THIS CAKE IS
IDEAL TO SERVE AS A DESSERT
ACCOMPANIED BY A LITTLE CREAM
OR FROMAGE FRAIS.

SERVES SIX TO EIGHT

INGREDIENTS
    450g/1lb cooking apples or tart dessert
        apples, cored and chopped
    115g/4oz/1 cup self-raising flour
    5ml/1 tsp baking powder
    115g/4oz/⅔ cup caster sugar
    90ml/6 tbsp milk
    50g/2oz/4 tbsp butter, melted
    3 eggs
    5ml/1 tsp fresh nutmeg, grated
For the topping
    75g/3oz/6 tbsp butter, softened, or
        sunflower margarine
    115g/4oz/½ cup caster sugar
    5ml/1 tsp vanilla essence
    sifted icing sugar, to dust

1 Preheat the oven to 160°C/325°F/
Gas 3. Grease and line the base of a deep
23cm/9in round cake tin.

2 Put the chopped apples into the base
of the cake tin.

3 Put all the remaining cake ingredients,
except 1 egg, into a bowl or food
processor. Beat to a smooth batter.

4 Pour the batter over the apples in the
tin, level the top then bake for
40–45 minutes until lightly golden.

5 Meanwhile, cream the topping
ingredients together with the remaining
egg. Remove the cake from the oven and
spoon over the topping.

6 Return the cake to the oven for a
further 20–25 minutes until it is golden
brown. Cool the cake in the tin, then turn
it out and finish with a light dusting of
icing sugar.

COURGETTE CROWN BREAD

Adding grated courgettes and cheese to a loaf mixture will keep it tasting fresher for longer. This is a good loaf to serve with a bowl of special soup.

SERVES EIGHT

INGREDIENTS
  450g/1lb courgettes, coarsely grated
  salt
  500g/1¼lbs/5 cups plain flour
  2 sachets fast-action yeast
  60ml/4 tbsp Parmesan cheese, freshly
    grated
  ground black pepper
  30ml/2 tbsp olive oil
  lukewarm water, to mix
  milk, to glaze
  sesame seeds, to garnish

1 Layer the courgettes in a colander and sprinkle them lightly with salt. Leave to drain for 30 minutes, then pat dry.

2 Mix the flour, yeast and Parmesan together and season with black pepper.

3 Stir in the oil and courgettes and add enough lukewarm water to give you a good firm dough.

4 Knead the dough on a lightly floured surface until it is smooth, then return it to the mixing bowl, cover it with oiled cling film and leave it to rise in a warm place.

5 Meanwhile, grease and line a 23cm/ 9in round sandwich tin. Preheat the oven to 200°C/400°F/Gas 6. When the dough has doubled in size, turn it out of the bowl, punch it down and knead it lightly. Break into eight balls, rolling each one and placing them in the tin as shown. Brush the tops with milk and sprinkle over the sesame seeds.

6 Allow to rise again, then bake for 25 minutes or until golden brown. Cool slightly in the tin, then turn out the bread to cool further.

ROSEMARY FOCACCIA

Italian flat bread is becoming increasingly popular and very easy to make using packet bread mix. Add traditional ingredients like olives and sun-dried tomatoes.

SERVES FOUR

INGREDIENTS
  450g/1lb pack white bread mix
  60ml/4 tbsp extra virgin olive oil
  10ml/2 tsp dried rosemary, crushed
  8 sun-dried tomatoes, snipped
  12 black olives, stoned and chopped
  200ml/7fl oz/¾ cup lukewarm water
  sea salt flakes

1 Mix the bread mix with half the oil, the rosemary, tomatoes, olives and water until it forms a firm dough.

2 Turn out the dough onto a lightly floured surface and knead thoroughly for 5 minutes. Return the dough to the mixing bowl and cover with a piece of oiled cling film.

3 Leave the dough to rise in a warm place until it has doubled in size. Meanwhile, lightly grease two baking sheets and preheat the oven to 220°C/425°F/Gas 7.

4 Turn out the risen dough, punch down and knead again. Divide into two and shape into rounds. Place on the baking sheet, and punch hollows in the dough. Trickle over the remaining olive oil and sprinkle with salt.

5 Bake the focaccia for 12–15 minutes until golden brown and cooked. Slide off onto wire racks to cool. Eat slightly warm.

BROWN SODA BREAD

*THIS IS VERY EASY TO MAKE —
SIMPLY MIX AND BAKE, AND AN
EXCELLENT RECIPE FOR THOSE
NEW TO BREAD MAKING.*

MAKES ONE 1KG/2LB LOAF

INGREDIENTS
  450g/1lb/4 cups plain flour
  450g/1lb/3 cups wholemeal flour
  10ml/2 tsp salt
  15ml/1 tbsp bicarbonate of soda
  20ml/4 tsp cream of tartar
  10ml/2 tsp caster sugar
  50g/2oz/4 tbsp butter
  up to 900ml/1½ pints/3¾ cups
    buttermilk or skimmed milk
  extra wholemeal flour, to sprinkle

1 Lightly grease a baking sheet. Preheat the oven to 190°C/375°F/Gas 5.

2 Sift all the dry ingredients into a large bowl, tipping any bran from the flour back into the bowl.

3 Rub the butter into the flour mixture, then add enough buttermilk or milk to make a soft dough. You may not need all of it, so add it cautiously.

4 Knead lightly until smooth then transfer to the baking sheet and shape to a large round about 5cm/2 in thick.

5 Using the floured handle of a wooden spoon, form a large cross on top of the dough. Sprinkle over a little extra wholemeal flour.

6 Bake for 40–50 minutes until risen and firm. Cool for 5 minutes before transferring to a wire rack to cool further.

CARDAMOM AND SAFFRON TEA LOAF

AN AROMATIC SWEET BREAD
IDEAL FOR AFTERNOON TEA, OR
LIGHTLY TOASTED FOR
BREAKFAST.

MAKES ONE 1 KG/2 LB LOAF

INGREDIENTS
    good pinch saffron strands
    750ml/1¼ pints/3 cups lukewarm milk
    25g/1oz/2 tbsp butter
    1 kg/2lb/8 cups strong plain flour
    2 sachets fast-action yeast
    30g/1½oz caster sugar
    6 cardamom pods, seeds extracted
    115g/4oz/⅔ cup raisins
    30ml/2 tbsp clear honey
    1 egg, beaten

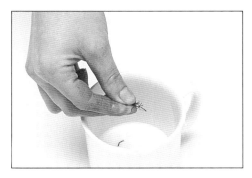

1 Crush the saffron into a cup containing a little of the warm milk and leave to infuse for 5 minutes.

2 Rub the butter into the flour then mix in the yeast, sugar and cardamom seeds (these may need rubbing to separate them). Stir in the raisins.

3 Beat the remaining milk with the honey and egg, then mix this into the flour along with the saffron milk and strands, stirring well until a firm dough is formed. You may not need all the milk: it depends on the flour.

4 Turn out the dough and knead it on a lightly floured board for about 5 minutes until smooth.

5 Return the dough to the mixing bowl, cover with oiled cling film and leave in a warm place until doubled in size. This could take between 1-3 hours.

VARIATION

For simplicity, leave out the saffron and cardamom and add 10ml/2 tsp ground cinnamon.

6 Turn the dough out onto a floured board again, punch it down, knead for 3 minutes then shape it into a fat roll and fit it into a greased loaf tin.

7 Cover with a sheet of lightly oiled cling film and stand in a warm place until the dough begins to rise again. Preheat the oven to 200°C/400°F/Gas 6.

8 Bake the loaf for 25 minutes until golden brown and firm on top. Turn out of the tin and as it cools brush the top with honey. Slice when cold and spread with butter. It is also good lightly toasted.

DINNER MILK ROLLS

MAKING BREAD ESPECIALLY FOR YOUR DINNER GUESTS IS NOT ONLY A WONDERFUL GESTURE, IT IS ALSO QUITE EASY TO DO. YOU CAN VARY THE SHAPES OF THE ROLLS TOO.

MAKES 12–16

INGREDIENTS
  750g/1½lb/4 cups strong plain flour
  10ml/2 tsp salt
  25g/1oz/2 tbsp butter
  1 sachet easy-blend fast action yeast
  about 450ml/¾ pint/scant 2 cups
    lukewarm milk
  extra cold milk, to glaze
  poppy, sesame and sunflower seeds or
    sea salt flakes, to garnish

1 Sift the flour and salt into a large bowl or food processor. Rub in the butter, then mix in the yeast.

2 Mix to a firm dough with the milk, adding it cautiously if the dough is a little dry in case you don't need it all.

3 Knead for at least 5 minutes by hand, or for 2 minutes in a food processor. Place in a bowl, cover with oiled cling film and leave to rise until doubled in size.

4 Turn out of the bowl, punch down and knead again, then break off into 12–16 pieces and either roll each one into a round or make into fun shapes.

5 Place on an oiled baking sheet, glaze the tops with extra milk and sprinkle over seeds or sea salt flakes of your choice.

6 Leave to start rising again, while you preheat the oven to 230°C/450°F/Gas 8. Bake the rolls for 12 minutes or until golden brown and cooked. Cool on a wire rack. Eat the same day, as homemade bread stales quickly.

INDIAN PAN-FRIED BREADS

INSTEAD OF YEAST, THIS DOUGH USES BICARBONATE OF SODA AS A RAISING AGENT. TRADITIONAL INDIAN SPICES ADD A TASTY BITE.

MAKES ABOUT 24

INGREDIENTS
  250g/8oz/2 cups wholemeal flour
  250g/8oz/2 cups plain flour
  5ml/1 tsp salt
  5ml/1 tsp sugar
  10ml/2 tsp bicarbonate of soda
  10ml/2 tsp cumin seeds
  10ml/2 tsp black mustard seeds
  5ml/1 tsp fennel seeds
  450g/1lb natural yogurt
  75g/3oz/6 tbsp vegetable ghee or
    clarified butter
  75ml/5 tbsp sunflower oil

1 Mix the flours with the salt, sugar, bicarbonate of soda and spices. Mix to a firm dough with the yogurt. Be sure to add the yogurt gradually, as you may not need it all.

2 If the dough is too dry, add cold water slowly until you achieve the correct consistency. Cover and chill for 2 hours.

3 Divide the dough into 24 pieces and roll each piece out to a thin round. Stack the rounds under a clean tea towel as you roll out the rest.

4 Fry the breads in the hot ghee or butter and oil; starting with a quarter and adding more ghee/butter and oil each time you fry. Drain the breads well on kitchen paper towel and store under the tea towel. Serve with curries and raitas.

INDEX

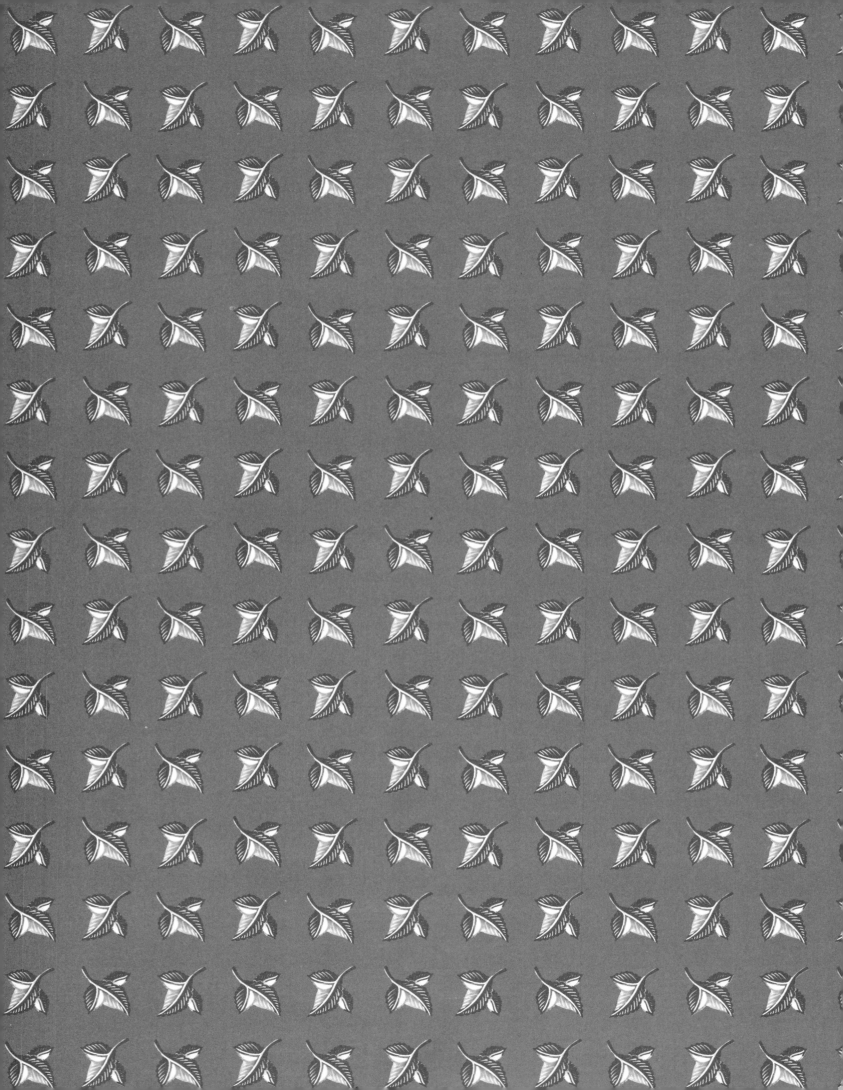